Lake Taihu, China

Monographiae Biologicae

VOLUME 87

Series Editor
H.J. Dumont
State University of Ghent, Institute of Animal Ecology, Ghent, Belgium
henri.dumont@rug.ac.be

Aims and Scope

The Monographiae Biologicae provide a forum for top-level, rounded-off monographs dealing with the biogeography of continents or major parts of continents, and the ecology of well individualized ecosystems such as islands, island groups, mountains or mountain chains. Aquatic ecosystems may include marine environments such as coastal ecosystems (mangroves, coral reefs) but also pelagic, abyssal and benthic ecosystems, and freshwater environments such as major river basins, lakes, and groups of lakes. In-depth, state-of-the-art taxonomic treatments of major groups of animals (including protists), plants and fungi are also elegible for publication, as well as studies on the comparative ecology of major biomes. Volumes in the series may include single-author monographs, but also multi-author, edited volumes.

For other titles published in this series, go to
www.springer.com/series/6304

Lake Taihu, China

Dynamics and Environmental Change

Edited by

Boqiang Qin

State Key Laboratory of Lake Science and Environment, Nanjing Institute of Geography and Limnology, Chinese Academy of Sciences, Nanjing, China

Springer

Editor
Boqiang Qin
State Key Laboratory of Lake Science
and Environment
Nanjing Institute of Geography and Limnology
Chinese Academy of Sciences
73 East Beijing Road
210008 Nanjing
China, People's Republic
qinbq@niglas.ac.cn

ISBN: 978-94-024-0493-7 ISBN: 978-1-4020-8555-0 (eBook)
DOI 10.1007/978-1-4020-8555-0

Softcover reprint of the hardcover 1st edition 2008

Cover Illustration: Lake Taihu, China. Photo by G. Zhu

Printed on acid-free paper

9 8 7 6 5 4 3 2 1

springer.com
Additional material to this book can be downloaded from http://extras.springer.com

Preface

In China, there are more than 2,759 lakes with surface area greater than 1 km^2, and the total lake area is 91, 019 km^2. One-third of these lakes are freshwater lakes, and the majority are situated in the middle and lower reaches of the Changjiang River or in eastern China's coastal areas. These lakes function as drinking water supplies, flood control systems, aquaculture and tourism resources, navigation channels, etc. Recently, many shallow lakes in China have been subject to rapid eutrophication and suffer from algal blooms. This issue has resulted in a shortage of drinking water and in degradation of their ecosystems. The control of eutrophication of shallow lakes is one of the main issues with which the local people and Chinese governments are concerned today.

Lake Taihu is the third largest freshwater lake in China, with an area of about 2338 km^2 and a mean depth of 1.9 m, a typical shallow lake located in the delta of Changjiang River, the most industrialized and urbanized area in China. Its main function is supplying drinking water for the surrounding cities, such as Wuxi, Suzhou, and Shanghai, but tourism, aquaculture, fisheries, and navigation are important as well. However, with economic development and increased population in the lake basin, Lake Taihu has suffered increasingly from serious eutrophication. The environmental issue of Lake Taihu is now a very common one, as most lakes from eastern China are confronted with it. Many efforts have been made to control the process of eutrophication and water quality deterioration. But so far, little improvement has been achieved. The hardness of a shallow eutrophic lake creates the need for deeply understanding the physical, geochemical, and biological processes associated with the eco-environmental issues within the lake. The work presented in this book was conducted in recent research activities. It is, therefore, a summarization of these studies about various aspects of Lake Taihu.

It is my pleasure to express my thanks to Prof. Henri Dumont for his interest in the manuscript and constructive suggestions during its preparation. He even helped with revising and editing some chapters. Without his concern and encouragement, this book would not have come into being. I also gratefully acknowledge the help of Yunlin Zhang for his hard work on text formatting, figure redrawing, and English editing. Many thanks to Xiaomei Liu for her excellent work in drawing the watershed map, which is the most detailed map so far of the lake. I am also grateful to Dr. Sarah Poyton and Mrs. Jennifer Purcell for effective and conscientious work in language polishing and supportive comments on successive drafts of the book's manuscript. I also would like to thank Dr. Zhijun Gong and Dr. Feizhou Chen for their help in English editing of Section 3 and Section 4 of Chapter 6.

Most of this book has been translated from its Chinese version, but chapters have been updated and revised, incorporating new research findings and data. Finally, I thank the "Science Press, China" for releasing the copyright of the Chinese version.

Nanjing, China Boqiang Qin

Contents

Contributors

Qiming Cai
State Key Laboratory of Lake Science and Environment, Nanjing Institute of Geography and Limnology, Chinese Academy of Sciences, 73 East Beijing Road, Nanjing 210008, P. R. China

Weimin Chen
State Key Laboratory of Lake Science and Environment, Nanjing Institute of Geography and Limnology, Chinese Academy of Sciences, 73 East Beijing Road, Nanjing 210008, P. R. China, e-mail: chenwm@niglas.ac.cn

Guang Gao
State Key Laboratory of Lake Science and Environment, Nanjing Institute of Geography and Limnology, Chinese Academy of Sciences, 73 East Beijing Road, Nanjing 210008, P. R. China, e-mail: guanggao@niglas.ac.cn

Junfeng Gao
State Key Laboratory of Lake Science and Environment, Nanjing Institute of Geography and Limnology, Chinese Academy of Sciences, 73 East Beijing Road, Nanjing 210008, P. R. China, e-mail: gaojunf@niglas.ac.cn

Wenyu Huang
State Key Laboratory of Lake Science and Environment, Nanjing Institute of Geography and Limnology, Chinese Academy of Sciences, 73 East Beijing Road, Nanjing 210008, P. R. China, e-mail: wyhuang@niglas.ac.cn

Wenchao Li
State Key Laboratory of Lake Science and Environment, Nanjing Institute of Geography and Limnology, Chinese Academy of Sciences, 73 East Beijing Road, Nanjing 210008, P. R. China, e-mail: wchli@niglas.ac.cn

Liancong Luo
State Key Laboratory of Lake Science and Environment, Nanjing Institute of Geography and Limnology, Chinese Academy of Sciences, 73 East Beijing Road, Nanjing 210008, P. R. China, e-mail: lcluo@niglas.ac.cn

Rui Mao
State Key Laboratory of Lake Science and Environment, Nanjing Institute of Geography and Limnology, Chinese Academy of Sciences, 73 East Beijing Road, Nanjing 210008, P. R. China

Boqiang Qin
State Key Laboratory of Lake Science and Environment, Nanjing Institute of Geography and Limnology, Chinese Academy of Sciences, 73 East Beijing Road, Nanjing 210008, P. R. China, e-mail: qinbq@niglas.ac.cn

Shuncai Sun
State Key Laboratory of Lake Science and Environment, Nanjing Institute of Geography and Limnology, Chinese Academy of Sciences, 73 East Beijing Road, Nanjing 210008, P. R. China

Qinglong Wu
State Key Laboratory of Lake Science and Environment, Nanjing Institute of Geography and Limnology, Chinese Academy of Sciences, 73 East Beijing Road, Nanjing 210008, P. R. China, e-mail: qlwu@niglas.ac.cn

Gang Xu
Wuxi Municipality, 631 South Jiefang road, Wuxi, 214001, P. R. China

Pengzhu Xu
State Key Laboratory of Lake Science and Environment, Nanjing Institute of Geography and Limnology, Chinese Academy of Sciences, 73 East Beijing Road, Nanjing 210008, P. R. China, e-mail: pzxu@niglas.ac.cn

Dingtian Yang
South China Sea Institute of Oceanology, Chinese Academy of Sciences, Guangzhou 510301, P. R. China, e-mail:dtyang@scsio.ac.cn

Liuyan Yang
Department of Environmental Science, School of the Environment, Nanjing University, 22 Hankou Road, Nanjing 210093, P. R. China, e-mail: yly@niglas.ac.cn

Lu Zhang
State Key Laboratory of Lake Science and Environment, Nanjing Institute of Geography and Limnology, Chinese Academy of Sciences, 73 East Beijing Road, Nanjing 210008, P. R. China, e-mail: luzhang@niglas.ac.cn

Yunlin Zhang
State Key Laboratory of Lake Science and Environment, Nanjing Institute of Geography and Limnology, Chinese Academy of Sciences, 73 East Beijing Road, Nanjing 210008, P. R. China, e-mail: ylzhang@niglas.ac.cn

Wanping Zhou
State Key Laboratory of Lake Science and Environment, Nanjing Institute of Geography and Limnology, Chinese Academy of Sciences, 73 East Beijing Road, Nanjing 210008, P. R. China

Guangwei Zhu
State Key Laboratory of Lake Science and Environment, Nanjing Institute of Geography and Limnology, Chinese Academy of Sciences, 73 East Beijing Road, Nanjing 210008, P. R. China, e-mail: gwzhu@niglas.ac.cn

Chapter 1
An Introduction to Lake Taihu

1.1 Hydrography and Drainage Basin

Shuncai Sun and Rui Mao

1.1.1 Lake Morphology and Drainage Basin

1.1.1.1 Morphological Characteristics

Lake Taihu is a well-known freshwater lake in the middle to lower valley of the Changjiang River, China. It is located in the broad Taihu plain, on the south side of the Changjiang delta, at 30°55′40″–31°32′58″ N and 119°52′32″–120°36′10″ E. Taking the mean water level of Lake Taihu as 3.0 m above sea level (a.s.l.), its total area is 2, 427.8 km^2, in which about 89.7 km^2 is occupied by 51 islands and islets, and the actual water area is 2, 338.1 km^2, making Taihu the third largest freshwater lake in China.

Lake Taihu is 68.5 km long in the north–south direction and on average 34 km wide from east to west. The maximum width, maximum depth, and mean depth are 56 km, 2.6 m, and 1.9 m, respectively. Therefore, it is a typical shallow lake. The topography of the lake bottom is flat, with a mean declivity of 0°0′19.66. The mean elevation of the lake bottom is 1.1 m above sea level (a.s.l.) (Fig. 1.1). The morphological characteristics of Lake Taihu are summarized in Table 1.1.

The map of the lake floor, and calculations derived from this map (Fig. 1.1, Table 1.2), show (i) areas of very shallow water ($<$1.5 m deep) along the shore zone and in Eastern Taihu Bay, total area 452.2 km^2, and occupying 19.3% of the total lake area; (ii) the area of deepest water ($>$2.5 m), mainly in the western Taihu region (maximum of 2.5–2.6 m, west and north of Pingtaishan Island in west Taihu), total area 197.3 km^2, and occupying 8.4% of the total area; and (iii) areas

Sun, S.C.
State Key Laboratory of Lake Science and Environment, Nanjing Institute of Geography and Limnology, Chinese Academy of Sciences, 73 East Beijing Road, Nanjing 210008, P. R. China

Qin, B. Q. (ed.), *Lake Taihu, China*,

Fig. 1.1 Bathymetric map of Lake Taihu

of moderate depth (1.5–2.5 m), which dominate the lake, total area 1, 688.6 km^2, occupying 72.3% of the total area. There is no deep trench in Lake Taihu, nor are there large-scale shoals. The lake is a submerged plain, on the floor of which are the remnants of buried rivers, evidence of its prior topographic character.

From natural and anthropogenic causes, Lake Taihu has five bays: Eastern Taihu Bay, Xukou Bay, Gonghu Bay, Meiliang Bay, and Zhushan Bay from east to west, with Dongdongtingshan (Dongshan Peninsula), Tanshan Mountain, Junzhang Mountain, and Guanzhang Mountain separating them. Currently, as a consequence of the slow exchange of streams, shallow water, and the action of prevailing winds, some bays have become seriously polluted or eutrophic.

1.1.1.2 Topographic Characteristics of the Drainage Basin

Based on topographic characteristics, the lake basin can be divided into four regions: the massif region in the western basin, the low plain region in the middle basin,

Table 1.1 Morphological characteristics of Lake Taihu[1,2]

Area (km^2)	Water area (km^2)	Catchment area (km^2)	Supply coefficient	Shoreline length (km)	Length (km)	Width (km)	Mean depth (m)	Maximum depth (m)	Volume ($\times 10^8 m^3$)	Annual transfer coefficient
2,427.8	2,338.1	36,895	15.8	405	68.5	34	1.9	2.6	44.28	1.18

[1] The data were obtained by direct measurement on the topographic map produced in the 1980s.

[2] Excluding Wuli Bay, which is 5.7 km^2.

Table 1.2 Distribution of different depth zones in Lake Taihu

Water depth (m)	< 1	1–1.5	1.5–2.0	2.0–2.5	> 2.5	Total
Area (km^2)	131.7	320.6	719.3	969.3	197.3	2338.1
Percentage (%)	5.6	13.7	30.8	41.5	8.4	100

the high plain region along Changjiang River shore, and the Lake Taihu region (Fig. 1.2).

(1) Massif region in the western basin. Based on a customary partition of the lake basin, the ground above 12 m elevation is defined as the low hill and massif region, situated at the west of the basin, with an area of 7,338 km^2. This region includes the Maoshan low hill, massif, and hummock region, and the Yili River low hill and massif region within Jiangsu Province, Xitiaoxi River vale, and plain region, and the Tianmu Mountain and massif region within Zhejiang Province. This area is the riverhead or upstream zone of most rivers feeding Lake Taihu (Fig. 1.2).

(2) The low plain region in the middle basin. This region lies below 5 m, is typified by a dense water web and a flat landform, and has an area of 19,350 km^2. It is divided into three parts. (i) The South Jiangsu province low plain region, mainly including the Tao-Ge (Lake Tao and Lake Ge) low plain around Lake Tao and Lake

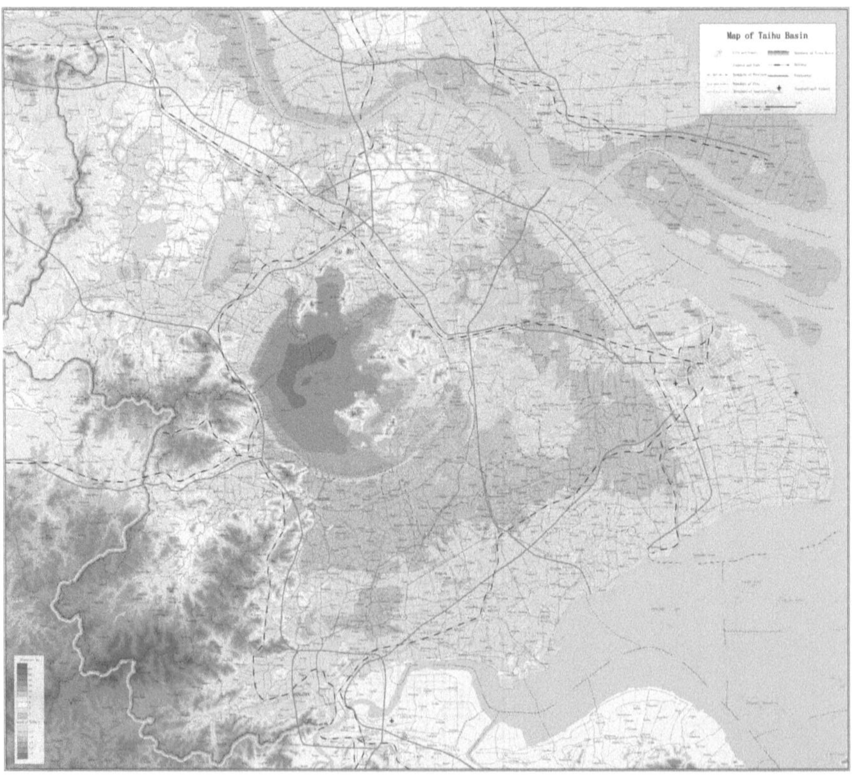

Fig. 1.2 Map of Lake Taihu. A detailed electronic version of this map can be viewed on the enclosed CD-Rom

Ge to the west of Taihu; Wujinggang-Xicheng (the area between Wujinggang River and Xicheng Canal) low plain with the Beijing-Hangzhou Canal (Grand Canal) as its axis, lying to the east of the Tao-Ge region and the northern suburbs of Wuxi City; and the Yangcheng-Dianliu low plain, around the lake cluster of Yangcheng and Dianliu Lake. (ii) From Hangzhou-Jiaxing-Huzhou (in brief, Hang-Jia-Hu) to West Zhejiang low plain; Hang-Jia-Hu Plain lies to the east of the Tiaoxi Diversion River, along both sides of the Beijing-Hangzhou Canal, at an altitude of 2.5–5.0 m; the plain is 110 km long from east to west, and 60 km wide from south to north; it is the largest low plain of the Taihu basin. Adjacent to it, in the west, is the West Zhejiang low plain, in the Changxing region, at an altitude of 4–5 m. The West Zhejiang low plain is vulnerable to mountain flood water. (iii) Along the Huangpu River low plain, including the Pudong and Puxi regions lying on east and west bank of the Huangpu River, respectively; the terrain is low, usually at an altitude of 2.5–3.5 m. The area is densely populated, and thus this low plain has been intensively used. The water level is high, and farmland (i.e., polders) is at low altitude, and for many years polders have been built. The polder area is currently 14, 542 km^2 in total, occupying 39% of the Taihu basin (Fig. 1.2).

(3) the high plain region along Changjiang River shore. Lying along the Changjiang River, the western end of this region lies at the east of Zhenjiang City, and the eastern end is Changshu City; the region is 135 km long from east to west, 30–50 km wide from south to north, and with an altitude of 6–12 m at its western end and 5 m at its eastern end. It is formed from sediment deposited by the Changjiang River and has an area of 7,015 km^2. The estuary high plain region extends from Hangzhou in the west, to Zhapu in the east, and is a narrow, discontinuous high plain, about 100 km long, with altitude of 6–7 m (5 m at the eastern end). The part of the estuary high plain in this region is far smaller than the part of the Changjiang River shore plain. The west massif region, the Changjiang River shore plain, and the estuary high plain region constitute the high margins around the Taihu basin.

(4) Lake Taihu region. This region includes the water surface, islands and islets, and belts between the lake regions. Lake Taihu is situated at the center of a dish-shaped basin, with shallow water, a flat bottom, and a mean altitude of only 1.1 m a.s.l. Its total area is 3,192 km^2 (occupying 8.6% of Lake Taihu basin), including 2,338.1 km^2 of water surface, 89.7 km^2 of islands and isles, and 764 km^2 of lowland along the shore.

The areas of the Lake Taihu basin situated at different altitudes are given in Table 1.3.

Table 1.3 Areas of the Lake Taihu basin situated at different altitudes

Height (m)	Distribution area (km^2)	Percentage (%)
<2	6,200	17.2
2–3	13,890	38.1
3–4	2,280	6.2
4–5	2,750	7.5
>5	11,766	31.8
Total	36,985	100

1.1.2 Formation and Development

As the Lake Taihu basin is in one of the most economically developed areas in China, with a long history of intense land and water use, research on its formation and development has long attracted attention from Chinese and international scholars. Within the last century, numerous publications, including more than 100 books, have focused on the lake, each advancing different theories and views. The principal literature (in chronological order) includes *Terrestrial Development of the Yangtze River Drainage Basin, Constitution and Degradation of Lake Taihu* (Wang & Ding, 1936), *Geographic Problems of the Lower Reach of the Yangtze River* (Wissman, 1941), *Geomorphic Development of the Estuary of the Yangtze River* (Chen et al., 1959), *Comprehensive Investigation Report of Lake Taihu* (Nanjing Institute of Geography, 1965), *Sea Level Change in Holocene and Formation and Development of Lake Taihu* (Yang et al., 1985), *Research on Modern Sediment in the Delta of the Yangtze River* (Yan & Xu, 1987), *Formation and Development of Lake Taihu and Modern Sedimentation Action* (Sun et al., 1987), *Lake Taihu* (Sun & Huang, 1993), and *Planning and Comprehensive Management of Lake Taihu Basin* (Huang, 2000).

Concerning the origin and formation of Taihu, opinions may be summarized as falling into three main groups.

ⓐ Lagoon formation hypothesis. Because it lies close to the mouth of Changjiang River and the East China Sea, and there are abundant sea sedimentary deposits below the Taihu Plain, it is believed that modern Lake Taihu originated from an ancient lagoon. During the early and middle Holocene, about 6,000–7,000 years ago, the Taihu Plain was a large bay connecting to the sea. The barrier spits formed at the south bank of Changjiang River, and the north bank of the Qiantang River stretched to the east, then enclosing the Taihu area, and as a result, the original bay gradually changed into a lagoon, and finally this lagoon became a lake isolated from the sea. Taihu and other lakes in the Taihu Plain evolved from this ancient lagoon (Wang & Ding, 1936; Chen et al., 1959).

ⓑ Tectonic origin hypothesis. It has often been considered that geological forces formed the big lake basins in the world, such as the Caspian Sea, Lake Baikal, Lake Victoria, and the fault depression lakes in Yunnan Province in China. Subsidence of the earth's crust, caused by tectonic movements, does indeed form lake basins. Large lakes such as Taihu may also owe their origin to fault-block differential movement in the Mesozoic or early Pleistocene, with the central subsiding part forming a lake basin, and water subsequently accumulating in it; in other words, the fault depression forms a lake.

ⓒ Flood water retention hypothesis. When dissemination of flood water is not immediate, waterlogging in lowlands can form a lake. Since the 1980s, the scientists from Nanjing Institute of Geography and Limnology, Chinese Academy of Sciences, have carried out systematic surveys around Lake Taihu and other lakes in the Taihu Plain, using a chain of deep and shallow drill cores in and around the lake, and a sub-bottom profiler, to detect the sediment as much as 40 m beneath the lake bottom. Their results indicate that the entire lake bed is composed of hard loess, except for local embedded rivercourses and hollows. Lake water directly lies over the loess formation, and the Taihu Plain is an alluvial plain, which became covered with loess before Lake Taihu came into being. As a consequence of inundation of this loess plain, waterlogging in the lowlands formed Lake Taihu. The age of formation of Lake Taihu, inferred from an ancient buried well of the Chunqiu and Zhanguo Period, should be about 2,000–2,500 years ago (Sun et al., 1987).

1.1.3 Influence of Lake Formation on the Aquatic Environment

As a result of the lack of immediate and direct flood drainage, water is retained in the lowland to form a lake. Thus, in the early stage of its formation in the Tang Dynasty (430–479 A.C.), Taihu was an expanse of water without clear boundaries. After the Tang and Song Dynasties, dams were built, opening up many inlets and outlets of waterways. According to literature from the Ming Dynasty (1368–1644 A.C.), there were 140 outlet channels from Wuloukou (at the outlet of the Wulougang River) to Liangxikou (at the outlet of the Liangxi River) in eastern Taihu, and 180 inlet channels in western Lake Taihu, giving a total of 320 rivercourses.

However, as a result of silting and anthropogenic activities, the number of rivercourses has markedly decreased. The sedimentation rate at the Wangyu River in the 1990s (measured using ^{210}Pb) was 3 cm/acre (a), and the sedimentation rate at the estuary of Dapugang River was 3–5 cm/a; silting thickness after 30 years exceeded 1 m. This sedimentation resulted in only 240 rivercourses remaining in the 1960s, 219 in the 1980s, and 97 in the 1990s. Recent investigation showed that there are 125 rivercourses. At the same time, the number of outlet channels declined from 150 to 90 from the 1960s to the 1990s, and now only 17 remain. The reduction in lake area, and in the number of inlet and outlet channels, has had two significant effects on the aquatic environment of Taihu; namely, an increase in water retention time and exacerbation of flood disasters.

1.1.3.1 Increase in Water Retention Time

Based on data from 1954–1988, the mean yearly inflow and outflow was approximately 57.37×10^8 m^3, and the retention period was about 281 days. After the 1990s, this period increased to 309 days, and this longer retention time accelerated the deterioration of the aquatic environment and the process of eutrophication. Both Lake Taihu and Lake Dianchi have relatively long retention periods and have become eutrophic. Therefore, dredging some primary inlets and outlets to enhance drainage capability, and to quicken renewal of the water in the lake, will be helpful for restoration of the aquatic environment. In contrast, other freshwater lakes such as Lake Dongting, Lake Poyang, and Lake Hongze in the middle and lower reaches of the Changjiang River have short retention periods, ranging from 20 to 168 days (Table 1.4), and have maintained a good water quality.

Table 1.4 Water retention time of six freshwater lakes in China

Lake	Taihu	Dongding	Boyang	Hongze	Chaohu	Dianchi
Water exchange time	309 d	20 d	57 d	35 d	168 d	2.5 a

d: days; a: year.

1.1.3.2 Exacerbation of Flood Disasters

In the 20th century, major flood disasters occurred in the Taihu basin in 1931, 1954, 1991, and 1999. These four floods, caused by "plum rain," each lasted for a long period and became a hazard throughout the drainage basin.

For the 1991 flood, the influence of change in retention time on lake level was calculated. As a result of the filling up of inlets and outlet channels and lakes, as well as lake reclamation, there was a decline in the capacity for flood water storage and drainage. According to the basic equation on regulation and storage of flood water by lakes:

$$H_i - H_0 = a_i \left(\sum P + \sum V - \sum Q\right) / A \tag{1.1}$$

here, H_i is water level of a certain day (in meters), H_0 is the water level before flood (m), P is daily rainfall (m^3), V is the flux entering the lake (m^3), Q is the flux leaving the lake (m^3), A is lake area (m^2), a is an uncertain integrative factor, including infiltration and leakage, which can be calculated by the following equation:

$$a = (H_{li} - H_0) \cdot A_{l1} / \left(\sum P_{li} + \sum V_{li} - \sum Q_{li}\right) \tag{1.2}$$

Evaporation volume during rainfall can be ignored. Here, subscript 1 shows the range of lake surface area. Based on 2, 338.1 km^2 of actual water area of Lake Taihu, and a capacity for flood regulation and storage of 47×10^8 m^3, it is simulated as follows:

ⓐ The influence of reduced lake area (as a result of reclamation and silt sedimentation) on capacity for flood regulation and storage. From 1954 to 1991, the reclamation area of Lake Taihu totaled 140 km^2, and the lost volume of flood storage was 4.7×10^8 m^3, occupying 12.9% of overall lake volume. According to the sedimentation rate (range, 0.16–2.99 mm/a) of each region, total siltation is approximately 1.56×10^8 m^3. Thus, 4.2% of storage volume was lost from 1954 to 1991, and this may cause a rise in water level. Under the circumstance that precipitation, and inlet and outlet flux, in 1991 were equal to that of 1954, the water level in Taihu would increase by 0.266 m.

$$\Delta H_{li} = H_0 + \Delta H_{2i} = 0.266 \tag{1.3}$$

where H_2 is the water level in 1954. This increase in water level is the result of reclamation and silting.

ⓑ Silting of inlets and outlets channels, and anthropogenic stoppage of flooding water discharge. In the 1950s, there were 84 outlet channels, whereas after the 1990s there were only 17; thus, one-third of the drainage capacity had been lost. In the flood of 1954, observed discharge volume of flooding water was $36 \times 10^8\,m^3$, whereas in the flood of 1991 it was only $21.53 \times 10^8\,m^3$. If it is presumed that the flux entering the lake ($\sum V$), precipitation ($\sum P$), and actual effective area of this lake were the same in 1991 as in 1954, and drainage capacity decreased by one-third compared with 1954, the rise of water level of Lake Taihu may be estimated as

$$H_3 = H_{3i} - H_0 - a_i\left(\sum P + \sum V - 1/3\sum Q\right)/A_1 \tag{1.4}$$

$$H_3 = H_0 + \Delta H = H_2 + 2a_i/3A_2\sum Q_2 \tag{1.5}$$

with A_1 the lake dimensions in 1991 and A_2 the lake dimensions in 1954. If other conditions remain unchanged, and drainage capacity decreases by one-third, then water level will rise to $2a_i/3A_2\sum Q_2$.

ⓒ Effects of reduction in volume and drainage capacity. Compared to the situation in 1954, storage volume in 1991was reduced by 17.1%, and drainage capacity by one-third, and thus water level will rise by

$$\Delta H_4 = H_{4i} - H_0 = a_i\left(\sum V_{4i} + \sum P_{4i} - 1/3\sum Q_{4i}\right)/A_4 \tag{1.6}$$

Assuming that $V_4 = V_2$, $P_4 = P_2$, and $V_4 = (1 - 0.171)A_2$ (due to silting, the volume of lake reduces by 17.1%), thus:

$$\Delta H_{4i} = a_i\left(\sum V_2 + \sum P_2 - 1/3\sum Q\right)/0.829A_2 = \Delta H_{2i} + 2a/3A_2\sum Q/A_2 \tag{1.7}$$

These results show that the water level would rise 1.5–1.8 m above the highest water point of 1954 (Fig. 1.3).

Figure 1.3 reveals that if the lake with the actual storage and drainage conditions of 1991 were subjected to the flood volumes of 1954, the water level would rise

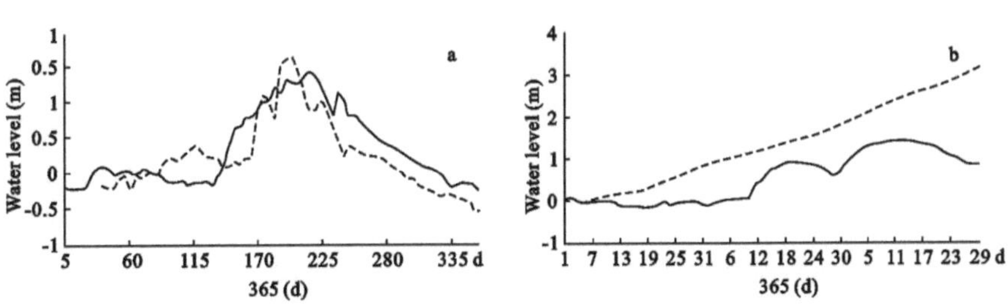

Fig. 1.3 Simulated lake level during flooding. (**a**) Increment of lake level in 1991 (*solid line* is measured; *dashed line* is calculated, assuming the flooding of 1954 were to occur in 1991). (**b**) Simulated lake level during flooding in 1954 and 1991 (*dashed line* is the 1954 flood; *solid line* is the flood in 1991)

1.6–1.8m above the highest level of 1954. Compared with 1954, water levels in 1991 would also rise more rapidly, and decrease more slowly.

A declining capacity for flood regulation and storage, and an increased retention period of lake water, will exacerbate flood calamities. In the flood of 1954, the maximum precipitation in 90 days was 890.5 mm, and the highest water level was 4.65 m; in 1991, the maximum precipitation across the drainage basin in 90 days was 681.2 mm, and the highest water level was 4.79 m; whereas in 1999, maximum precipitation was 672.8 mm, and the highest water level 5.08 m. Clearly, the precipitation in 90 days in the 1990s was much less than that in 1950s, and the maximum lake level during flooding in 1990s was much higher than that in the 1950s. It is evident that lake reclamation and silting of rivers and lake area have direct effects on the limnetic environment.

1.2 Hydrography and Water Resources

Rui Mao, Pengzhu Xu, and Junfeng Gao

1.2.1 Hydrography

Lake Taihu basin is a depression of land, in which water from all sides gathers to the center and then diffuses in all directions, forming a complex hydrosystem that contains interlaced rivers, dense water nets, and dotted depression lakes of different sizes.

As a result of human activities, the structure of the water system in the basin has changed dramatically in the past decades, especially after the floods of 1991. The present system mainly consists of regional backbone watercourses and small interior rivers. The backbone watercourses, most of which had taken shape before 1991, are designed to meet irrigation and transportation requirements. After the floods, central and local governments spent more than 10 billion Chinese yuan to have the watercourses dredged, riverbanks reinforced, and dams built. As a result, the whole area attained higher standards of flood prevention and gained a greater diversion and outlet capacity. There are still 125 inflowing and outflowing watercourses.

The western parts of the basin are hilly and the eastern parts are lowland plains; thus, rivers tend to flow from west to east. The northern boundary between upstream and downstream is Zhihugang River in Wuxi City, and the southern boundary is Wulougang River in Wujiang County. Rivers in the west mainly flow into Lake Taihu and belong to the upper reaches. In the upper reaches, rivers form an arborescence, and in the lower reaches, the drainage system is fan shaped. The total length of the rivers in the basin is 12,000 km, and the river net density is 3.24 km/km^2; thus, a river or other watercourse is encountered at 600- to 800-m intervals. There are also 189 lakes of different sizes, covering an area of 3,159 km^2.

1.2.1.1 The Upper Reaches

According to their sources, the water systems in the upper reaches can be classified into four systems: Tiaoxi Rivers (including Dongtiaoxi River and Xitiaoxi River), Yili River, Lake Tao and Lake Ge, and Wujinggang–Zhihugang Rivers.

(1) Tiaoxi river (Xitiaoxi River and Dongtiaoxi River) system

The Tiaoxi river system originates from the Tianmu mountainous area, southwest of Lake Taihu, and has a catchment area of 5,931 km^2. This system can be further divided into Xitiaoxi (West Tiaoxi River), Dongtiaoxi (East Tiaoxi River), and rivers around Changxing County and Huzhou City, which all flow into Lake Taihu (Fig. 1.2).

Mao, R.
State Key Laboratory of Lake Science and Environment, Nanjing Institute of Geography and Limnology, Chinese Academy of Sciences, 73 East Beijing Road, Nanjing 210008, P. R. China

The Xitiaoxi River has its source at the foot of Tianmu Mountain, in the west of Zhejiang Province. The river collects waters in Anji and Huzhou and forms a 145-km-long main stream. The Dongtiaoxi River has its source at the southern foot of Tianmu Mountain and collects water in Lin'an, Yuhang, and Deqing (see Fig. 1.2). Its 165-km-long main stream meets the Xitiaoxi River at the Daqiangang River (i.e., Daqiankou) and the Xiaomeigang River (i.e., Xiaomeikou), and then flows through various rivers and canals (such as Xiaomeikou) and finally flows into Lake Taihu. Part of the currents flow east into Youtang River, merging with the surface flow of the Hangzhou-Jiaxing-Huzhou plain (Fig. 1.2). Thereafter, some of the converged current runs along Lake Taihu's southern bank and finally flows into the lake. Part of the currents flows east into Huangpujiang River, by the way of Grand Canal. In all, about 70% of the amount of water in the Tiaoxi River system flows into Lake Taihu. The Xitiaoxi and Dongtiaoxi Rivers converge at the west of Huzhou and then travel into Lake Taihu by way of the Changdougang and Xiaomeigang Rivers (Fig. 1.2). The planned water diversion project at the middle reaches of the Xitiaoxi River will be able to discharge water from the Xitiaoxi River into Lake Taihu. Rivers in the Changxing area carry the water yield from the Changxing County plains, and its western hilly areas, to Lake Taihu through several rivers and canals, such as Jiapugang River, Hexixingang River, Changxinggang River, and Yangjiapugang River. Among these rivers, the Jiapugang River is a separate river system that opens into the lake. Hexixingang and Changxinggang chiefly discharge the incoming flows from the upper reaches watercourses and rivers, while Yangjiapugang discharges the incoming flows from Siantang River, which meets the Xitiaoxi River in its low reaches, so part of its flow runs into the latter.

(2) Yili river system

The Yili, also called the Nanhe River, has a basin of $3,091\,\mathrm{km}^2$ (calculated from a 1:10000 relief map). It originates from the western part of the Lake Taihu basin, at the common boundary of Jiangsu, Anhui, and Zhejiang Provinces (see Fig. 1.2). The Yili system gathers water from the Guodi mountainous area, the hilly areas at the south of Maoshan Mountain (Fig. 1.2), and in the plains along the Yili River. Most water produced in the west of the basin flows first into the Nanhe and Beixi rivers, then into three seasonal lakes (Lakes Xijiu, Tuanjiu, and Dongjiu) (Fig. 1.2), and finally into Lake Taihu, passing the three channels to Dapugang River, Chengdonggang River, and Hongxianggang River (Fig. 1.2).

(3) Lake Tao and Lake Ge river system

This system covers an area of $4,480\,\mathrm{km}^2$. It contains water from the Lake Tao and Lake Ge districts, and the Beijing-Hangzhou Canal area, including the east of Mao mountain and the south of the Changjiang River shore high plain (see Fig. 1.2). The system also collects water from Lake Tao, Lake Ge, and the Taige Canal. The backbone watercourses flowing into the lake here are the Taige Canal, Caoqiao River, Yincungang River, and Shaoxianggang River. Lakes Tao and Ge, which are in the middle of the area, are regulation lakes (Fig. 1.2). The south part of the area exchanges water with the Yili River via watercourses such as the Danjinlicao, Mengjin, and Wuyicaohe Rivers (Fig. 1.2).

(4) Wujinggang-Zhihugang river system

This system covers an area of 1,000 km^2 around Wuxi city, and contains water from the low plain of Wujinggang-Xicheng, west of the Xicheng Canal. Wujinggang River and Zhihugang River are the two main watercourses leading to Lake Taihu (see Fig. 1.2). They are the most heavily polluted parts of the Lake Taihu area.

1.2.1.2 Lake Taihu Water System

Lake Taihu is the largest natural reservoir in the basin and serves to regulate floods in the basin. From the west, the lake receives water of several sub-basins, such as the Tiaoxi and Yili Rivers and the Taige Canal. To the east, the water goes sequentially through the Liangxi River, Wangyu River, and Eastern Taihu, Taipu River, and Guajingkou (outlet of Guajinggang River at the east end of Eastern Taihu Bay) (see Fig. 1.2).

1.2.1.3 River System in the Low Reaches

The low reaches can be divided into the following three systems: Changjiang River (Yangtse River) northward drainage system, Huangpujiang River system, and Hangzhou Bay southward drainage system. Each system drains surface runoff from two-thirds of the plains of the Lake Taihu basin, in addition to the Lake Taihu floods. Connected by regional backbone watercourses, these three water systems are interconnected.

(1) The Changjiang River northward drainage system is composed of a dozen watercourses between Jiangyin and Shanghai, including the Baiqugang River (Dongqing River), Wangyu River, Qiputang River, and the Xinliu River. Wangyu River, one of Lake Taihu's two main outlets (besides the Taipu River), discharges Lake Taihu's floods. The Wangyu River, in addition to the Grand Canal, which goes through the east of Lake Taihu, and four districts of Lake Yangchenghu, Lake Chenghu, Lake Dianshan, and Hangzhou-Jiaxing-Huzhou plain, belongs to the Changjiang River watercourse system, as it intersects with the interlaced rivers and lakes within the Changjiang River basin.

(2) The Huangpujiang River lies between the Changjiang River northward drainage and the Hangzhou-Jiaxing-Huzhou water system. Located in Lake Taihu's lowest reaches, its sources are the Wusongjiang River, Lake Cheng, Lake Dianshan, Taipu River, and part of Dongtiaoxi River flow in the Hangzhou-Jiaxing-Huzhou region. As one of Lake Taihu's main drainage watercourses, the Huangpujiang River can drain more than 60% of flood waters. The Taipu River, with a drainage capacity of half of Lake Taihu's water collection, is located at the Huangpujiang River's upper reaches. The Wusongjiang River is also in the Huangpujiang's upper reaches and mainly discharges waterlogging in drainage areas; it accepts only a small amount of water from Lake Taihu. There are three branches of flow that converge in the Huangpujiang River's upper reaches: the north branch comprises Xietang River, Xihe River (not shown in Fig. 1.2), and Lanlugang River; the middle branch comprises the Yuanxiejing and Yuhuitang Rivers (not shown in Fig. 1.2);

and its south branch is formed by the Liugang River (not shown in Fig. 1.2) which receives incoming water from Hangzhou-Jiaxing-Huzhou. Because the water level of the Huangpujiang River is affected by estuary tides as far as its source, water levels will consequently reach high levels during floods, and the flood water cannot then discharge efficiently; conversely, in winter, the water level is low and the water from Lake Taihu discharges quickly.

(3) The Hangzhou Bend south drainage system consists of the Haiyantang River, Changshan River, a few backbone watercourses of the Yanguan River (Yanguanshanghe River, Yanguanxiahe River) basin, and several other regions. All the backbone watercourses carry surplus runoff from Hangzhou-Jiaxing-Huzhou area into Hangzhou Bay, by way of Nantaitou gate, Changshan gate, and Yanguan gate. Dongtiaoxi River offshoots, Lake Taihu and Taipu river, are connected to the west and the north part of the local river nets, and provide emergency drainage in years of great floods.

1.2.1.4 Constitution of Inflowing and Outflowing Rivers and their Distribution

As the center of the basin, Lake Taihu is of extreme importance in flood control, water resource regulation, and environment protection. To a large extent, it is Lake Taihu that has brought the rapid regional economic growth to areas such as the Suzhou-Wuxi-Changzhou (in brief, Su-Xi-Chang) and Hangzhou-Jiaxing-Huzhou (in brief, Hang-Jia-Hu) districts, and Shanghai. Lake Taihu will continue playing an important role as the processes of irrigation construction and environment improvement accelerate.

Lake Taihu is a typical lowland shallow lake. It is characterized by shallow water, a fast water cycle, and numerous incoming rivers. There were 225 rivers and canal entrances around the lake before the 1980s, and as the shoreline is 405 km long, there was 1 entrance approximately every 2 km. When the embankment around Lake Taihu was built, 54 original entrances were blocked, although 27 locations have 1.5-m-aperture culvert gates. Among the other 171 entrances, 45 were kept open; the remainder are controlled by building gates.

A total of 37 rivers leading to Lake Taihu are found along the Yixing shoreline (see Fig. 1.2). The Shaoxianggang River, the entrances of which were all kept open, is a watercourse dug in 2002 to connect Lake Taihu with Lake Ge Basin. Regional backbone rivers into Lake Taihu are the Taige Canal and 6 rivers: Dapugang River, Chengdonggang River, Hongxianggang River, Shaoxianggang River, Shatanggang River, and Yincungang River (Fig. 1.2). Other smaller rivers are the Dagang, Wuxigang, Huangdugang, Zhudugang, Kandugang, and Xindugang Rivers.

There are 23 rivers along the Wuxi shoreline (see Fig. 1.2), all of which, except for the Wujinggang and Zhihugang Rivers, run into Lake Taihu. All are passages from Lake Taihu to the Grand Canal of Wuxi section with added or reduced flows. Gates have been fixed for these 23 rivers. The backbone rivers here are Liangxi River, Lihe River, Daxi River, Wujinggang River, Zhihugang River, and Wuli Bay. Wuli Bay used to be an arm of Lake Taihu until, in 1991, a gate was built to separate them. Thus, Wuli Bay should be considered as a "watercourse" connected to Lake Taihu.

There are 75 watercourses leaving Lake Taihu along the Suzhou shoreline (see Fig. 1.2). The Wangyu River and Taipu Rivers serve as two major drainage and diversion channels. If there is insufficient water in Lake Taihu, the Wangyu River carries supplemental water from the Changjiang River. When there is a water shortage in Shanghai, or water quality deterioration in Huangpujiang River, the Wangyu River delivers water from Lake Taihu to the Huangpujiang River (Fig. 1.2).

Altogether, there are 37 watercourses flowing along the Huzhou shoreline (see Fig. 1.2). The Changdugang River is the largest incoming river in the Lake Taihu basin. Its upper reaches are the Dongtiaoxi River and Xitiaoxi River, which collect more than half of the water entering Lake Taihu catchment's upper reaches. All the rivers run east to Lake Taihu. Of those rivers that run west to Changdugang River, some flow into the lake, while others flow out. Nine of the rivers running into the lake are kept open; the others are controlled by sluice gates.

Although Lake Taihu has many inflowing and outflowing rivers, they are all short, with low flow volumes (Table 1.5).

1.2.2 Water Resources and Water Balance

The Lake Taihu basin is located in the fan-shaped delta of the main Changjiang River. The Changjiang drainage basin has the most water of all river basins in China. With a capacity of $9.61 \times 10^{11}\,m^3$, it accounts for 34.1% of China's total water resources. There is 2,358 m^3 of water per capita in the Changjiang basin and 480 m^3 per capita in the Lake Taihu basin.

In 2001, the whole Lake Taihu basin had an annual precipitation of $4.41 \times 10^{10}\,m^3$. Annual runoff was $1.78 \times 10^{10}\,m^3$, of which $1.52 \times 10^{10}\,m^3$ was from surface water and $2.6 \times 10^9\,m^3$ was from ground water. The runoff yielding coefficient is $4.8 \times 10^5\,m^3$ per kilometer squared.

1.2.2.1 Rainfall

Atmospheric precipitation is the main source of surface water in the Lake Taihu basin. The local distribution of annual precipitation is influenced by factors such as climate, terrain, and moisture sources. In general, the south of the basin receives more precipitation than the north, and the mountainous west receives more precipitation than do the plains to the east. Annual precipitation in the whole basin in 2000 was 1,182 mm.

The basin can be classified as humid (annual precipitation, 800–1,600 mm). In its southwest, because of orographic rain, the Tianmu mountain area has the highest annual precipitation, 1,300–1,900 mm. In the northwest and north of the basin (Maoshan Mountain, Jintan, Changzhou, Kunshan, Wusongjiang River), annual precipitation is less than 1,050 mm. Across the basin, the highest annual precipitation

Table 1.5 Major inflowing and outflowing rivers of Lake Taihu[1]

River	Inflow				Outflow				Annual runoff ($\times 10^8$ m^3)
	Day (d)	Maximal daily mean flux (m^3/s)	Mean flux (m^3/s)	Inflow total runoff ($\times 10^8$ m^3)	Day (d)	Maximal daily mean flux (m^3/s)	Mean flux (m^3/s)	Outflow total runoff ($\times 10^8$ m^3)	
Xitiaoxi	235	273.0	35.0	7.10	124	−57.8	−12.89	−1.38	5.72
Sanliqiaogang	102	52.0	9.0	0.79	263	−19.2	−5.25	−1.19	−0.40
Dongtiaoxi	119	400.0	71.3	7.33	246	−122.0	−45.39	−9.65	−2.32
Changxinggang	301	61.0	7.0	1.82	64	−10.4	−2.21	−0.12	1.70
Gulougang	343	52.0	3.9	1.14	18	−1.9	−0.73	−0.01	1.13
Chendonggang	338	147.0	34.2	9.98	26	−38.7	−10.07	−0.23	9.76
Caoqiaohe[1]	347	72.9	18.5	5.56	18	−14.8	−5.47	−0.09	5.47
Yincungang									
Wujinggang	339	49.8	6.6	1.95	26	−7.1	−2.89	−0.06	1.88
Zhihugang	244	78.1	8.3	1.75	27	−14.1	−5.56	−0.13	1.62
Liangxi	174	60.7	11.8	1.78	60	−23.3	−9.58	−0.50	1.28
Wuli Bay	6	71.1	47.3	0.25	0	0.0	0.00	0.00	0.25
Daxigang[1]	230	53.1	11.4	2.27	135	−35.8	−10.66	−1.24	1.03
Yuantangqiaogang									
Wangyu	116	240.0	82.8	8.30	249	−177.0	−44.10	−9.49	−1.19
Xinyunhe	32	33.1	15.4	0.43	333	−183.0	−31.47	−9.05	−8.63
Dapugang	14	4.2	2.7	0.03	348	−14.2	−5.78	−1.74	−1.71
Xigang	7	16.5	8.7	0.05	358	−44.3	−18.72	−5.79	−5.74
Taipu	8	73.2	27.5	0.19	357	−296.0	−119.47	−36.85	−36.66

[1]If the number of days of inflow and outflow total less than 365, the difference represents number of days of no-flow.

recorded was 1,593 mm (in 1999), and the lowest was 680.5 mm (in 1978); the ratio of the highest to the lowest precipitation is 2.34.

Because the basin is affected by the monsoon, both monthly and yearly precipitation is subject to strong variation. When comparing precipitation of periods of 4 consecutive months, the highest amount occurs from June to September, some 45–55% of the annual total (Table 1.6); the second highest amount occurs from May to August. June and July tend to get the most monthly precipitation and January or February the least. Flood season comes earlier in the south of the basin than in the north, and earlier in the mountainous regions than in the plains.

The coefficient of variation of annual precipitation, C_v, is an indicator of precipitation perennial change: the greater the value of C_v, the more variable the precipitation. Normally C_v fluctuates between 0.17 and 0.3 in the basin; in the northwest to Danyang, C_v is higher (0.26–0.30), and in Hangzhou-Jiaxing-Huzhou district and Shanghai, it is lower (0.18–0.20). Precipitation was 1,130 mm in a median frequency (50% frequency of reoccurrence), 1,030 mm in a low frequency (25% frequency of reoccurrence), and 1,280 mm in a high frequency (20% frequency of reoccurrence).

1.2.2.2 Runoff

Subtracted by surface and land evaporation and evapotranspiration, the remaining amount of precipitation is surface and subsurface runoff. Because both climate and the underlying surface vary across the Taihu basin, each region has a distinctly different yearly runoff. The highest yearly runoff, $3.98 \times 10^9\,m^3$, occurred in west Zhejiang Province, and the lowest was $5.35 \times 10^9\,m^3$ in the region around the lake. The mean yearly runoff depth, which depends greatly on precipitation and runoff generation, is highest in west Zhejiang Province (533 mm/a), and lowest in lake around areas (156 mm/a). Within the Lake Taihu basin, the highest yearly runoff depth occurs in the Tianmu mountainous region (800–1000 mm), followed by the Hangzhou-Jiaxing district. There is a relatively small yearly runoff (only about 300 mm) in the south of Jiangsu Province.

Most of the runoff (33%) occurs in summer (from June to August); the least (11%) occurs in winter (December–February). Runoff volume varies greatly from year to year, with a maximum difference (eightfold) recorded at the Huxisha River monitoring station.

Yearly runoff, and the precipitation distribution within a year, are at maximum in the flood season and vary in different seasons. There is considerable regional variation in the 4 successive months when the maximum occurs; for example, in Zhejiang province, it is from April to July; in Jiangsu province, from May to August; and in Shanghai, it is from June to September. Runoff shows strong fluctuations, both within and between years.

1.2.2.3 Evaporation

There are numerous factors affecting lake evaporation rate, making it hard to directly measure the actual value and to calibrate the various measured or calculated results.

Table 1.6 Seasonal precipitation at six locations in the Lake Taihu basin

Rainfall gauge location	Province/ municipal region	No. of years	Spring (Mar.–May) (% of year)	Summer (Jun.–Aug,) (% of year)	Autumn (Sep.–Nov.) (% of year)	Winter (Dec.–Feb.) (% of year)	Annual pre-cipitation (mm)	Flooding season (Jun.-Sep.) (% of year)
Xujiahui	Shanghai	107	24.8	39.7	22.6	12.9	1143.1	51.7
Changzhou	Jiangsu	41	25.1	42.4	19.3	13.2	1058.4	52.3
Wuxi	Jiangsu	40	26.4	41.2	18.9	13.5	1066.3	51.0
Suzhou	Jiangsu	50	27.2	38.5	20.1	14.2	1057.0	49.9
Hangzhou	Zhejiang	44	26.7	35.7	22.3	15.3	1290.3	49.0
Jiaxing	Zhejiang	52	27.5	35.5	22.0	15.0	1155.4	48.8

By examination of many years of monitoring data from the hydrological experimental station in Yixing, the yearly surface evaporation of Lake Taihu varies from 800 to 1,000 mm, with a mean of 950 mm. The evaporation capacity in the western Zhejiang's mountainous region (i.e., west Zhejiang Province district, with fluctuations between 800 and 900 mm) is lower than that in the plain areas (rather constant at around 1,000 mm). The area along the lake has higher evaporation, as much as 1,100 mm per year.

Water surface evaporation variation within a year is unimodal. Evaporation in January and February is low, accounting for only 3.5–4.1% of that for the whole year; evaporation in June and August is the highest, accounting for 13.5–16.4% of that for the whole year (Table 1.7). For seasonal evaporation, winter has the lowest, just 11.3–12.1%, and summer the highest, about 39.0–42.1%, of the yearly amount. Spring and summer have about 22.6–26.4% of the year's total, respectively.

Table 1.7 Monthly distribution of surface evaporation at three locations in the Lake Taihu basin (in percentage)

Month	1	2	3	4	5	6	7	8	9	10	11	12
Yixing	3.6	3.5	6.5	6.8	9.3	11.4	13.5	14.5	11.4	8.6	6.4	4.5
Shanghai	3.8	4.1	6.3	8.3	10.1	11.6	13.5	13.9	10.0	8.4	5.8	4.2
Hangzhou	3.5	3.8	5.8	8.2	10.0	10.7	16.4	15.0	9.5	7.7	5.4	4.0

Land evaporation includes water surface (except for lake surface) evaporation, soil evaporation, and evapotranspiration. The land annual evaporation ranges from 600 to 800 mm, which is about 70–80% of that of the Lake Taihu surface. The general trend is that the south has higher evaporation rate than the north, and mountainous areas have a higher rate than do the plains. For the entire basin, the gross evaporation can be computed in terms of the evaporation from land and water surface. During the period 1956–1979, the gross evaporation in Lake Taihu basin was $2.77 \times 10^{10}\,m^3/a$, an amount equivalent to 63% of the normal precipitation in the same period.

1.2.2.4 Storage Capacity of the Lake

Lake storage capacity is the volume of water storage, an important component in the lake's water resources. The actual measured value of Lake Taihu at lowest water level was 1.78 m a.s.l. (in 1934), and the highest water level was 5.08 m a.s.l. (in 1999). The normal water level is 3.07 m (derived from the years 1954–2001, in Xishan Island station). During the flood of 1999, Lake Taihu had its highest water level, 5.08 m with a gross storage of $9.3 \times 10^9\,m^3$. During the flood in 1991, its highest water level was 4.78 m and its gross storage was $8.6 \times 10^9\,m^3$. Table 1.8 shows the water level and water capacity volume.

Table 1.8 Water level, area, and capacity of Lake Taihu in 1984

Water level	0.39	0.5	1.0	1.5	2.0	2.5	3.0	3.5	4.0	4.5	5.0
Area (km^2)	0	197.3	1,166.6	1,886.0	2,126.0	2,233.1	2338.1	2338.1	2338.1	2338.1	2338.1
Volume ($\times 10^8\ m^3$)	0	0.1	3.3	11.0	21.9	33.2	44.3	56.0	67.7	79.4	91.1

1.2.2.5 Water Balance of Lake Taihu

There are many rivers and canals around Lake Taihu; those in the upper reaches flow into the lake, while those in the low reaches flow from it. The field survey conducted in 2001 revealed 125 watercourses around the lake, 33 of which flowed into it. Most of the river network had become blocked, and only 13 watercourses discharged efficiently. Furthermore, because the terrain around the lake is lower and flat, rivers run in all directions and counterflow (reverse flow) happens rather frequently, especially in dry years.

In view of the aforementioned situation, conventional hydrometry is unreliable for calculating the incoming and outgoing flow of the lake. Since 1956, Jiangsu Province and Zhejiang Province have made collaborative efforts to monitor the hydrology of Lake Taihu. Because there are numerous watercourses running in two directions, it was necessary to choose a limited number of locations for fixed-point observation and undertake tour gauging around the lake from one watercourse to another. This survey was performed from the bridges of the highway around Lake Taihu. The area between the lake and the highway was included; the area of this land was about $2,264\,km^2$, the outflow of which, W, should be added when calculating the lake's inflow V.

Lake Taihu water balance formula:

$$P + V = E + Q + q \pm \Delta V \tag{1.8}$$

$$V = V' \pm W \tag{1.9}$$

In the formula, P represents lake surface precipitation of a given period, V and Q are the lake's incoming and outgoing runoffs, respectively, of that period, E is lake surface evaporation, q is industrial and agricultural water consumption of that period, ΔV is lake basin's pondage variation of that period, V' is the gauged runoff, and W is the runoff of the area between the lake and the highway (subsurface runoff is excluded). The data for the Lake Taihu water balance in different years are given in Table 1.9.

In recent decades, the Lake Taihu water regime has changed as sluice gates have been built to control the outflowing watercourses. Presently, Lake Taihu has the characteristics of a reservoir as the water balance can be brought under human control. The characteristics of the new water balance of Lake Taihu have been analyzed and specified as follows:

Table 1.9 Water balance of Lake Taihu during 1966–1988

Year	Hydrological frequency (%)	Input (10^8 m^3)				Output (10^8 m^3)				
		Inflow river runoff	Precipitation	Change of pondage	Total	Outflow river runoff	Evaporation	Change of pondage	Total	Water used
1969	50	61.51	25.17	5.04	91.72	67.04	22.87		89.91	0.14
1977	10	75.32	33.91		109.2	87.84	20.97	5.13	113.94	0.34
1978	95	30.55	15.06	7.88	53.49	30.67	24.16		54.83	0.46
1987	17	67.18	32.56		99.74	74.94	19.95	4.35	99.24	3.6
1988	70	42.47	24.26	6.14	72.87	40.90	22.71		63.61	4.0
1966–1988					52.74				52.29	

(1) Multiyear averaged annual inflow and outflow of Lake Taihu is 5.25×10^9 m^3, and its gross storage is 4.43×10^9 m^3. Lake Taihu water exchange time (the time for all lake water to be replaced) is 300 days, which is much longer than that of the other freshwater lakes in China; this means that Lake Taihu has a slower water exchange and that this will impede its self-purification and accelerate eutrophication.

(2) The Lake Taihu water recharge coefficient is 7.0, which is the ratio of the basin's catchment area to lake area. Poyang Lake (the largest freshwater lake in China) and Dongting Lake (the second largest freshwater lake in China) also belong to the Changjiang River basin, yet have markedly different water recharge properties. Poyang Lake's water recharge coefficient is 56 and Dongting's is 42. Furthermore, every water balance element of these two lakes differs from that of Lake Taihu. In Taihu, the incoming and outgoing flow is only 60% of total inflow and outflow; for Poyang and Dongting Lakes, the incoming and outgoing flow is 98% of the total inflow and outflow.

(3) Lake Taihu has approximately the same amount of inflow and outflow in flooding years as in normal years. However in 1978 and 1988, which were relatively dry years, water consumption was hard to estimate, and the flow ran in changing directions. So, survey monitoring alone resulted in much deviation in measuring incoming and outgoing flows. There tends to be more evaporation in dry years than in flood years. In 1987, the lake's evaporation was higher than its surface precipitation. The amount of incoming flow in dry years is only about 40% of that in wet years.

1.3 Pollution and Eutrophication

Wenyu Huang, Gang Xu, Qinglong Wu, and Boqiang Qin

1.3.1 Development of the Water Quality in Lake Taihu and Its Rivers

In 1960, water quality in Lake Taihu was good; total inorganic nitrogen concentration (TIN) was only 0.05 mg/L, phosphorus (PO_4^{3-}-P) was 0.02 mg/L, and the organic pollution index, COD_{Mn}, was 0.9 mg/L [Nanjing Institute of Geography and Limnology, Chinese Academy of Sciences (NIGLAS), 1965]. However, by 1981 the water quality had deteriorated, TIN had increased to 0.894 mg/L (nearly 18 times higher than in 1960), and COD_{Mn} increased to 2.83 mg/L; phosphorus had not markedly increased. By 1988, TIN and total nitrogen (TN) were 1.115 mg/L and 1.84 mg/L, respectively (Sun & Huang, 1993); in 1998, levels rose to 1.582 mg/L and 2.34 mg/L, respectively (according to the investigation of the Taihu Laboratory for Lake Ecosystem Research, Chinese Academy of Sciences). Total phosphorus (TP) and COD_{Mn} also increased markedly from 0.032 mg/L and 3.30 mg/L in 1988 to 0.085 mg/L and 5.03 mg/L in 1998, increases of 2.66 and 1.53 fold, respectively (Table 1.10).

In 2000, the mean concentration of total P in Lake Taihu was 0.1 mg/L, and that in Meiliang Bay and Wuli Bay was also high, 0.26 and 0.2 mg/L, respectively. The mean concentration of TN was 2.3 mg/L, and in Meiliang and Wuli Bays the values

Table 1.10 Changes in water quality indices of Lake Taihu from 1960 to 1999 (mg/L)[1,2,3,4]

Year	TIN	TN	PO_4^{3-}-P	TP	COD_{Mn}
1960[1]	0.05		0.02		1.90
1981[2]	0.894	0.9	0.014		2.83
1988[2]	1.115	1.84	0.012	0.032	3.3
1994[3]	1.135	2.05	0.010	0.086	5.77
1995[3]	1.157	3.14	0.011	0.111	5.53
1998[4]	1.582	2.34	0.007	0.085	5.03
1999[4]	1.79	2.57	0.004	0.105	4.99

[1] Nanjing Institute of Geography, Chinese Academy of Sciences, 1965.

[2] Sun & Huang, 1993.

[3] The editorial group of "the '9th 5-year-period' and the year of 2010 plan of water pollution protection for Lake Taihu," 1998.

[4] The mean of 14 samples of Taihu Laboratory for Lake Ecosystem Research, Chinese Academy of Sciences.

Huang, W.Y.
State Key Laboratory of Lake Science and Environment, Nanjing Institute of Geography and Limnology, Chinese Academy of Sciences, 73 East Beijing Road, Nanjing 210008, P. R. China
e-mail: wyhuang@niglas.ac.cn

Table 1.11 Water quality in five regions of Lake Taihu in 2000

Region	COD_{Mn} (mg/L)	TP (mg/L)	TN (mg/L)	TLI_c[1]
Wulihu Bay	8.1	0.2	6.6	64.8
Meiliang Bay	7.8	0.26	5.6	55.9
Western littoral zone	7.0	0.17	3.1	55.7
Lake center	5.5	0.10	1.6	55.9
Eastern littoral zone	5.9	0.11	2 .4	54.5

[1]TLIc, an index of weighted sum based on concentration of Chl a, TN, TP, and transparency.

were 5.6 and 6.6 mg/L, respectively. The highest biomass of algae in 2000 was 2.1×10^8 cell/L, much higher than that in 1999. The water quality in the different regions in 2000 is shown in Table 1.11.

According to the current national environmental quality standard of surface water (GB3838–2002), Lake Taihu was a class I–II water body in the 1960s, and then severely deteriorated, becoming class II in the 1970s, class II–III at the beginning of the 1980s, class III in the late 1980s, and class IV in the middle 1990s (with one-third of the lake area being class V). By the year 2000, of the total surface area of the lake, class III water accounted for 6.7% (156.7 km^2), class IV water accounted for 85% (1, 989.7 km^2), and class V and worse accounted for 8.2% (191.6 km^2). Some 83.5% of Lake Taihu had a high level of nutrients, and only 16.5% had a moderate level of nutrients (east Lake Taihu and the lake centre). These data demonstrate that, every 10 years, the water quality in Lake Taihu dropped by one grade (State Environmental Protection Bureau, 2000). Furthermore, after 1990, the decline markedly accelerated.

Before the 1980s, deteriorating water quality, mainly evident as increases in total nitrogen and COD_{Mn}, was closely related to the development of agricultural production. The 1980s were a turning point for the environment of Lake Taihu, and although increases in total nitrogen diminished, the levels of phosphorus and chlorophyll *a* (Chl a) increased. During the 1980s and 1990s, deterioration of the water quality was mainly evident as increases of phosphorus and chlorophyll, resulting from urbanization, secondary and tertiary industry, and the consequent increase in living standards in the basin.

In the beginning of the 1970s, blooms of blue-green algae first appeared in Wuli Bay of Wuxi, and subsequently their scale and frequency constantly increased. In the mid- to late 1980s, algal blooms occurred two to three times every year, and expanded to Meiliang Bay. In the middle and late 1990s, algal blooms occurred four to five times every year and gradually expanded to most of the lake. In 2000, the center of Lake Taihu also suffered a bloom of blue-green algae. Because Meiliang Bay area is the main water source for Wuxi City, the algal bloom seriously affected the water supply, and directly influenced productivity and people's lives, leading to great economic losses. At present, even though the trend of deterioration of the water environment has been contained to some extent, the ecosystem is still degrading (Table 1.12).

Since the 1980s, the river inflow to Lake Taihu has been increasingly polluted, and this situation continues to date. In 1998, water quality monitoring results in

Table 1.12 Comparison of the main pollutant parameters in Meiliang Bay, Lake Taihu between December 1998 and December 2000[1]

Year	TIN	TN	PO_4^{3-}-P	TP	COD_{Mn}	Chl a
1998	1.180	1.892	0.002	0.0714	4.270	10.829
1999	1.241	2.654	0.002	0.260	5.267	11.097
2000	1.156	1.990	0.006	0.038	4.100	7.210

[1]Concentrations are given as mg/L, except for Chl a, which is given as μg/L; data are monthly observation results from the Taihu Laboratory for Lake Ecosystem Research, Chinese Academy of Sciences.

23 sections on the border of the lake loop in Jiangsu Province indicated that the sections with class V, IV, and III water quality accounted for 22%, 48%, and 30% of the total, respectively. Among the 23 river sections were 10 (some 43.5%) in which water quality did not reach surface water quality. There were five rivers in which water quality was worse than class V, namely, the Liangxi and Zhihugang in Wuxi City, Wujingang and Taige Canal in Wujing County, Changzhou, and the Yuecheng in Suzhou City.

The inflow, over a total length of 1,598 km, was inspected by the Taihu Basin Authority (TBA) of the Ministry of Water Conservancy in December 2000, and the water quality was appraised according to national surface water environmental quality standard (GB3838–2002). Some 80% of sites had water quality of class IV or worse, and 53% of the sites were class V or worse; only 19.5% of sites reached class III or better. Within the 20 provincial boundary river sections monitored, 85% of sections had water quality of class IV or worse. Within the 35 sections monitored along the river mouth that enters the lake, 80% had water quality of class IV or worse (Table 1.13). In July 2000, the Nanjing Institute of Geography and Limnology, Chinese Academy of Sciences, investigated the Dapu River in Yixing and determined that levels of the main pollutants TN, TP, and COD_{Mn} were at the highest levels ever recorded, exceeding 4.5 mg/L, 0.2 mg/L, and 7.5 mg/L, respectively. These data demonstrate that the Lake Taihu river water pollution has not markedly improved, despite the State Council's "zero point action" (the so-called zero point action is a movement aiming at industrial pollution control at the end of 1998, which was organized by the State Council) of 1998.

1.3.2 Pollutant Emissions and Amounts Flowing into the Lake

Studying the budget of the inflow and outflow of nitrogen, phosphorus, and the chemical oxygen demand (COD) of the lake help determine the main nutrient pathways and pollutant sources. The data presented here were collected in 1998, and concern the whole Lake Taihu basin, including 33 cities such as Shanghai, Hangzhou, Suzhou, Wuxi, Changzhou, Jiaxing, Huzhou; 17 counties and cities are in Jiangsu, 13 counties and cities are in Zhejiang, and the remaining 3 in Shanghai.

Table 1.13 Water quality assessment for categorization of rivers in Lake Taihu basin in 2000

Time	II		III		IV		V		> V	
	River length (km)	Percentage (%)	River length (km)	Percentage (%)	River length (km)	Percentage (%)	River length (km)	Percentage (%)	River length (km)	Percentage (%)
Dry period	4	0.2	329	20.6	363	22.7	505	31.6	397	24.9
Wet period	21	1.3	260	16.3	513	32.1	465	29.1	339	21.2
All year	12	0.7	299	18.7	434	27.2	484	30.3	369	23.1

1.3.2.1 Pollutant Emission and the Amounts Entering Rivers in Lake Taihu Basin

(1) Amount of wastewater entering the lake from major industrial pollution sources:

Based on an investigation by the environmental protection administration, the amount of wastewater pollutants from important industrial sources entering the Lake Taihu basin in the dominant Jiangsu Province in 1998 were COD_{Cr}, 93,822.1 t/a, TN, 2,686.0 t/a, and TP, 191.4 t/a. When all three provinces affecting the basin were considered, the emissions into the whole basin were COD_{Cr}, 160,478.2 t/a, TN, 4,041.8 t/a, and TP 264.74 t/a (Table 1.14).

(2) Domestic pollution sources

ⓐ Amount of pollutant emission

The domestic pollutant discharge coefficients were COD_{cr} 21.9 kg per·capita for the urban population, TP 0.62 kg per·capita, and TN 3.65 kg per·capita; and COD_{Cr} 21.9 kg per·capita, TP 0.55 kg per·capita, and TN 3.65 kg per·capita for the rural population. In the study district, the urban and rural populations were 7.7083 and 17.1854 million, respectively, in 1998. Thus, the amount of pollution produced by the urban population was COD_{Cr} 168,811.8 t/a, TP 4,779.15 t/a, and TN 28,135.3 t/a, respectively; and by the rural population, COD_{Cr} 376,360.3 t/a, TP 9,451.97 t/a, and TN 62,726.7 t/a, respectively. In total, the quantities of domestic pollutants produced were COD_{Cr}, 54 5,172.1 t/a, TP, 14,231.12 t/a, and TN, 90,862.0 t/a.

ⓑ Amount of pollution entering the lake

Because most urban sewage enters a water body through primary treatment tanks, there is deposition and loss of solid matter, estimated at 90%. Pollutants from rural residents were dispersed, and the percentage loss of both COD and TN was estimated as 12.1% while that of TP was 33.6%. Thus, the pollutants from domestic sewage that entered the lake were CODCr 197,470.2 t/a, TP 7,477.10 t/a, and TN 32,911.7 t/a.

(3) Amount of pollution from agriculture production

According to the Nanjing Institute of Environmental Science, State Bureau of Environmental Protection, farmland (rice paddy and dry land) yielded the pollution

Table 1.14 The regional distribution of industrial wastewater pollutants from major sources into the Lake Taihu basin in 1998

Province	City	COD_{Cr} (t/a)	TN (t/a)	TP (t/a)
Jiangsu	Suzhou	46, 942.3	1, 051.0	82.58
	Wuxi	16, 259.2	1, 193.2	77.71
	Changzhou	26, 460.0	348.4	28.10
	Zhenjiang	4, 160.6	93.4	3.01
	Total	93, 822.1	2, 686.0	191.40
Zhejiang	Hangzhou	14, 893.0	286.8	24.10
	Jixing	33, 846.3	499.1	23.11
	Huzhou	15, 881.7	250.9	17.32
	Total	64, 621.0	1, 052.5	64.53
Shanghai	Three counties	2, 035.1	287.6	8.81
Total		160, 478.2	4, 041.8	264.74

Table 1.15 Pollutant concentrations from different agricultural sources in the Lake Taihu basin (1998)

Agricultural source	COD_{Cr} (mg/L)	TN (mg/L)	TP (mg/L)
Paddy fields	62.82	13.45	1.030
Dry fields	30.30	4.30	0.695
Runoff from land	58.12	5.90	0.850
Seepage from rice fields	16.66	3.26	0.204

loads shown in Table 1.15, and from this we derive the resulting runoff pollution shown in Table 1.16. The total pollutants from farmland passed to Lake Taihu were COD_{Cr}, 272,805.7 t/a, total nitrogen, 53,970.0 t/a, and total phosphorus, 3,990.41 t/a. Drainage and seepage were the primary and secondary ways by which these pollutants left the paddy fields.

(4) Soil erosion

According to calculations by the Jiangsu Provincial water and soil conservation office, the areas in the basin with slight soil erosion reached about 1,292.86 km^2, those with a moderate degree of soil erosion 206.96 km^2, and those with a severe degree of soil erosion 40.13 km^2. The annual loss of soil was 0.4–2.0 mm in the "slight area," 2.0–4.0 mm in the "moderate area," and 4.0–6.0 mm in the "severe area." Loss of soil was about 2.97 million t/a, and approximately 0.4 million t silt entered the lake. In the upstream regions of the Taihu basin, the mean concentration of total phosphorus in soils was 0.032–0.048% and total nitrogen was 0.18–0.23%. Thus, through soil erosion, the silt entering the lake basin carried about COD_{Cr} 8,585 t/a, 800 t/a total nitrogen, and total phosphorus 192 t/a.

(5) Amount of pollution from livestock and poultry production

The Nanjing Institute of Environmental Science estimated total pollution loads originating from livestock and poultry in 1998 (Table 1.17), and from this, the percentage and load flowing into the Taihu basin were calculated (Tables 1.18, 1.19). The pollutant load that flowed into the Lake Taihu basin from livestock and poultry production was calculated as COD_{Cr} 37,850 t/a, total nitrogen 6,710 t/a, and total phosphorus 2,820.00 t/a. The COD_{Cr} derived primarily from poultry excrement and secondarily from pig excrement; total nitrogen derived primarily from poultry excrement, secondarily from pig urine; and total phosphorus derived primarily from poultry and secondarily from sheep excrement.

Table 1.16 Runoff pollutant load of agricultural non-point sources in Lake Taihu basin (1998)

Agricultural source	COD_{Cr} (t/a)	TN (t/a)	TP (t/a)
Runoff from land	28168.9	2859.6	410.53
Dry fields	28803.6	4143.5	664.89
Paddy fields	132696.1	30698.8	1896.98
Seepage from rice fields	83137.1	16268.1	1018.01
Total	272805.7	53970.0	3990.41

Table 1.17 Total pollution load originating from livestock and poultry in 1998 (10^4 t/a)

Item	Cattle excrement	Cattle urine	Hog excrement	Hog urine	Poultry excrement and urine	Sheep excrement and urine	Total
COD_{Cr}	1.564	0.151	27.121	7.745	19.936	1.237	57.754
TN	0.220	0.202	2.806	2.840	4.545	2.003	12.616
TP	0.060	0.010	1.778	0.447	2.507	0.694	5.496

Table 1.18 Percentage of pollution from livestock and poultry production flowing into the Taihu basin in 1998 (in %)

Item	Cattle excrement	Cattle urine	Hog excrement	Hog urine	Poultry excrement and urine	Sheep excrement and urine
COD_{Cr}	6.16	5.00	5.58	5.00	8.59	5.50
TN	5.68	5.00	5.34	5.00	8.47	5.30
TP	5.50	5.00	5.25	5.00	8.42	5.20

Table 1.19 Pollution load from livestock and poultry production flowing into the Taihu basin in 1998 (10^4 t/a)

Item	Cattle excrement	Cattle urine	Hog excrement	Hog urine	Poultry excrement and urine	Sheep excrement and urine	Total
COD_{Cr}	0.096	0.008	1.513	0.387	1.713	0.068	3.785
TN	0.013	0.010	0.015	0.142	0.385	0.106	0.671
TP	0.003	0.001	0.009	0.022	0.211	0.036	0.282

(6) Amount of pollution from aquaculture

The area with production fishponds in the basin was estimated at $53,704\,hm^2$ for 1998. Taking the pollutant discharge coefficients/hm^2 as COD_{Cr} 745 t/a, TN 101 t/a, and TP 11 t/a, the pollutant discharge from fishponds was COD_{Cr} 40,009.9 t/a, total nitrogen 5,424.2 t/a, and total phosphorus 590.72 t/a.

In addition, the aquaculture area in Lake Taihu itself was about $3,300\,hm^2$ in 1998, 95% of which was for the cultivation of river crab, and the remaining 5% for fish culture. About 8,500 t food for these species is put into the lake every year, equal to about 782.0 t/a COD_{Cr}, 279.7 t/a total nitrogen, and 44.2 t/a total phosphorus.

The total pollutant discharge from aquaculture in the Taihu basin and lake are, therefore, COD_{Cr} 40,791.9 t/a, total nitrogen 5,703.9 t/a, and total phosphorus 634.92 t/a.

(7) Amount of pollution from lakeside tourism

According to statistics for the main hotels and domestic pollution sources around the lake in Wuxi and Suzhou City, about 20,000 people visit Lake Taihu every day. The pollutant discharge coefficients, including those of the night soil, are COD_{Cr} 300.4 t/a, total nitrogen 162.2 t/a, and total phosphorus 11.20 t/a. If it is assumed that 50% of the pollution from tourism enters the water body, the discharge loads into the lake are 150.2 t/a COD_{Cr}, 81.1 t/a total nitrogen, and 5.60 t/a total phosphorus.

(8) Atmospheric pollution from rainfall and dust on the lake surface

According to calculations by the Nanjing Institute of Environmental Science, State Bureau of Environmental Protection, COD_{Cr} entering the lake from rainfall amounts to 23,595.0 t/a, total nitrogen is 2,759.5 t/a, and total phosphorus is 60.10 t/a. Dust on the lake surface adds a further 420.9 t/a total nitrogen and 33.0 t/a total phosphorus.

(9) Shipping pollution

Nearly 2,200 fishing boats work in Lake Taihu. If it is assumed that the fishermen live on their boats for half a year, the sanitary sewage, excrement, and urine discharged into the lake amounts to COD_{Cr} of 98.2 t/a, total nitrogen 38.9 t/a, and total phosphorus 3.26 t/a. In addition, shipping in Lake Taihu and its rivers involve some 87,000 boats per year. Assuming two persons as crew per ship, their pollution discharge amounts to COD_{Cr} 3,981.9 t/a, total nitrogen 1,577.1 t/a, and total phosphorus 128.8 t/a. Thus, the total pollutants entering the water body from fishing and shipping are COD_{Cr} 4,080.1 t/a, total nitrogen 1,616.0 t/a, and total phosphorus 132.1 t/a.

(10) Total pollution

The foregoing calculations show that, in 1998, the total pollutants entering the Lake Taihu basin were COD_{Cr} 745,609.9 t/a, total nitrogen 108,937.1 t/a, and total phosphorus 15,609.89 t/a (Table 1.20). The principal pollution sources varied for each of these three main pollutants. For COD_{Cr}, the main pollution source was agriculture (36.59%), followed by sewage (26.47%) and industrial waste (21.51%). For total nitrogen, the main source was agriculture (49.54%), followed by domestic sewage (30.22%) and livestock and poultry production (6.16%). For total phosphorus, the main source was domestic sewage (47.89%), followed by agriculture (25.56%) and livestock and poultry production (18.06%).

1.3.2.2 Estimates of Pollutants Entering Lake Taihu from the Area of Jiangsu Province

Some nutrients enter Lake Taihu via rivers that flow into the lake, whereas other nutrients, such as those from soil erosion, aquaculture, tourism, rainfall, dust precipitation, and shipping, enter the lake directly. The pollutants entering the lake via the main rivers of the dominant Jiangsu Province in 1998 were COD_{Cr} 117,646.3 t/a, total nitrogen 19,241.4 t/a, and total phosphorus 961.2 t/a, and via other rivers were COD_{Cr} 53,404.7 t/a, total nitrogen 6,099.7 t/a, and total phosphorus 355.66 t/a. Thus, the total pollutants entering Lake Taihu by the rivers were COD_{Cr} 171,051.0 t/a, total nitrogen 25,341.1 t/a, and total phosphorus 1,316.86 t/a.

The pollutants transported by surface runoff were, according to the Nanjing Institute of Environmental Science, as follows: from lakeside vegetated areas, COD_{Cr} 1,521.4 t/a, total nitrogen 231.5 t/a, and total phosphorus 17.8 t/a; from lakeside urban areas (towns and villages) COD_{Cr} 1,287.4 t/a, total nitrogen 26.3 t/a, and total phosphorus 2.84 t/a; and from the lakeside highway COD_{Cr} 38,215.7 t/a, total nitrogen 7,693.0 t/a, and total phosphorus 1,072.20 t/a.

Table 1.20 Sources of pollutant emission in Lake Taihu basin in 1998

Sources of pollutants	COD_{Cr}		TN		TP	
	Emission (t/a)	Percentage (%)	Emission (t/a)	Percentage (%)	Emission (t/a)	Percentage (%)
Industry	160,478.2	21.51	4,041.8	3.71	264.74	1.70
Sewage	197,470.2	26.47	32,911.7	30.22	7,477.10	47.89
Agriculture	272,805.7	36.59	53,970.0	49.54	3,990.41	25.56
Soil erosion	8,585.0	1.15	800.0	0.73	192.00	1.22
Poultry production	37,850.0	5.08	6,710.0	6.16	2,820.00	18.06
Aquaculture	40,791.9	5.47	5,703.9	5.24	634.92	4.07
Tourism	150.2	0.02	81.1	0.07	5.60	0.04
Precipitation	23,595.0	3.16	2,759.5	2.53	60.10	0.39
Dust fall	–	–	420.9	0.39	33.00	0.21
Shipping	4,080.1	0.05	1,616	1.41	132.1	0.86
Total	745,806.3	100.0	109,014.9	100.0	15,610.0	100.00

The actual loads and percentages of pollutants entering Lake Taihu from each of ten different sources in Jiangsu Province are shown in Table 1.21. The dominant source of COD_{Cr} was rivers (71.13% of the total), followed by the lakeside highway (15.89%), and precipitation (9.82%). A similar pattern pertained for total nitrogen, which was mainly brought in by rivers (64.69% of total), followed by the lakeside highway (19.64%) and precipitation (7.04%). Although total phosphorus was also mainly brought in from rivers (45.96% of total) and the lakeside highway (37.15%), an additional source, namely soil erosion, was also important (6.65%).

The pollutant load entering the lake from each of the 13 main rivers in Jiangsu Province is shown in Table 1.22. The COD_{Cr} was mainly received from the Taige Canal and the Caoqiao, Tiaoxi (Xitiaoxi and Dongtiaoxi), and Zhihugang Rivers (each of which contributed approximately 13% of the total). Total nitrogen was mainly received through the Zhihugang River (23.64% of the total), the Taige Canal, and the Wujingang, Caoqiao, and Liangxi Rivers (each of which contributed approximately 8% of the total). Total phosphorus was mainly received through the Wujingang and Zhihugang Rivers (21.15% and 15.07% of the total, respectively), followed by the Shedugang and Liangxi Rivers (9.76% and 7.46%, respectively).

Among the pollutants entering the lake during 1987–1988, the amount of COD_{Cr} was 145,419.8 t/a (COD_{Cr} was converted to COD_{Mn} by 2.98 times), TN was 28,106.0 t/a, and TP was 1,988.53 t/a (Huang, 2001). Compared with the amount in 1998, the pollutants of COD_{Cr}, TN, and TP entering the lake increased by 95,067 t/a, 11,065 t/a, and 897.8 t/a, respectively in 10 years; that is, COD_{Cr}, TN, and TP increased 65.4%, 39.4%, and 45.1%, respectively. The rapid increases obviously worsened water quality and accelerated eutrophication of the lake. High demands on the management of the environment of Lake Taihu, especially in pollution control, would be required.

Analysis of pollutant composition of COD, TN, and TP from different sources in 1994 and 1998 will be able to show the changes in contributions from various pollutant sources. According to the sources of pollutants in 1994 (State Environmental Protection Bureau, 2000), for COD_{Cr}, industrial effluent accounted for 39%, urban domestic sewage for 42%, and agriculture and rural resident sewage for 10%; for total nitrogen, industrial effluents accounted for 16%, agricultural and breeding discharges for 10%, urban domestic sewage for 38%, and agriculture for 42%; and for total phosphorus, industrial effluents accounted for 10%, turban domestic sewage for 15%, and agricultural sewage for 69%. In 1998, the COD_{Cr} discharged within the basin was partitioned as follows: industry, domestic, and agriculture, 21.5%, 26.5%, and 42%, respectively; for TN, the contributions were industry, domestic, and agriculture 3.7%, 30.2%, and 55%, respectively; and for TP, the contribution rates were 1.7%, 48%, and 30%, respectively (Table 1.23).

The proportion of pollutant sources in 1994 and 1998, as documented by the State Environmental Protection Bureau in 2000, are compared in Table 1.23. The dominant source of each of the three main pollutants changed during this 4-year period: both COD and TN mainly originated from domestic sewage in 1994 and from agricultural non-point sources in 1998; for TP, the reverse pertained. Thus,

Table 1.21 Loads of pollutants entering Lake Taihu from different sources in the catchment of Jiangsu Province in 1998

Source of pollutant	COD_{Cr}		TN		TP	
	Total (t/a)	Percentage (%)	Total (t/a)	Percentage (%)	Total (t/a)	Percentage (%)
River	171,051.0	71.13	25,341.1	64.69	1,326.56	45.96
Aquaculture	782.0	0.34	279.7	0.71	44.20	1.53
Tourism	150.2	0.06	81.1	0.21	5.60	0.19
Soil erosion	8,585.0	3.58	800.0	2.04	192.00	6.65
Precipitation	23,595.0	9.82	2,759.5	7.04	60.10	2.08
Dust fall	–	–	420.9	1.08	33.00	1.14
Shipping	3,883.7	1.61	1,538.2	3.93	132.02	4.58
Discharge from forest	1,521.4	0.63	231.5	0.59	17.81	0.62
Discharge from urban area	1,287.4	0.54	26.3	0.07	2.84	0.10
Uncontrolled beach area	38,215.7	15.89	7,693.0	19.64	1,072.20	37.15
Total	240,486.4	100.0	3,9171.3	100.0	2,886.33	100.0

Table 1.22 Loads of pollutants entering Lake Taihu from the 13 main rivers in Jiangsu Province 1998

River	COD_{Cr}		TN		TP	
	Input (t/a)	Percentage (%)	Input (t/a)	Percentage (%)	Input (t/a)	Percentage (%)
Liangxi	8,941.1	5.23	2,112.3	8.34	99.0	7.46
Zhihugang	21,646.8	12.66	5,991.3	23.64	280.6	21.15
Wujinggang	9,867.6	5.77	2,176.3	8.59	199.9	15.07
Caoqiao	22,754.0	13.30	2,162.8	8.53	61.1	4.61
Taige Canal[1]	22,875.4	13.38	2,221.2	8.77	88.6	6.68
Shedugang	2,459.8	1.44	311.6	1.23	129.5	9.76
Dapugang	8,592.8	5.02	916.0	3.61	22.0	1.66
Wuxigang	2,861.9	1.67	469.4	1.85	8.3	0.63
Other rivers in Jiangsu	17,646.9	10.32	2,880.5	11.37	144.2	10.87
Changxinggang	11,570.0	6.76	1,715.3	6.77	77.46	5.84
Xitiaoxi[2]	11,348.5	6.63	981.1	3.87	46.28	3.49
Dongtiaoxi[2]	10,756.1	6.29	1,345.1	5.31	66.37	5.00
East to Lake	19,730.1	11.53	2,058.1	8.12	168.23	12.68
Total	171,051.0	100.0	25,341.1	100.0	1,326.56	100.0

[1]Connecting Lake Taihu and Lake Ge.

[2]The Xitiaoxi and the Dongtiaoxi Rivers together are called the Tiaoxi River.

Table 1.23 Proportion of pollution sources to Lake Taihu in 1994 and 1998

Year	Pollution source	COD_{Mn}		TN		TP	
		Output (t)	Percentage (%)	Output (t)	Percentage (%)	Output (t)	Percentage (%)
1994	Industrial sewage	11,106	39	12,544	16	591	10
	Domestic sewage	119,029	42	19,948	38	3,394	15
	Agricultural non-point	28,138	10	29,842	25	852	60
	Aquaculture			13,195	17	533	9
1998	Industrial sewage	160,478	21.5	4,042	3.7	264.7	1.7
	Domestic sewage	197,470	26.5	32,912	30.2	7,477	47.9
	Agricultural non-point	272,806	36.6	53,970	49.5	3,990	25.6
	Aquaculture	40,792	5.5	5,704	5.2	635	4.0

industrial pollution control could not yet create the desired and expected result for lake eutrophication control in Lake Taihu (Qin et al., 2002).

1.3.3 Causes of Worsening Water Quality

The Lake Taihu basin, extending across the provinces of Jiangsu, Zhejiang, Anhui, and the city of Shanghai, occupies only 0.4% of the area of the country, yet accounts for 14% of the country's gross industrial and agricultural output and 14% of revenue. The basin has become one of the most economically developed areas in China, and has the highest degree of urbanization, with the urban population of 0.76 million in 1980 growing rapidly to 1.12 million in 1999.

The rapid economic development in the basin has aggravated contradictions among the population, resources, the environment, and further economic development. The economic development, mainly driven by increasing resource input and workforce, has resulted in overconsumption of natural resources. The natural environment of Lake Taihu has already deteriorated sharply, and water pollution and eutrophication have become serious, as evidenced by indices such as total phosphorus, total nitrogen, biological oxygen demand $(BOD)_5$, CODMn, and volatile phenols that do not meet national standards, and water quality that has dropped by one grade in each period of 5–10 years (see Section 3.1, above). Because the serious pollution of the water in the basin now seriously restricts the sustainable development of the regional economy, it is necessary to discuss the causes of the water problems from the social and economic angle to redefine the targets for comprehensive restoration.

The main reasons for the continued deterioration of the environment of Lake Taihu are increased water use and discharge; changes in agricultural practices and in fisheries and aquaculture; insufficient wastewater treatment; and an unsuitable management system. Each of these aspects is now considered in detail.

(1) With the development of industry and urbanization, the water requirements and wastewater discharge have increased, and the amount of pollution entering the rivers and the lake has increased. Although the rapid development of the regional economy has facilitated increases in income and living standards, the per capita water requirements have increased very rapidly. For example, water consumption in Wuxi City has increased from just 105 L/d in 1980 to 175 L/d in 1990, and reached 284 L/d in 1999 (a nearly threefold increase in just 20 years). Meanwhile, the peak daily water delivery amount has increased by 11% annually. The total amount of domestic water use increased from 80.3 thousand t/d to 0.32 million t/d between 1980 and 1999. If the discharge coefficient is taken at 0.8, the daily domestic sewage discharged to the river system increased from 64 thousand t/d to 0.254 million t/d from 1980 to 1998. On average, nearly 100 t CODMn enters the network of waterways with sewage every day.

In parallel with development of the economy, water transport has shown unprecedented growth. There are now thousands of ships traveling on the canals and rivers in the Lake Taihu basin every day, discharging a large amount of domestic sewage and other pollutants directly into the water [as discussed in Section 3.2, part (9), above].

The rapid development of industry is also accompanied by increased demands for water. The water supplied from the urban water system for industrial consumption increased from 41.16 million t/a in 1980 to 122.46 million t/a in 1998 in Wuxi City. Although technological developments have improved the water recycling utilization ratio, the total use is increasing, as is the total amount of industrial sewage.

Industries vary in their water consumption, and in the volume and nature of their wastewater discharge. For example, in Wuxi City, the industry is dominated by textiles, metal smelting, chemical industry, food and tobacco processing, beverage manufacturing, papermaking, and machinery production (Table 1.24). These industries together used 73% of the total volume of freshwater (excluding that used in water processing and supply), and their CODMn discharges accounted for 86% of industrial wastewater for Wuxi City. Although some of these core industries have recently decreased their pollution output, total industrial pollution output is increasing very fast, along with increasing water demand. Because the output of pollutants is particularly high in certain industries, such as leather and petroleum refining, effective management of their water use and pollution discharge should have a great impact on water pollution control. As the core industries of Wuxi City, namely machinery, electrical equipment, and electronic manufacturing, they contribute less to wastewater discharge than do the textile and chemical industries. Electricity, gas, and water production and ferrous metal metallurgy produced fewer pollutants but they consumed great amounts of water in Wuxi City (Tables 1.24, 1.25).

(2) Changes in agricultural production and land use have had an increasingly important impact on the aquatic environment. Historically, the land in the Taihu area was mainly used for growing mulberry bushes (as food for silkworms) and rice, and for fish breeding. However, since the policy changes in the 1980s, there have been great changes in land use. Large amounts of former agricultural land are now used for industry; as city size has expanded, there is competition for land use, and the amount of cultivated land per capita has reduced to 0.045 ha. To compensate for the large loss of cultivated land, a guaranteed stable agricultural output (grain) has been achieved by adopting new techniques. For example, use of chemical fertilizers has increased rice and wheat output by 10.3% and 34.9%, respectively, over the decades. This practice increased the yield of grain per unit area in the Lake Taihu basin to 7,272 kg/hm^2 at the end of the 1990s, which is approximately 1.4 times the mean for the whole country.

Cultivated land has mainly been lost from the plain of the Taihu basin. Although there is less space for agricultural development, investment and economic growth have increased. Although agricultural land has an output value of only 260,000 yuan (US $32,500)/hectare, industrial land has an output value of about 3 million yuan (US $375,000)/hectare. To improve output, agricultural land use has been adjusted, with the proportion of the economic cropping area to all agricultural production areas changing from 7:3 in the early 1980s to 6:4 in the late 1990s. Some low-lying land and rice terraces have been excavated or modified for use as fish ponds for aquaculture.

Table 1.24 Water consumption and wastewater discharge of 20 industries in Wuxi city (1997) (data of 770 industrial enterprises)

Industry	Wastewater (10^4 t)	COD (10^4 t)	Water consumption (10^4 t)	Wastewater/10^4 yuan production (t/10^4 yuan)	COD discharge/10^4 yuan production (t/10^4 yuan)
Mining	30	144	23	24.8	0.012
Food, tobacco	1,923	5,220	3,421	86.5	0.024
Textile	4,645	25,782	15,927	37.3	0.021
Leather, pelage, down	110	1,135	123	350.7	0.15
Papermaking	1,826	7,790	2,820	350.7	0.15
Printing and media	0.87		1.93	2.9	
Oil and petrochemical	266	1,755	3,953	35.5	0.023
Chemical industry	4,457	12,946	20,160	65.7	0.019
Pharmacy	176	908	2,364	25.4	0.013
Chemical fiber manufacture	135	85	4,882	10.6	0.001
Rubber industry	116	170	427	12.1	0.002
Plastic industry	36	76	184	5	0.001
Non-metal manufacture	313	219	706	17.3	0.001
Ferrous metal metallurgy	4,499	2,219	8,116	10.9	0.001
Rare metal metallurgy	291	17	298	83.8	0.004
Metal product manufacture	494	409	875	39.5	
Mechanical and electrical manufacture	1,555	1,645	9,726	18.2	0.002
Electricity, gas, and water production	5,081	848	64,852	12.1	0.001
Other industries	297	919	666	416.2	0.007

Table 1.25 Correlation coefficients of output value and waste water discharge of different industries in Wuxi city (1996–1999)

Industry	Coefficient of wastewater production
Mining	0.0868
Food, tobacco	0.9081
Textile	0.7924
Leather, pelage and down	0.8915
Papermaking	0.9207
Printing and media production	−0.7758
Oil and petrochemistry	0.9200
Chemical industry	0.942
Pharmacy	0.7605
Chemical fiber manufacture	0.3139
Rubber industry	0.7324
Plastic industry	0.7211
Non-metal manufacture	0.7752
Ferrous metal metallurgy	0.9481
Rare metal metallurgy	0.7840
Metal product manufacture	0.7221
Mechanical and electrical manufacture	−0.1679
Electricity, gas and water production	0.9985
Other industries	0.7552

During the conversion of agricultural land to nonagricultural use, some rivers have been filled in while others became separated from the main river system, thus losing or changing their original ecological functions, influencing nutrient circulation between rivers, reducing absorption and dilution of pollutants, and indirectly causing deteriorating water quality.

Changes in agricultural technology have also influenced the aquatic environment. The main arable production in the Lake Taihu basin is two varieties of rice and wheat (or rice and rape). Because of the development of rural township enterprises, the comparative economic benefit from agriculture has been low, and labor productivity has largely turned to secondary and tertiary industries. In addition, some traditional techniques such as crop rotation and use of domestic organic fertilizer were gradually abandoned, which in turn indirectly affects the water quality of the lake. The traditional technique of using sediments from rivers, ponds, or lakes as a fertilizer no longer exists. This change has caused river and lake deposits to increase significantly, soil fertility has decreased, nutrient return and consumption are unbalanced, and chemical fertilizers have to be used, with their consumption increasing yearly. The proportion of organic to chemical fertilizer has changed from 3:7 in the mid-1980s to 1:9 in the mid-1990s. The amount of chemical fertilizer used (N, P, K) was 25 kg/ha in the 1980s and increased to 45 kg/ha in the 1990s. Nitrogen application now exceeds 30 kg/ha, accounting for more than 70% of chemical fertilizers. Because the utilization ratio of the chemical fertilizer is currently relatively low, fertilizer loss is serious. Similar problems also exist with extensive pesticide use. The increased use of chemical fertilizers and chemicals in agriculture has been accompanied by increased content of nitrogen, phosphorus, and potassium in the water bodies in or adjacent to farmland. Fertilizers and chemicals flow into rivers directly with surface runoff, thus causing not only nutrient loss, but also water pollution. The extensive, dispersive, seasonal, and flow characteristics of the chemical fertilizers and agricultural chemicals make them a significant source of eutrophication.

Since the mid-1990s, new cropping systems and agricultural structures have been introduced, namely grain farming, developing fruit orchards and poultry culture, innovative cultivation techniques, and concomitant use of fertilizer technology. Fertilizer loss and impact on the environment has, however, been reduced through cropping rotation, changing the base manure and application method of fringe fertilizer, cultivation using water-saving methods, fertilizing the land, and improving the utilization ratio of chemical fertilizer. At the same time, with the internal agricultural structural adjustment, the consumption of chemical fertilizer and agriculture chemicals has been reduced (Table 1.26). These developments have had new influences on the aquatic environment. For example, with the development of extensive feeding, numerous centralized poultry breeding facilities have been built around Lake Taihu, and the direct discharge of untreated waste became one of the major pollution sources of the basin. A further significant concern is the large amount of unconsumed fish food in aquaculture facilities, which settles on the lake bottom, increasing the organic content in water and sediment.

Nitrogen pollution, from grain and oil crop production, has a great impact on the aquatic environment. Some nitrogen (about 25%) is absorbed by crops, some is absorbed by the soil, and some enters the air (thus contributing to "greenhouse effects"), while another part enters the water with farmland runoff. In 1996, the nutrient content of rice seedling beds, fields, and irrigation canals and ditches was determined, and the mean total nitrogen content of 33 samples was 8.26 mg/L, which is 69% higher than the content of total nitrogen in rivers around the lake. The nitrogen loss with surface runoff per mu (15 mu = 1 ha) paddy field per year is 2.7–3.4 kg, which accounts for 12–17% of the total nitrogen applied. Taking Wuxi as an example, with 133,000 ha of grain fields, more than 5,400–6,800 t nitrogen enters the water body every year.

(3) Increasingly serious impacts of the fishery on the lake environment. Net enclosure fish production is concentrated in a small area in the east of Lake Taihu, with a production area of $5,400\,hm^2$. The development of closed net breeding has had many adverse effects on the aquatic environment, including reduced the purifying ability of the lake and contributing to eutrophication. Such aquaculture requires the input of a large quantity of feed, the amount of which is closely correlated with

Table 1.26 Chemical fertilizer use in Wuxi city (1993–1997)

		1993	1994	1995	1996	1997
Total sales	Carbonic ammonia (10^4 t)	13.59	14.61	8.99	9.92	8.42
	Carbamide (10^4 t)	9.27	8.39	8.22	9.10	7.75
Usage/hectare	Carbonic ammonia (kg)	532.2	590.7	362.55	398.55	338.55
	Carbamide (kg)	362.55	338.4	331.35	365.55	311.7

increases in nitrogen and phosphorus load in the water body. For the per mu yield of 7,500–10, 125 kg/hm^2 from intense fish breeding, each 1 t fish produced adds 141.5 kg nitrogen and 14. 4 kg phosphorus into the lake. For example, below a 2-year-old enclosed fish-pen-culture net area, the organic matter, organic carbon, total nitrogen, and organic nitrogen in the top layer of the sediment are 190.7%, 141.4%, 87.5%, and 86.2% higher, respectively, than before (Fig. 1.4). From calculating the mean data in the 1990s, in the enclosed net area of east Lake Taihu, the nitrogen from aquaculture accounted for 2.8% of the total amount in Lake Taihu, and the phosphorus for 10.1%, thus demonstrating that fish breeding contributes significantly to the eutrophication of Lake Taihu. The increased nutrient content in the water body has resulted in nearly doubling the amount of algae, compared with the preaquaculture levels; increased diversity and biomass of protozoa; a three- to fourfold increase in aerobic bacteria and colon bacilli; and increases in aquatic plants, mollusks, and aquatic insect larvae.

(4) Insufficient wastewater treatment. Currently the wastewater treatment capacity in the Lake Taihu basin area is insufficient. The handling of industrial effluent still relies mainly on treatment in the factory. Urban sewage treatment needs massive investment, and the operating cost is high. According to the investment and operation cost of the urban sewage treatment plants that have already been built, to build a plant with capacity of 10,000 t/d needs about 5 million yuan (US $660,000), and to operate such a plant, about 520,000 yuan per year (US $70,000) is needed. For reasons of these great expenses, urban sewage treatment has developed slowly; for example, the municipal sewage handling rate of the urban area in Wuxi City is only 44% (1998 data), and if the suburbs are included, the rate was only 28%.

(5) The current water management system is unsuitable. Under the current system, the environmental protection bureaus, which have supervisory and management authority, are within the water environmental protection agency. This management system has effectively imposed the waste discharge fee and a fine for failing

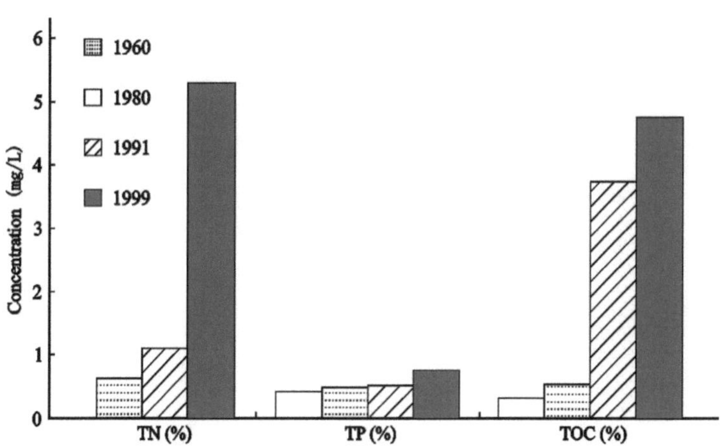

Fig. 1.4 Temporal changes in TN, TP, and TOC in surface sediment in East Lake Taihu, 1960–1999

to meet standards. However, the management system failed to manage the municipal sewage effectively and also failed to manage the rivers that bear and receive the sewage discharge. A large amount of municipal sewage has entered the rivers directly. Although river management lies with the water administration authorities, there is no legal basis from which the water administration authorities could manage the water environment and prevent water pollution. Water administration authorities cannot order enterprises to process waste within a defined time period, cannot collect the waste discharge fee from enterprises, and cannot invest in building the sewage treatment plants; thus, the water management authorities can only allow the rivers to bear and receive sewage.

If considering non-point source pollution, land resource management is closely related to water resource management. Thus, the current situation of no unified management of water quality and the water resource is hindering the process of aquatic environmental improvement and protection.

1.4 Swamping of Eastern Taihu Bay

Wenchao Li

Eutrophication of Lake Taihu, while manifesting as blue-green algal blooms in west and north Taihu, is manifest as swamping, or macrophyte-dominated eutrophication, in east Taihu, with plants covering more than 50% of the lake surface. Between the 1960s and 1990s, the biomass of aquatic plants has increased by one order of magnitude (Yang & Li, 1996b), corresponding with increasing nutrients in the surface sediment. Eastern Taihu Bay is of considerable economic importance for the region because it plays a key role in flood prevention and alleviation, supplies water for Shanghai and the east Zhejiang Province, and is one of the most developed areas of aquaculture production in China. Lying in the southeast of Lake Taihu, this bay is 27.5 km long and up to 9.0 km wide, and over its 131.25 km^2 area the water is now less than 1 m deep on average (see Fig. 1.2).

Swamping is the process of deposition within a lake, causing it to become shallower and finally to develop into a marsh. During this process, large amounts of plant material accumulate on the lake bed, rapidly choking the lake basin, blocking rivers, hindering shipping, diminishing storage capacity, and impeding flood drainage and water supply. In addition, decomposition of the plants releases organic pollutants and noxious substances that compromise water quality, further influencing the water supply and fishery (Li, 1997a).

The apparent cause of lake swamping is simply overgrowth of the aquatic plants, such that they progressively cover the lake surface. However, the actual cause is deposition, causing the lake basin to become shallower, combined with increased nutrient levels; both of which may originate beyond the lake itself. Only when these conditions occur can there be overgrowth of macrophytes.

Some human activities, such as nutrient loading and inappropriate dredging and water level management, can accelerate lake swamping (Yang et al., 1995; Yang & Li, 1996a). Conversely, proper measures such as control of nutrient sources, removal of silt, increasing the water level, and cutting and utilizing the aquatic plants can prevent or delay lake swamping (Li, 1998).

Aquaculture has both delayed and accelerated swamping in Eastern Taihu Bay (Li, 1998). From the first aquaculture along the lakeside in the 1970s, to the extensive production from 3,300 hm^2 in the 1990s, aquatic plants from Eastern Taihu Bay, principally *Zizania latifolia* Turcz, were used as the main feed source, with a harvest of 600,000 t in the 1990s. This intensive harvesting eliminated *Z. latifolia* from the entire northwest bank and from most of the southeast bank, thus initially delaying swamping. However, aquaculture now occupies almost all of eastern Taihu, making it impossible to harvest and use the aquatic plants. Furthermore,

Li, W.C.
State Key Laboratory of Lake Science and Environment, Nanjing Institute of Geography and Limnology, Chinese Academy of Sciences, 73 East Beijing Road, Nanjing 210008, P. R. China
e-mail: wchli@niglas.ac.cn

net enclosures block water flow during the flood season, and the stagnant conditions encourage growth of floating-leaf plants. In addition, excessive planting of *Z. latifolia* has accelerated swamping in some lake regions. Thus, aquaculture has recently accelerated swamping.

1.4.1 Current Status of Swamping

More than 50% of Eastern Taihu Bay is swampy, with plants and plant debris floating on the surface and depositing on the bottom, and the lake bed is now elevated by more than 2 m (Li, 1997a). The most seriously swamped region is Dongjiaozui, where a 2-m-high beach formed from deposits now reaches towards the opposite bank, leaving only a 1-km-wide channel open, seriously impeding the sluicing function of Eastern Taihu Bay. The largest swamped area, along the southeast bank of the lake, blocks the mouths of more than 20 rivers, including the Taipu, rendering water supply insufficient.

1.4.1.1 The Topography and Depositional State of the Lake Bed

(1) Lake bed topography

Lake bed topography was measured in Eastern Taihu Bay in 1991, and again in 1998 between the mouth of the Taipu River and Dongjiaozui (a 50 km^2 area with active deposition) (Fig. 1.5). The comparatively simple lakebed is 1.24–3.24 m a.s.l., with a shoal in the Dongjiaozui lake area. Only 10% of the lake bed is more than 2.00 m a.s.l., including a small beach more than 2.50 m high at the mouth of the Taipu river, the *Z. latifolia* growing area of the southeast bank, which is 2.0–2.5 m high, and a large shallow beach in the reeded area in the west of the Lujiagang River mouth where about 7 km^2 is more than 2.50 m a.s.l. Part of the lake bed is less than 2.00 m a.s.l., including the north of Eastern Taihu Bay center area where some 12 km^2 lies at 1.50–1.70 m a.s.l., a deep trough in the west of the Miaogang River mouth at 1.24 m, and the narrow deep-water zone less than 1.8 m a.s.l. that joins the deep-water region in the north of the bay center and west Taihu, forming the main passage for the entry and exit of water in Eastern Taihu Bay.

Southeast winds usually prevail, and long-term erosion by the wind-induced waves has resulted in the special topography of the deep-water zone in the lake centre near the northwest bank. West or northwest winds, which prevail in winter, generate storm waves leading to strong erosion near the southeast bank at Eastern Taihu Bay mouth and forming a deep trough in the lake bottom. The constant extension of Dongjiaozui results in the shrinking of the mouth of the Eastern Taihu Bay, limiting the sluicing capacity of Eastern Taihu Bay.

(2) Depositional state

Sediment deposition in Eastern Taihu Bay is very serious (Figs. 1.6, 1.7), with nearly 1 m (0.96 m) of soft deposits (hardness $< 5\,kgf/cm^2$), and about 15×10^7 t being deposited each year, more than 98% of which is inorganic matter (silt) originating from outside the lake (Li, 1997b). Deposition is most serious in the

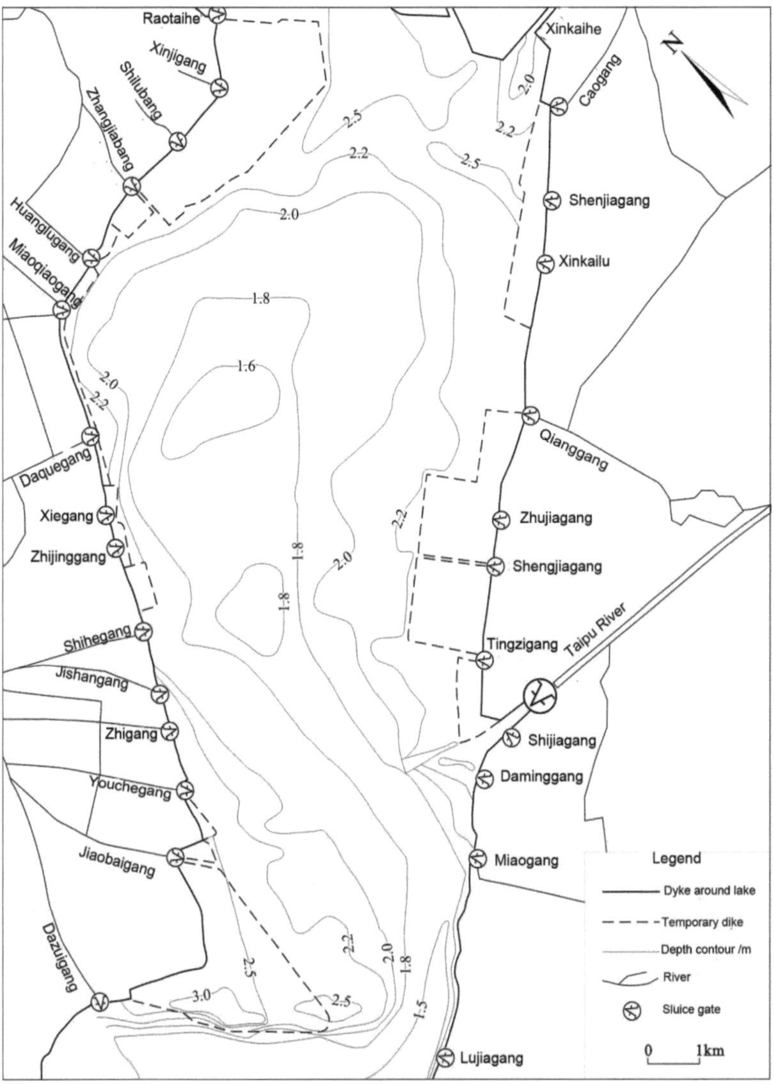

Fig. 1.5 Underwater topography of Eastern Taihu Bay in the 1990s

Z. latifolia cultivation area along the southeast bank, at the mouth of the Taipu River, and at Dongjiaozui, where some deposits are more than 2.0 m deep. In contrast, in open lake regions, deposits are less than 1.0 m deep, and at the centre and mouth of Eastern Taihu Bay deposits are less than 0.5 m deep (Fig. 1.6). This deposition pattern indicates that the silt comes principally from west Lake Taihu and is mainly transported by the inflow and outflow.

Analyzing the lake bed elevation of the Eastern Taihu Bay mouth in 1991 and 1998, it was found that deposition in this region was very active (Fig. 1.7). Except

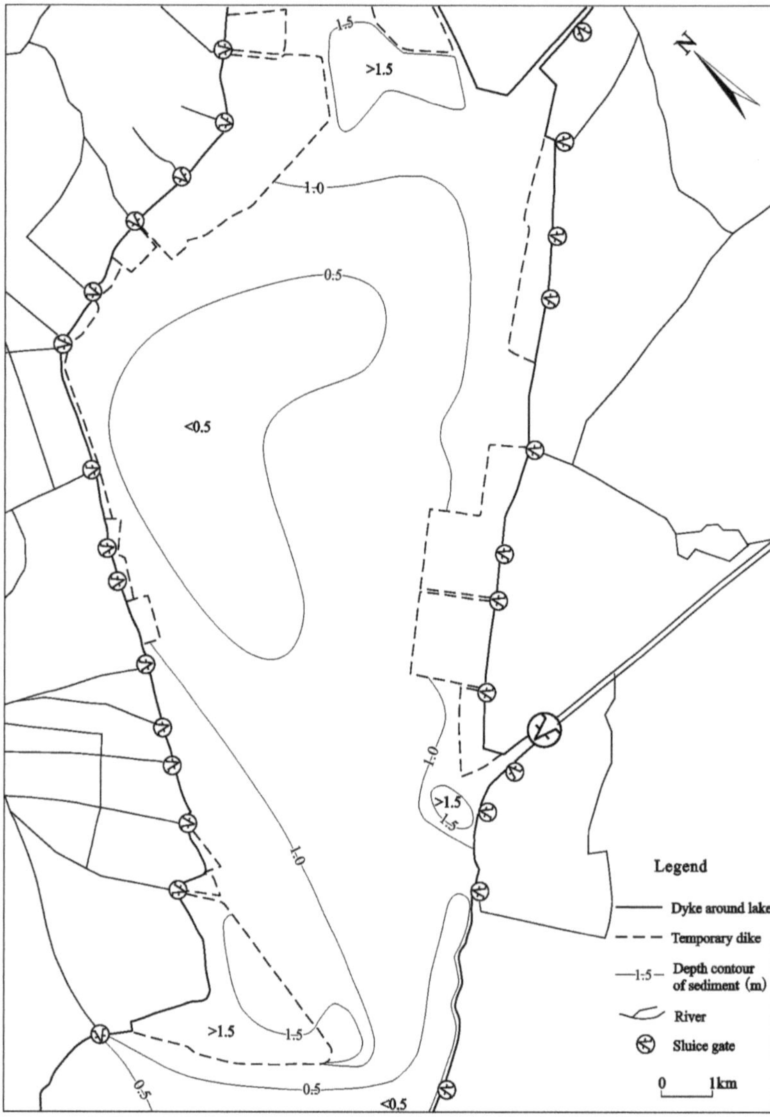

Fig. 1.6 Depth contour of soft sediment in Eastern Taihu Bay (hardness $< 5\,\mathrm{kgf/cm^2}$)

for slight erosion in the lake bed along the southeast, there was deposition to various degrees on all other lake bed areas. Among them there is an ox tongue-shaped deposition zone extending from east of Dongjiaozui to the Taipu river mouth, of about $6\,\mathrm{km}^2$, with recently deposits more than 0.20 m deep. There is also a strip-shaped zone near Dongjiaozui, of $1\,\mathrm{km}^2$, where new deposits are more than 0.30 m deep. In the *Zizania latifolia* Turcz growing area of the Taipu River mouth, new deposits are more than 0.30 m deep (Table 1.27).

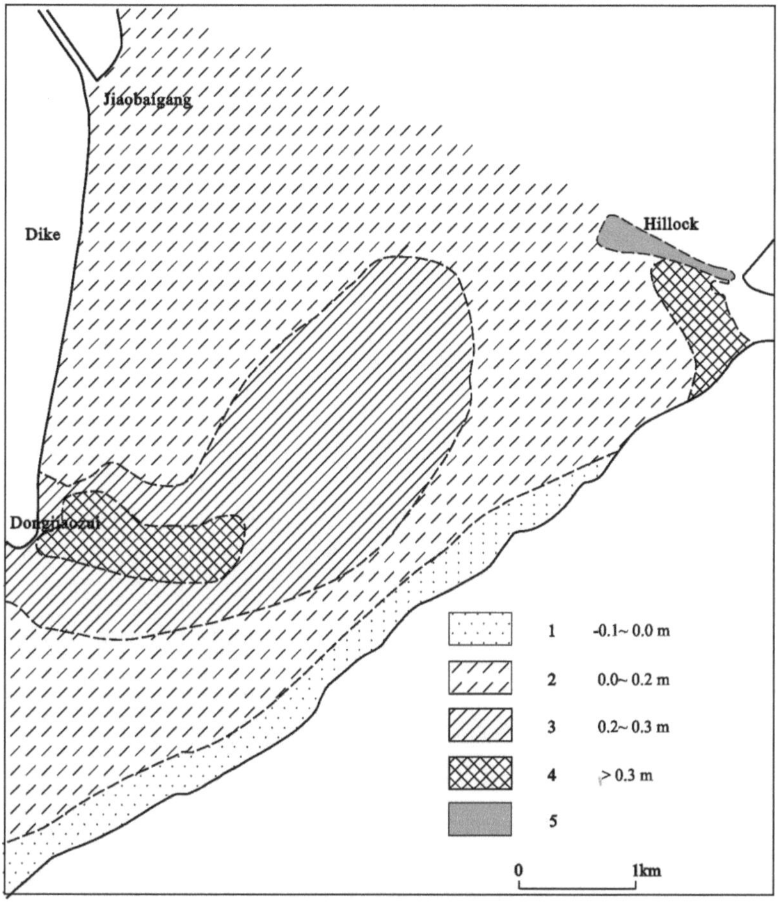

Fig. 1.7 Spatial pattern of deposition in the mouth of Eastern Taihu Bay in 1990s. 1, Slight erosion; 2, slight deposition; 3, marked deposition; 4, serious deposition; 5, hillock

1.4.1.2 Status of Aquatic Vegetation

More than 95% of the area of Eastern Taihu Bay was covered with aquatic plants, the exception being the enclosed net-pen aquaculture areas in the northeast (Fig. 1.8). The distinct vegetation was, in descending order of area: (i) in the center of the bay, *Potamogeton maackianus* A. Benn. (44% of the vegetated area); (ii) in the southeast, *Z. latifolia* (26 km^2); (iii) along the northwest bank, the community of cultured *Z. latifolia* and *Nymphoides peltatum*, *Trapa incisa*, *Hydrocharis dubia*, and some submerged plants (19 km^2); (iv) in the Dongjiaozui and southwest side of Eastern Taihu Bay, the reed growing area (8 km^2); and (v) the deep trough near Lujiagang, *Vallisneria natanus* (0.5 km^2). In addition, there was a transition zone between the *V. natanus* community and the *Potamogeton maackianus* community

Table 1.27 Characteristics of the aquatic vegetation of Eastern Taihu Bay in August 1997

Communities	Location	Coverage (km^2)	Percentage (%)	Biomass (g/m^2)	Total biomass (t)
Total	All the lake	132.38	100	3, 816	505, 200
Phragmites communis Trin.	Dongjiaozui around areas	8.25	6.2	4, 800	39, 600
Zizania latifolia Turcz	Along southeast bank	26.37	19.9	5, 600	147, 700
Zizania latifolia Turcz + *Nymphoides peltatum* + *Hydrocharis dubia*	Along northwest and northeast end	19.15	14.5	3, 500	67, 000
Nymphoides peltatum + *Trapa incisa* + *Hydrilla verticillata*	From Dongjiaozui to Zhigang, outer zone of *Zizania latifolia* Turcz growing area in the southeast area	16.07	12.1	1, 800	28, 900
Potamogeton maackianus	Center of bay	58.56	44.3	3, 590	210, 200
Potamogeton maackianus + *Vallisneria natanus*	From Miaogang to Lujiagang beach	3.31	2.5	3, 200	10, 600
Vallisneria natanus	Lujiagang	0.67	0.5	1, 760	1, 200

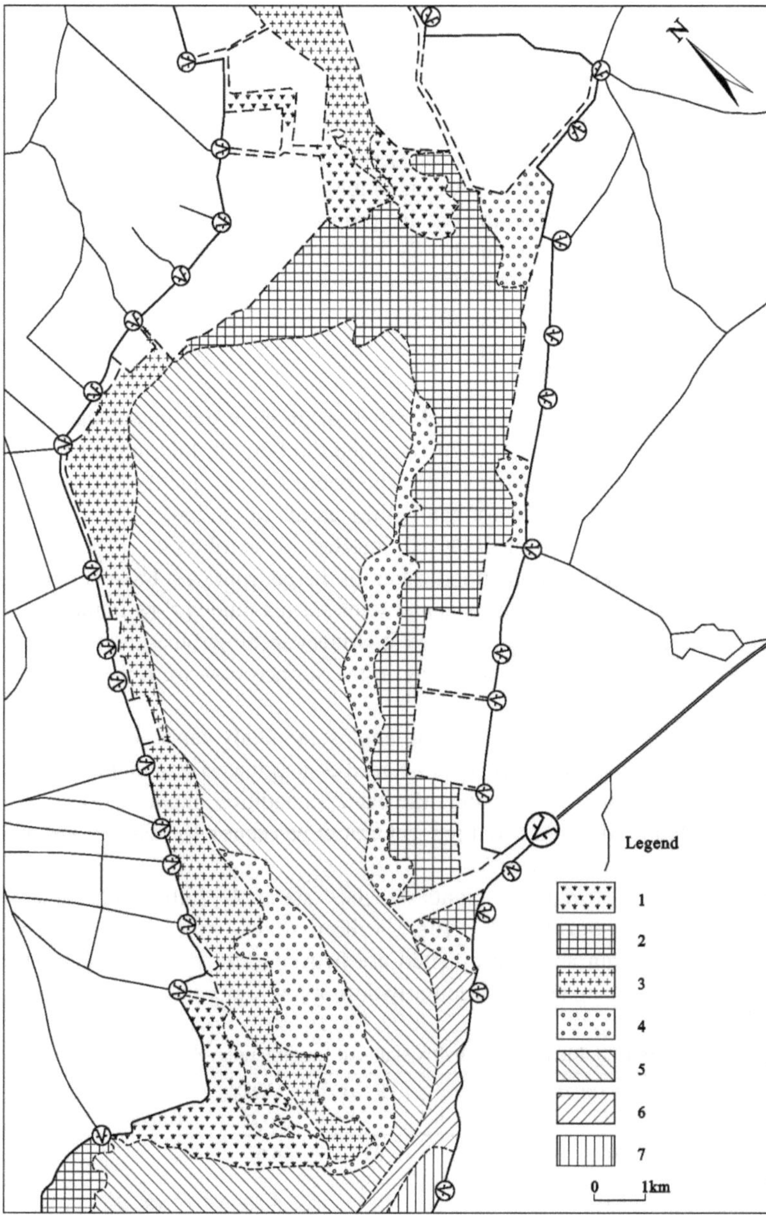

Fig. 1.8 Spatial pattern of aquatic vegetation in Eastern Taihu Bay (1997). Data derived from a 1997 survey and reference Spot-TM satellite image: 1, *Phragmites communis* Trin; 2, native *Zizania latifolia* Turcz; 3, newly cultured *Zizania latifolia* Turcz; 4, floating-leaved vegetation; 5, *Potamogeton maackianus*; 6, *Potamogeton malaianus*; 7, *Vallisneria natanus*

($3\,km^2$). Between Dongjiaozui on the northwest bank and Daquegang, *Spirogyra* was abundant, with biomass reaching $3\,kg/m^2$.

The abundant biomass of aquatic vegetation in Eastern Taihu Bay was estimated at 500,000 t in the 1990s. For most communities, biomass was 2–5 kg/m^2, whereas that of *Z. latifolia* reached 5.6 kg/m^2, and that of the floating-leaf plants and the *V. natanus* community was lower. Among the aquatic plant communities, the *P. maackianus* community had the highest proportion of total coverage (42%), followed by the *Z. latifolia* community (29.5%); the *V. natanus* community had the lowest proportion of coverage (0.2%).

1.4.1.3 Water Quality Status

Eastern Taihu Bay has the characteristics of a macrophyte-dominated shallow lake, namely, water clarity that is so deep as to see the bottom, midrange TP and Chl a, and high TN and COD (Table 1.28). The poorest water quality is found where wastewater enters at the northwest bank of the mouth of the bay.

The sources of pollution for Eastern Taihu Bay are as follows:

(1) The wind-driven flow and flood water discharge from west Lake Taihu mainly carry organic pollutants and nutrients and often form a muddy flow in the mouth of Eastern Taihu Bay. In addition, northwest winds can blow blue-green algae from west Lake Taihu into eastern Taihu Bay in the surface flow from the north or the west.

(2) The rivers from the east mountain peninsula mainly carry organic pollutants and nutrients, which usually form a fan-shaped contaminated area in the river mouth and influence the water quality along the bank. In addition, input of sewage, especially in the Daquegang area with high water flow, can cause decreased transparency of the lake water over a large area.

(3) Production of aquatic plants reaches some 1 million t/a, and most decompose as "*Zizania latifolia* Turcz yellow water," releasing nitrogen and phosphorus. The main areas so affected are between the southeast bank and the Dongjiaozui-Jishangang zone, particularly during low water in summer.

Table 1.28 Mean water quality status of Eastern Taihu Bay in 1997

Location	TN (mg/L)	NH_4^+–N(mg/L)	TP (μg/L)	COD_{Mn} (mg/L)	BOD_5 (mg/L)	Chl a (μg/L)
Mouth of Eastern Taihu	1.31	0.12	31	4.81	–	4.38
Estuary along northeast bank	1.94	0.50	182	5.16	3.18	14.12
Aquaculture area in northwest beach	1.48	0.22	33	5.93	–	3.92
Center of bay	1.38	0.24	35	6.04	–	5.15
Southeast	1.24	0.15	24	4.97	–	3.25
Outlets of southeast area	1.22	0.16	24	5.28	–	3.42

(4) Aquaculture releases fish excrement and uneaten food containing abundant organic matter and nutrients. This pollution is mainly distributed along the northwest bank of the lake, and when mixed with sewage and decomposing aquatic plants, forms a contaminated zone along the lake bank.

1.4.1.4 Division of Areas According to the Degree of Swamping

(1) Quantification of degree of swamping

Lacustrine bogs are defined according to the type of aquatic vegetation: those with emergent plants are bogs, those with submerged plants are natural ponds, and those with no large-scale growth of aquatic plants are lakes. Water depth is only a reference point, although the type of vegetation does reflect water depth.

In contrast, the degree of swamping of different areas of a specific water body has not yet been classified. However, Eastern Taihu Bay provides an opportunity to develop a classification system. Here, the aquatic vegetation is well developed and naturally distributed, and there are marked differences within the lake in the rates of decomposition and accumulation of different species of aquatic plants, thus representing development from potential to actual swamping. Therefore, the degree of swamping can be expressed by the structure of the aquatic vegetation, combined with the lake bed deposit factor. *ZZHv*, the vegetation index of the degree of swamping, is divided into four grades according to type of vegetation (Table 1.29); let *ZZHv* = {0, 1, 2, 3}.

ZZHs is the deposition index of the degree of swamping. The hard lake bed of Eastern Taihu Bay is rather flat, and silt accumulation is evident as differences in height of the lake bed silt. Supposing lake bed elevation range is H = [h1, h2], the deposition index of swamping degree of a location in a specific vegetation type is

$$ZZH_s = (H - h1)(h2 - h1)$$

The index of swamping degree is therefore *ZZH* = *ZZHv* + *ZZHs*. For Eastern Taihu Bay, this is a nonoverlapping index, with continuous numerical values from 1 to 4, reflecting the quantification principle used, namely, taking vegetation as the core and deposition as a supplement.

The swamping degree of the whole bay can be expressed with the area-weighted mean degree of swamping index *ZZHmean*:

$$ZZH_{mean} = \sum ZZH \cdot Pa \qquad (1.10)$$

Table 1.29 Vegetation index for degree of swamping

Type of aquatic vegetation	Emergent	Floating-leaf	Submerged	No macrophytes
Swamping degree index, *ZZHv*	3	2	1	0

$ZZHmean = [1, 2]$ is the early marsh stage, $ZZHmean = [2, 3]$ is the peak marsh stage, and $ZZHmean = [3, 4]$ is the later marsh stage.

(2) Division of Eastern Taihu Bay according to degree of swamping

Taking the swamping index *ZZH* in ranks of 0.5 increments, we can divide Eastern Taihu Bay into six zones representing habitats ranging from extensive marsh to no swamping (Table 1.30, Fig. 1.9):

ⓐ Extensive marsh: there are two such zones, at the Dongjiaozui and at the southeast of Eastern Taihu Bay, together accounting for less than 10% of the area of Eastern Taihu Bay. The aquatic vegetation here is mainly a reed community, with decomposing *Phragmites communis* forming a layer more than 0.5 m thick. The poor water quality known as "yellow water due to *Z. latifolia*" is the most serious in this zone, and few fish and shrimps survive. The lake bed elevation is greater than 2.5 m, and during the low water season, the lake bed appears as a beach.

ⓑ Limited marsh: this area accounts for 35% of the bay area. The aquatic vegetation is dense *Z. latifolia* or a mixed community of sparse *Z. latifolia*, with *N. peltatum*, *T. incisa*, *H. verticillata*, *H. dubia*, and *Myriophyllum*; the layer of decomposing plants is some 0.1–0.5 m thick. In summer, water quality often deteriorates at low water levels because of "yellow water caused by *Z. latifolia*," killing numerous fish. Lake bed elevations are 1.9–2.5 m, and the lake bed near the bank is exposed at extremely low water levels.

ⓒ Severe swamping: this is present in the Xinkaihe River and outside Qianggang River, accounting for approximately 2% of the area of Eastern Taihu Bay. This zone is characterized by small floating-leaf plants. Lake bed elevation is 2–3 m, and a beach is often visible in low water seasons.

ⓓ Moderate swamping: this floating-leaf plant zone occupies some 10% of the bay. Lying outside the *Z. latifolia* area, this community comprises *N. peltatum*, *T. incisa*, *H. verticillata*, *H. dubia*, and *Myriophyllum*. Because the lake surface is covered with vegetation, ship passage is impossible, and lack of oxygen often kills fish and shrimps. Lake bed elevation is 2.0–2.5 m.

ⓔ Slight swamping: the peripheral area of submerged aquatic plants occupies nearly 30–40% of the bay area and comprises the *Potamogeton maackianus* community. Water quality is good, the layer of decomposing plant material on the lake bed is less than 0.1 m deep, and fish and shrimp grow well. Some tall submerged aquatic plants, such as *Myriophyllum*, begin to invade, a sign of early swamping. Because the surface of the water is unobstructed, small ships can pass at times of high water level. The lake bed elevation is 1.7–2.2 m.

ⓕ No swamping: the submerged aquatic vegetation zone in the bay center and near west Lake Taihu accounts for 17.7% of the total bay area. Vegetation is the *P. maackianus* community, the *Vallisneria natanus* community, or a transitional community lying between the two; there is no decomposing material on the lake bed, oxygen is sufficient, and fish and shrimp grow well. The water quality is good, water flows well, and small ships can usually pass. The lake bed elevation is less than 1.7 m.

Table 1.30 Degrees of swamping in Eastern Taihu Bay

Degree of swamping	*ZZH*	Area (km^2)	Percentage (%)	Dominant plant communities	Bottom elevation (m)
Marsh					
Extensive	(3.5, 4.0]	10.60	7.8	Dense *Phragmites communis* Trin	2.5–3.1
Limited	(3.0, 3.5]	47.38	35.0	Sparse *Zizania latifolia* Turcz	1.9–2.5
Swamping					
Severe	(2.5, 3.0]	2.38	1.8	*Nymphoides peltatum* + *Trapa incisa* Sieb et Zucc var. *quadricaudata* Glukc	2.5–3.0
Moderate	(2.0, 2.5]	12.46	9.2	Sparse *N. peltatum*, *Trapa incisa* Sieb et Zucc var. *quadricaudata* Glukc	2.0–2.5
Slight	(1.5, 2.0]	38.51	28.5	*Potamogeton maackianus* A. Benn.	1.7–2.2
No swamping	[1.0, 1.5]	23.90	17.7	*Potamogeton maackianus* A. Benn. community + *Vallisneria natanus* (Lour.) Hara	1.2–1.7

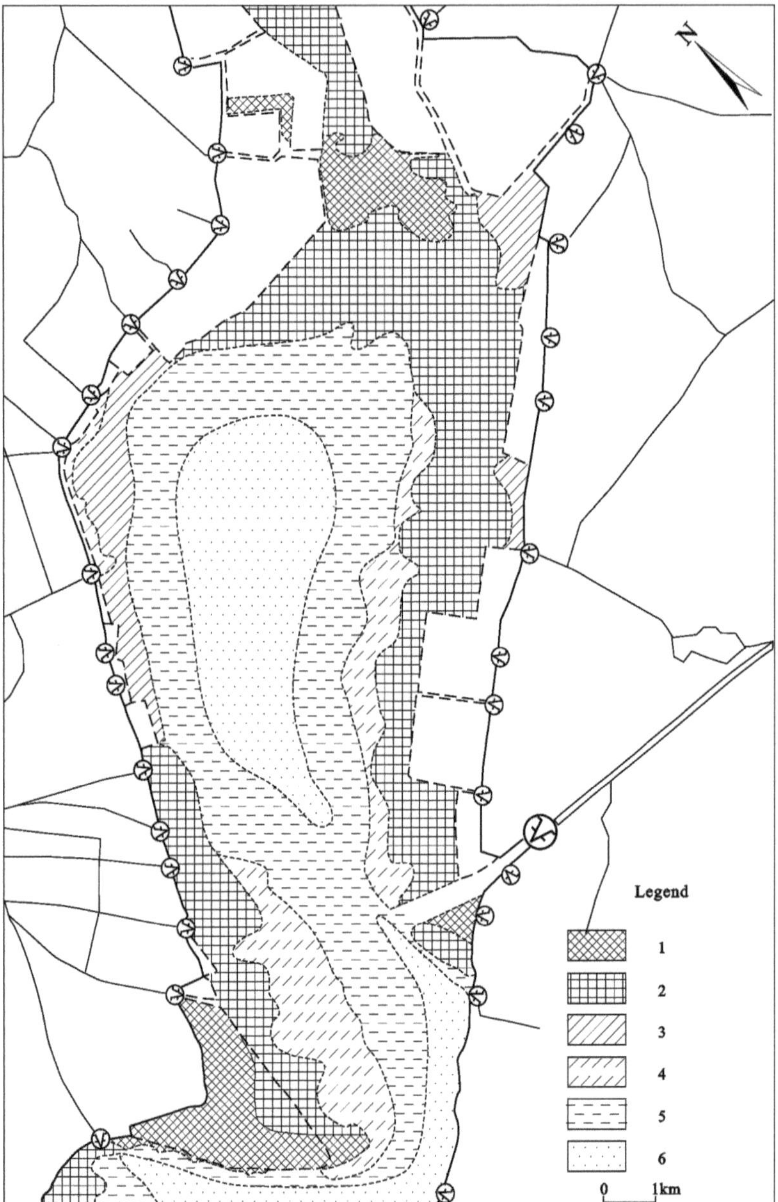

Fig. 1.9 Degree of swamping in Eastern Taihu Bay: 1, extensive marsh; 2, limited marsh; 3, severe swamping; 4, moderate swamping; 5, slight swamping; 6, no swamping

1.4.2 The Development of Swamping in Eastern Taihu Bay and Its Driving Forces

The swamping of Eastern Taihu Bay is rapidly accelerating. Both inside and outside the bay, human activity has intensified, and although some of these activities can delay the swamping process, most of them accelerate swamp development. A comparison of the conditions in 1997 and the 1960s illustrates this. In 1997, 42.8% of Eastern Taihu Bay was covered with aquatic plants such as reeds (7.8%) and *Z. latifolia* (35%) and had developed into a marsh. A further 11% was dominated by floating-leaf plants and had entered the late swamping stage (severe and moderate), which was a markedly expanding trend. Some 30% had submerged vegetation, representing an earlier swamping stage. The mean swamping index of the whole bay in 1997 was 2.41, representing active swamping.

1.4.2.1 Development of the Marsh Vegetation

Changes in the aquatic vegetation indicate the direction and speed of marsh formation.

(1) The types of aquatic vegetation and changes in their distribution

The earliest investigation is the "Aquatic Biological Survey Report in Eastern Taihu" published in 1959 by the biological department of East China Normal University (ECNU) (ECNU, 1959). At that time, *P. communis* and *Z. latifolia* formed a narrow strip along the bank of the bay, and only a sparse submerged plant community of *P. malaianus* and *V. natanus* was present on the open lake surface, reaching a mean density of biomass of $504\,g/m^2$ in summer. The lake bed could not be seen, and water transparency was 0.3–0.8 m. This kind of vegetation and environmental state is similar to that which now exists (2006) in west Lake Taihu near Xishan Island.

By 1980, the structure of the vegetation, its species composition, and the biomass had greatly changed (Fig. 1.10). The emergent plant community, principally *Z. latifolia*, had expanded and occupied the whole bank, covering 29.1% of the bay. The submerged plant community of *P. malaianus*, *H. verticillata*, and *V. natanus*, although still occupying 33.3% of the lake area, had doubled its peak biomass to $1,088\,g/m^2$. The *V. natanus* community in the southwest center of the bay and the *P. maackianus* community in the northeast occupied 21% and 4.6% of the surface, respectively. The mean density of biomass of aquatic plants in the whole bay was $891\,g/m^2$ in summer, an increase of 51% compared with 1959. The water transparency had markedly improved in 1980 compared to 1959, the lake bed could be seen in more than two-thirds of the lake area, and the transparency in the center of the lake was 0.2–0.8 m.

In 1986, the *P. maackianus* community around the *Z. latifolia* community occupied 31.4% of lake surface, and the *P. malaianus* community area had shrunk. The mean density of biomass of aquatic plants had increased markedly to $2,722\,g/m^2$ in summer. In more than 90% of the bay area, the lake water was clear and the lake bed could be seen (Li, 1993).

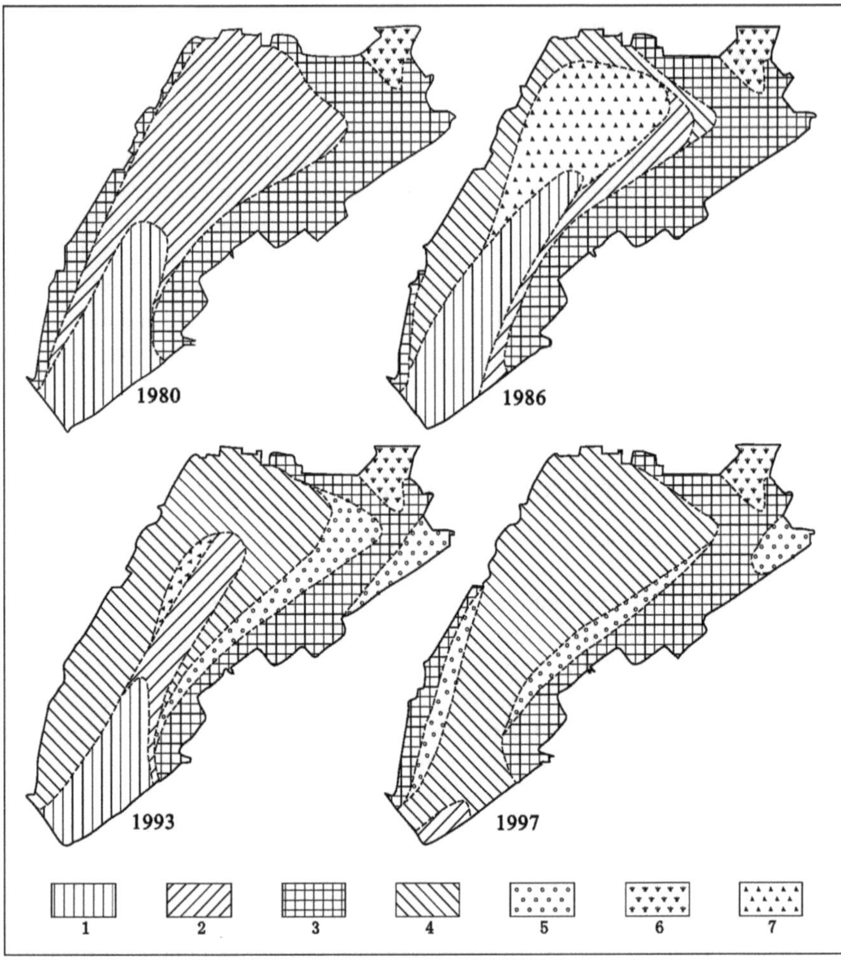

Fig. 1.10 Changes in the distribution of dominant macrophyte species in Eastern Taihu Bay between 1980 and 1997. 1, *Vallisneria natanus*; 2, *Potamogeton malaianus*; 3, *Zizania latifolia* Turcz; 4, *Potamogeton maackianus*; 5, floating-leaf macrophyte; 6, *Phragmites communis* Trin; 7, *Eleocharis palustris*

In 1993, the mean biomass of aquatic plants had further increased to 4, 990 g/m^2 in summer, and the transparency of the lake water had further improved. The *Z. latifolia* community along the northwest bank had disappeared as a result of intensive harvesting and an unusually high water level in 1991, and the *P. maackianus* community had expanded to occupy 42.9% of the lake surface. The *Z. latifolia* community in the southeast of the lake shrank about 0.5 km, and a floating-leaf plant zone developed, made up of *N. peltatum* and *T. incise*, occupying 8.7% of the bay surface. The *P. malaianus* community was reduced to a narrow strip, and the *V. natanus* community occupied only half the area it did in 1997.

Between 1993 and 1997, succession of the aquatic vegetation in Lake Taihu has comprised a 1-km-wide secondary vegetation zone forming along the northwest bank, composed of *Z. latifolia*, floating-leaf plants, and floating plants; a large area of floating leaf plants forming outside the *Z. latifolia* region between Dongjiaozui and Zhigang; and disappearance of the *V. natanus* and *P. malaianus* communities, with their replacement by the *P. maackianus* community.

(2) Succession of vegetation

Succession of the main kinds of vegetation in Eastern Taihu Bay, using area as a key indicator, is shown in Table 1.31. The *P. malaianus* and *V. natanus* communities were replaced by *P. maackianus* and, in turn, *P. maackianus* was replaced by *Z. latifolia* and floating-leaf plants. These changes represent accelerating swamping in Eastern Taihu Bay, especially since the 1990s, when emergent plants and floating-leaf plants have developed very rapidly.

Because leaves of emergent plants and floating leaf plants are exposed to air, they derive sufficient illumination and CO_2 from the atmosphere for photosynthesis and have higher productivity than submerged macrophytes. However, the harvest-utilization ratios of these emergent and floating leaf plants are lower than that of the submerged plants, and great quantities of the former decompose in the lake, often causing the serious "yellow water due to *Z. latifolia*" in summer. When the plant fragments that are difficult to decompose (20% of the dry weight of the plants) sink to the bottom, this accelerates deposition on the lake bed. In 1993, even though some 600,000 t aquatic plants was harvested, 520,000 t remained in the lake, about 56% of which were emergent plants and floating-leaf plants. The enclosed net-pen aquaculture is an obstacle to efficient harvest of the aquatic plants, and thus the amount remaining in the lake increases; it is estimated that this residuum may reach 1 million t/year.

(3) Increasing human activity and impact on the aquatic vegetation

In the 1950s, *Z. latifolia* grew naturally in the Dongjiaozui region and the neighbouring area in Eastern Taihu Bay. By the 1960s, it had extended on a large scale, and by the 1970s it occupied the whole lakeshore and had begun spreading towards the centre of the bay. In the early 1980s, it was harvested and used as food for carp breeding in the enclosed reclamation area and enclosed-net fish breeding areas, suppressing its development and reducing the area of *Z. latifolia* grass along the northwest bank. By 1991, *Z. latifolia* had disappeared from the northwestern bank area. It has been estimated that the enclosed net-pen fish breeding area was responsible for shrinking the area of *Z. latifolia* by about one-third between the 1980s and the 1990s.

After 1993, *Z. latifolia* was planted along the northwest bank to provide feed for fish culturing. The plants flourished and spread rapidly as a result of the high levels of nutrients in the wastewater from adjacent aquaculture and from rivers entering the lake, increasing the fertility of the sediment. Within 4 years, a dense zone of *Z. latifolia* was formed, which severely restricted the net-baffled water flow and wind waves, formed a stagnant environment, and limited access for harvesting aquatic plants in the fish-pen culture area. Around the edge of the *Z. latifolia* zone, a mixed zone of floating-leaf plants and floating plants formed, and this developed to the centre of the bay with the rapid expansion of enclosed net-pen aquaculture.

Table 1.31 Temporal changes in area, density, and biomass of aquatic vegetation in Eastern Taihu Bay (1959–1997)

Year	Type of communities	Distribution area		Mean density of biomass	Biomass	
		Area (hm^2)	Percentage (%)	(g/m^2)	Biomass (t)	Percentage (%)
1959	*Potamogeton malaianus*	14, 500	90.0	504	81, 194	––
1980	Total	14,600	100.0	891	443,000	100.0
	Phragmites communis Trin	800	5.5	5,625	45,000	10.1
	Zizania latifolia Turcz	4,667	32.0	6,724	310,000	70.0
	Potamogeton maackianus	733	5.0	2,046	14,000	3.2
	Potamogeton malaianus	5,333	36.5	1,088	58,000	13.1
	Vallisneria natanus	3,067	21.0	522	16,000	3.6
1986	Total	12,967	100.0	2,722	353,000	100.0
	Phragmites communis Trin	567	4.4	3,750	21,000	6.0
	Vallisneria natanus	3,400	26.2	5,241	178,000	50.4
	Potamogeton maackianus	4,067	31.4	2,149	87,400	24.7
	Eleocharis palustris	1,133	8.7	840	9,500	2.7
	Potamogeton malaianus	733	5.6	1,410	10,300	2.9
	Vallisneria natanus	3,067	23.7	1,527	46,800	13.3
1993	Total	12,807	100.0	4,990	641,650	100.0
	Phragmites communis Trin	726	5.7	5,610	40,700	6.3
	Zizania latifolia Turcz	2,822	22.0	7,632	215,300	33.6

Table 1.31 (continued)

Year	Type of communities	Distribution area		Mean density of biomass	Biomass	
		Area (hm^2)	Percentage (%)	(g/m^2)	Biomass (t)	Percentage (%)
	Nymphoides peltatum and *Trapa incisa*	1,113	8.7	6,517	72,600	11.3
	Potamogeton maackianus	5,489	42.9	4,683	257,100	40.1
	Eleocharis palustris	140	1.1	3,000	4,200	0.7
	Potamogeton malaianus	1,145	8.9	3,904	44,700	7.0
	Vallisneria natanus	1,372	10.1	514	7,050	1.1
1997	Total	13,238	100.0	3,816	505,200	100.0
	Phragmites communis Trin	825	6.2	4,800	39,600	7.8
	Zizania latifolia Turcz	2,637	19.9	5,600	147,700	29.2
	Zizania latifolia Turcz + *N. peltatum* + *Hydrocharis dubia*	1,915	14.5	3,500	67,000	13.3
	N. peltatum + *Trapa incise* + *Hydrilla verticillata*	1,607	12.1	1,800	28,900	5.7
	Potamogeton maackianus	5,856	44.3	3,590	210,200	41.7
	Potamogeton maackianus + *Vallisneria natanus*	331	2.5	3,200	10,600	2.1
	Vallisneria natanus	67	0.5	1,760	1,200	0.2

P. malaianus and *V. natanus* both have the characteristics of mud tolerance and resistance to stormy waves, preferring water flow, whereas *P. maackianus* needs limpid water; the enclosed fish-pen net in the bay centre has weakened the disturbance of the lake bed and has reduced bottom mud resuspension, improving the transparency of the lake water. This is the main reason that *P. maackianus* can replace *P. malaianus* and *V. natanus*. The secondary vegetation zone between Dongjiaozui and Jishangang, made up of *Z. latifolia* and floating-leaf plants, has developed rapidly, reaching to the opposite bank, and has almost closed the Taipu river mouth. These plants can seriously block the water flow and promote deposition, thus threatening the sluice activities of the Taipu River.

(4) Aquatic vegetation succession and swamping

The main natural factors promoting development of the aquatic vegetation in Eastern Taihu Bay are deposition and accumulation of nutrients. Human activities such as reclamation, planting and harvesting of macrophytes, and aquaculture can change the direction and speed of aquatic vegetation development. Harvesting results in the residual amount of emergent plants and floating-leaf plants greatly exceeding that of the submerged plants. For example, in the large harvest in 1993, the dry residue of Z. *latifolia* and floating-leaf plants reached $1,774\,g/m^2$ and $2,119\,g/m^2$, respectively, which was four to five times that of the submerged plants ($418\,g/m^2$).

Aquatic vegetation can be beneficial because there is accelerated deposition of debris and improved clarification of the water, and rivers flowing out from the lake are not blocked. However, aquatic vegetation can be disadvantageous, because deposition of fragments of aquatic plants can increase lake silting, accelerating the swamping process. Since *Z. latifolia* developed, deposition of plant material has markedly increased in Eastern Taihu Bay, as evident in the changes in total organic carbon (TOC) in the sediment at different depths (Fig. 1.11). The *Z. latifolia* residue can be distinguished in the top 30 cm of the sediment, where the TOC content reaches 5.6–17.3%, representing 10–30% of the organic matter.

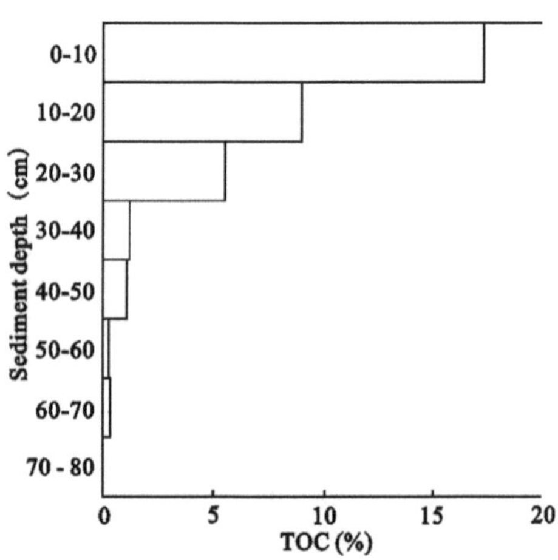

Fig. 1.11 Changes in the total organic carbon (*TOC*) content in different depths of sediment in *Zizania latifolia* Turcz growing area in Eastern Taihu Bay

The rapid expansion of the area of emergent plants and floating-leaf plants is the main characteristic of aquatic vegetation succession in Eastern Taihu Bay; this increases deposition of biological material, accelerating the swamping process.

1.4.2.2 Change in Hydrology and Swamping in Eastern Taihu Bay

Eastern Taihu Bay was originally a shallow lake bay, with a hard lake bed at an elevation of about 1 m above sea level (a.s.l.). The bay served as the flushing passage of flood water for Lake Taihu, with the Guajingkou River to the north being the main river mouth exiting the lake. Silty flood waters from west Lake Taihu enter Eastern Taihu Bay, causing extensive deposition nearly 1 m deep of soft deposits. Silt accounts for 98.46% of the content and organic matter only 1.54%, thus demonstrating that the external silt source is the main cause of deposition in Eastern Taihu Bay.

Before the Taipu river water conservancy construction was built in the 1990s, the lake water flushing through the river along the southeast bank of the bay formed a relatively even deposition zone of incoming and outgoing flow between Miaogang and Guajinggang. According to data from 1954, the silt deposited in Eastern Taihu Bay was about 143,650 t/a, the mean amount per unit area was 1.1 kg/m^2, and the rate of deposition was less than 0.1 cm/a. After the Taipu River water conservancy construction was built, silty lake water entered Eastern Taihu Bay, then flowed via the Taipu River to the East China Sea. The amount of silt entering the lake north of the Taipu River decreased, and the rate of deposition was also reduced. However, the region from the mouth of the Taipu River to Dongjiaozui became the main deposition location and deposition markedly accelerated here; between January 1991 and February 1998, there was 20–30 cm of new deposits, with 3–5 cm being deposited each year. Such rapid deposition created conditions for expansion of *Z. latifolia* and provided nutrition for the extensive growth of floating-leaf plants and submerged plants, thus promoting further silt deposition and acceleration of swamping. The development of the aquatic plants further promoted silt deposition, thus accelerating the swamping process in the bay region, and seriously jeopardized the function of the Taipu River water supply and floodwater discharge.

1.4.2.3 Reclamation of the Beach and Swamping

Between 1949 and the late 1970s, the area of enclosed reclaimed beach in Eastern Taihu Bay reached 112 km^2, which was approximately 0.85 times that of the existing water surfaces and larger than the natural beach area. Within the embankment north of Lujiagang, the reclaimed area is nearly 182 km^2; within the embankment encircling the bay there are ten dykes within a reclaimed area of 51 km^2, and the remaining area of bay surface is only 131 km^2. The remaining reclaimed area is in the northeast bay where the water surface is only 1 km wide, and the swamping problem is very serious, with reeds and *Z. latifolia* covering the bay surface.

The enclosed reclamation areas reduce the lake area and accelerate deposition. Development of marsh in Eastern Taihu Bay is accompanied by reduced wind fetch and wave action, which in turn promotes growth of aquatic plants and accelerates swamping.

1.4.2.4 Aquaculture Development and Swamping

Eastern Taihu Bay is one of the most developed areas of intensive aquaculture in the region, with three kinds of production: enclosed reclaimed areas with ponds for fish breeding, large lake surface net-pen culturing, and natural catch in enclosures. As yield from the wild catch fishery has fallen sharply, aquaculture has developed rapidly. In the 1990s, the total production area (including the reclaimed ponds) was already 6, 600 hm^2 in the 1990s, yielding an annual production of 28,000 t, with a value of 0.2700 billion yuan.

(1) Fisheries in reclaimed areas

The main development of the reclaimed pond area in Eastern Taihu Bay took place in the 1970s and was stabilized by the 1990s, with an area of 3, 300 hm^2, annual production of 24,000 t (Figs. 1.12, 1.13), and a value of 0.1500 billion yuan. The principal species were grass carp and bream, which were fed plants collected from Eastern Taihu Bay. The maximum harvest, in 1993, was 600,000 t, representing more than half the total for Eastern Taihu Bay.

This form of aquaculture is efficient at reducing secondary pollution and delaying swamping. By harvesting the aquatic plants, some 2,000 t nitrogen and 300 t phosphorus were removed from Eastern Taihu Bay every year, equivalent to 28% and 57% of the external load, respectively, thus delaying eutrophication. Thus, providing food for the herbivorous fish in the enclosed reclamation area has had a positive impact on the environment in Eastern Taihu, and this activity should be protected and supported. However, in recent years, most of the lake surface has been enclosed with nets, obstructing plant harvest and limiting the positive impacts just described. Further work is needed to achieve sustainable development of aquaculture in Eastern Taihu and to protect its environment.

(2) Enclosed net-pens

When enclosed net-pens began in 1984, they were limited to the Jiaobaigang and Jishangan regions along the northwest bank of the bay and concentrated on production of grass carp and bream. In the next decade of development, the net-pens was found irregularly around the bank of the bay, occupying less than 660 hm^2, and yielding a total output of nearly 3,000 t valued at 0.024 billion yuan in 1993. There was a diversification of species, with the addition of crabs, which accounted

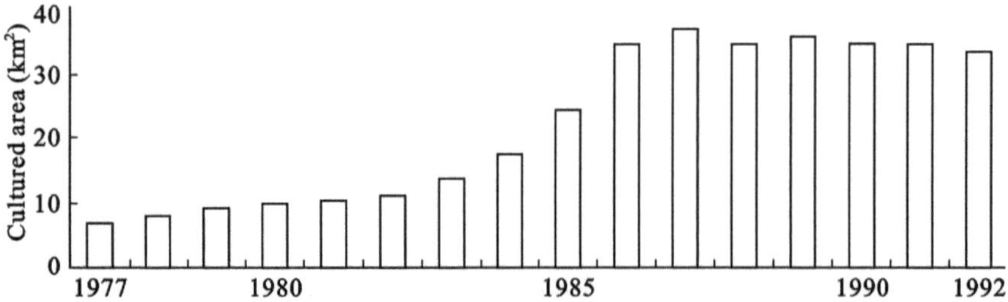

Fig. 1.12 Growth of the area for cultured aquatic products in the reclaimed pond area in Eastern Taihu Bay, 1977–1992

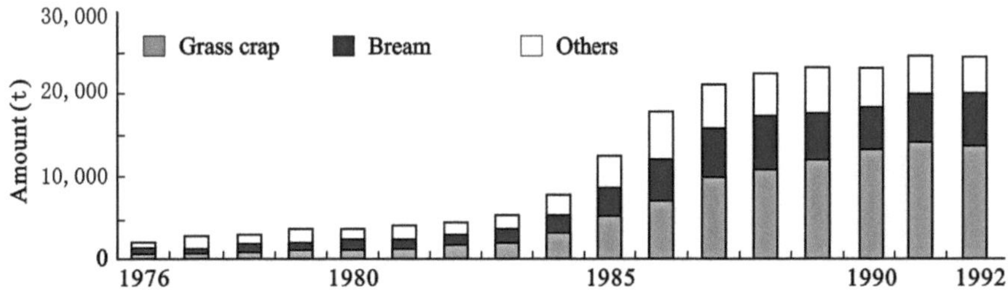

Fig. 1.13 Growth of annual aquaculture production in the reclaimed pond area in Eastern Taihu Bay, 1977–1992

for nearly 40% of the area. After 1993, there was extensive development of the net-pens, and by 1997 they had spread over the lake, covering an area of 3, 200 hm^2 (24% of Eastern Taihu Bay); output exceeded 4,000 t, which was valued at nearly 0.12 million yuan (more than that of the natural fishery harvest from all of Lake Taihu). Crab had become a very important product, representing 2, 660 hm^2 and an output value of 96.6 million yuan, more than 80% of the area and value of the net-pen aquaculture.

Enclosed net-pen aquaculture has produced significant economic and social benefits for the local fisherman, peasants, and aquaculturists around the Taihu region. However, a lack of systematic planning and management has had numerous negative consequences, as described next.

ⓐ Serious hindering of the harvest and utilization of aquatic plants has accelerated lake swamping. Taking the year 1993 as an example, the amount of the water plants growing every year was 1,120,000 t, which absorbed nitrogen at 3916 t and phosphorus at 496 t, equivalent of 58% and 95% of the outside source load in Eastern Taihu Bay separately; among them, 600,000 t water plants was reaped and utilized, mainly used as the fodder of pond breeding fish in the enclosed reclamation area, taking 1891 t nitrogen and 296 t phosphorus away from the lake, the equivalent of 28% and 57% of the outside source load in Eastern Taihu Bay separately. However, the enclosed net-pen breeding divides up most of the lake surface, hindering harvest, and thus a significant means of pollutant removal from Eastern Taihu Bay is lost. Furthermore, extensive *Z. latifolia* has been planted by some aquaculturists, and the pens that are more than 3 years old are covered by weeds such as *Spirogyra*, *T. incisa*, *N. peltatum*, and *H. dubia*. These plants block water flow and storm waves, resulting in a stagnant environment, in which plants decompose and accumulate on the lake bed, giving rise to episodes of yellow water caused by decay of *Z. latifolia*, all of which accelerates the depositing and swamping process.

ⓑ Influence on floodwater drainage and water supply. The main sluicing rivers along the southeast bank in Eastern Taihu Bay are already seriously silted. Furthermore, the nets on most of the lake hamper construction of water conservancy and environmental improvement projects, and the mouth of the Taipu, the biggest sluicing river, is surrounded by nets, which can collapse, allowing fish to escape.

ⓒ Disrupting and blocking shipping. Eastern Taihu Bay is now the main artery for agricultural shipping and small-scale cargo vessels. Originally small ships

Table 1.32 Water quality of 13 river mouths in Eastern Taihu Bay in winter, 1997

River name	TN (mg/L)	NH_4^+–N (mg/L)	TP (μg/L)	COD_{Mn} (mg/L)	BOD_5 (mg/L)	Chl a (μg/L)
Jiaobaigang	2.10	0.18	110	6.11	2.97	10.23
Youchegang	0.74	0.05	120	3.20	1.96	0.93
Zhigang	1.02	0.05	90	3.27	1.97	4.19
Jishangang	2.04	0.11	120	5.98	3.86	23.99
Shihegang	1.35	0.05	80	4.28	2.29	8.37
Zhijinggang	7.45	4.28	560	11.11	9.53	56.79
Xiegang	4.20	1.29	450	10.01	7.33	52.08
Dongdaquegang	1.55	0.05	330	4.63	1.72	7.53
Miaoqiaogang	1.66	0.19	140	4.40	1.54	13.39
Huanglugang	1.23	0.05	100	4.13	2.40	4.56
Zhangjiabang	0.70	0.14	80	3.15	1.68	0.50
Shilubang	0.61	0.08	80	3.54	2.36	0.50
Raotaihe	0.60	0.03	110	3.26	1.70	0.50
Mean	1.94 ± 1.91	0.50 ± 1.18	182 ± 157	5.16 ± 2.60	3.18 ± 2.46	14.12±19.09
Inflow from Dongjiaozui	1.31	0.12	31	4.81	–	4.38

Table 1.33 Trend of mean water quality in Eastern Taihu Bay from 1980 to 1997

Sampling number and sampling frequency	TN (mg/L)	NH_3^+–N (mg/L)	TP (μg/L)	COD_{Mn} (mg/L)	BOD_5 (mg/L)	Chl a (μg/L)	Water colour	Transparency (m)
2 sites, and 3 times 1980	0.65	0.05	30	2.87	0.98		14.7	0.53
11 sites, and 9 times 1990–1991	1.01	0.14	43	5.50	1.25	3.5	15.0	0.80
43 sites, and 4 times 1993–1994	1.04	0.20	75	5.57	1.07	4.3	–	Clear to bottom
20 sites, and 4 times 1997	1.39	0.19	31	5.51	–	4.4	–	Clear to bottom

navigated some 48 rivers along the lake; however, currently only one route between Daquegang to Qianggang and one along the south bank are navigable. The remainder have been obstructed by enclosure nets or blocked by weeds.

ⓓ Aggravated water pollution. Sewage from the Dongshan peninsula enters Eastern Taihu Bay through more than ten rivers. Before the installation of the enclosure nets, the sewage was purified by strong dilution after entering Eastern Taihu Bay. Abundant nets and weeds now render the northwest bank stagnant, and the sewage cannot be diluted. In addition, the lack of oxygen in the water and mud means that organic pollutants cannot be oxidized, and a 1-km-wide zone of contamination has developed. The "yellow water" caused by *Z. latifolia* along the northwest bank has negatively impacted the fishery, with disappearance of shrimp along the northwest bank, spread of epidemics affecting fish and crab in nets, increased mortality, and decreased growth rate and quality. Expansion of the enclosed nets towards the lake centre is accompanied by expansion of the contaminated zone along the northwest bank towards the lake center, directly jeopardizing the supply water quality from Taipu River to Shanghai.

1.4.2.5 Outside Pollution and Swamping

Thirteen rivers entering the lake between the Jiaobaigang and Yaotai Rivers along the northwest bank have been polluted with domestic sewage, village and urban industrial sewage, and farmland and fish pond drainage water (Table 1.32). After entering the bay these river waters pollute the river mouths and adjacent banks. Phosphorus is the most significant pollutant, minimally 80 μg/L; the TP content of 9 rivers exceeds 100 μg/L, with that in the seriously polluted areas of Zhijinggang and Xiegang reaching 330–560 μg/L. The most serious TN and BOD_5 pollution also occurs in Zhijinggang and Xiegang.

Since the 1980s, eutrophication in Eastern Taihu Bay has been rapid (Table 1.33), with TN, TP, and COD content continuously increasing. In contrast, BOD and Chl a concentrations are steady, and transparency is increasing as a consequence of numerous water plants and improved water quality protection. Swamping of Eastern Taihu Bay is accelerating, driven by excessive economic activity with associated pollution, land reclamation, and fishery development. The swamping of Eastern Taihu Bay and eutrophication of north Taihu have now become two of the biggest simultaneous environmental problems affecting the lake.

References

Chen, J. Y., Z. Y. Yu & C. X. Yun, 1959. Morphology of Yangtze River Delta. Acta Geographica Sinica 25: 201–220 (In Chinese).

East China Normal University (ECNU), 1959. Investigation report of hydrobiology in East Lake Taihu (In Chinese).

Huang, X. W., 2000. Layout and integrated management of Taihu basin. Beijing: China Water Resources and Hydropower Press, 1–20 (In Chinese).

Huang, Y. P., 2001. Contamination and control of aquatic environment in Lake Taihu. Beijing: Science Press, 231–241.
Li, W. C., 1993. Study on the succession trend of submerged vegetation in East Taihu Lake, China and its ecological cures. Memoirs of Nanjing Institute of Geography and Limnology Academia Sinica. Beijing: Science Press, 83–93 (In Chinese with English abstract).
Li, W. C., 1997a. "Yellow water" in East Taihu Lake caused by *Zizania latifolia* and its prevention. Journal of Lake Sciences 9: 364–368 (In Chinese with English abstract).
Li, W. C., 1997b. Silting up effect of aquatic plants in East Taihu Lake and accumulation of phosphorous in the sludge. Chinese Journal of Environmental Science 18: 9–12 (In Chinese with English abstract).
Li, W. C., 1998. Utilization model of aquatic macrophyte for fishery in Chinese shallow lakes. Ecological Engineering 11: 61–72.
Nanjing Institute of Geography, Academia Sinica, 1965. Investigation of Lake Taihu. Beijing: Science Press, 1–84 (In Chinese).
Qin, B. Q., Q. L. Wu, J. F. Gao, C. X. Fan, G. Xu, W. M. Chen, R. Mao & Y. W. Chen, 2002. Water environmental issues in Taihu Lake of China: Problems, causes and management. Journal of Natural Resources 17: 221–228 (In Chinese with English abstract).
State Environmental Protection Bureau, 2000. Water protection in the three rivers and three lakes of China. Beijing: Chinese Environmental Press, 151–175 (In Chinese).
Sun, S. C. & Y. P. Huang, 1993. Lake Taihu. Beijing: China Ocean Press, 23–89 (In Chinese with English abstract).
Sun, S. C., Y. F. Wu & B. F. Dong, 1987. The bottom configuration and recent deposition of Lake Taihu. In: Nanjing Institute of Geography and Limnology, Academia Sinica (ed.), Memoirs of Nanjing Institute of Geography and Limnology, Academia Sinica (4). Beijing: Science Press, 1–14 (In Chinese with English abstract).
Wang, H. Z. & W. J. Ding, 1936. Formation and shrinkage of Lake Taihu. Monthly Report of Water Resources 11: 8–13 (In Chinese).
Wissman, F., 1941. Geographical problem in the lower delta of Yangtze River (translated by Wang Deji). Geography 1: 74–77.
Yan, Q. S. & S. Y. Xu, 1987. Recent Yangtze Delta Deposits. Shanghai: East China Normal University Press, 68–75 (In Chinese with English abstract).
Yang, H. R., Z. R. Xie & D. Y. Yang, 1985. Sea-level change in Holocene and evolution of Lake Taihu. In: Yang Huairen (ed.), Contribution to the Quaternary Glaciology and Quaternary Geology. Beijing: Geological Publishing House, 49–64 (In Chinese).
Yang, Q. X. & W. C. Li, 1996a. Environmental changes since foundation of pen-fish-farming in East Taihu Lake. China Environmental Science 16: 101–106 (In Chinese with English abstract).
Yang, Q. X. & W. C. Li, 1996b. Effect of intensive pen fish farming on aquatic vegetation and its ecological strategy. Chinese Journal of Applied Ecology 7: 83–88 (In Chinese with English abstract).
Yang, Q. X., W. C. Li, L. Yu & J. Wei, 1995. Pen-fish-farming development in East Taihu Lake and its effects on lake environment. Journal of Lake Sciences, 7: 256–262 (In Chinese with English abstract).

Chapter 2
Optical Properties of Lake Taihu and Radiative Transfer Simulation

2.1 Optical Properties and Their Ecological Significance

Yunlin Zhang

Study of the optical properties of water includes quantitative descriptions of the inherent and apparent optical properties (IOPs and AOPs, respectively), the relationship between them, and their relationship to concentrations of optical constituents. The optical constituents of water bodies include pure water, chromophoric dissolved organic matter (CDOM), phytoplankton, and nonalgal particulates (Kirk, 1994).

Sunlight is both absorbed and scattered by pure water, phytoplankton, and nonalgal particulates; in contrast, sunlight is only absorbed, but not scattered, by CDOM (Kirk, 1994). If the sunlight entering water is all direct light, its attenuation is called the beam attenuation coefficient, and represents the sum of absorption and scattering; these three properties are all IOPs. For diffuse light, the diffuse attenuation coefficient describes light attenuation in different water layers, which is an AOP.

The AOPs are related to the constituents of water bodies and vary with the ratio of direct to diffused light. Generally, the absorption and scattering coefficients of pure water are constants, and their values have frequently been measured (Kirk, 1994); the coefficients of other constituents are obtained through field measurements, experiments, or calculation.

The IOPs are also related to the constituents of water bodies, but in contrast to AOPs, they do not vary with changes in incident light field. Parameters used to describe the IOPs of pure water, CDOM, phytoplankton, and nonalgal particulates include the beam attenuation coefficient, absorption coefficient, scattering coefficient, specific absorption coefficient, and volume scattering functions. According to the Lambert-Beer law, the total beam attenuation coefficient, absorption coefficient, and scattering coefficient of water bodies can be described as the linear sums of the beam attenuation coefficient, absorption coefficient, and scattering coefficient of the

Zhang, Y.L.
State Key Laboratory of Lake Science and Environment, Nanjing Institute of Geography and Limnology, Chinese Academy of Sciences, 73 East Beijing Road, Nanjing 210008, P. R. China
e-mail: ylzhang@niglas.ac.cn

Qin, B. Q. (ed.), *Lake Taihu, China*,

optical constituents (the beam attenuation coefficient is the sum of the absorption coefficient and the scattering coefficient) (Kirk, 1994).

$$c(\lambda) = c_w(\lambda) + c_{CDOM}(\lambda) + c_{ph}(\lambda) + c_d(\lambda) \tag{2.1}$$

$$a(\lambda) = a_w(\lambda) + a_{CDOM}(\lambda) + a_{ph}(\lambda) + a_d(\lambda) \tag{2.2}$$

$$b(\lambda) = b_w(\lambda) + b_{ph}(\lambda) + b_d(\lambda) \tag{2.3}$$

where $c(\lambda)$, $c_w(\lambda)$, $c_{CDOM}(\lambda)$, $c_{ph}(\lambda)$, and $c_d(\lambda)$ are the total beam attenuation coefficient and the beam attenuation coefficients of pure water, CDOM, phytoplankton, and nonalgal particulates, respectively; $a(\lambda)$, $a_w(\lambda)$, $a_{CDOM}(\lambda)$, $a_{ph}(\lambda)$, and $a_d(\lambda)$ are the total absorption coefficient and the absorption coefficients of pure water, CDOM, phytoplankton, and nonalgal particulates, respectively; and $b(\lambda)$, $b_w(\lambda)$, $b_{ph}(\lambda)$, and $b_d(\lambda)$ are the total scattering coefficient and the scattering coefficients of pure water, phytoplankton, and nonalgal particulates, respectively (it is thought that CDOM does not scatter, but only absorbs, light).

The parameters used to describe the AOPs of water include downward and upward irradiance, downward and upward radiance, irradiance ratio, water-leaving radiance, remote sensing reflectance, and the diffuse attenuation coefficients of these parameters. In studies on optical environments and their ecological effects, the diffuse attenuation coefficient and euphotic depth are of particular importance, whereas in remote sensing of water color, the irradiance ratio, water-leaving radiance, normalized water-leaving radiance, and remote sensing reflectance are of particular importance.

The AOPs represent the interaction between incident light and IOPs. In a specific solar radiation field, in which there are given angles of sun elevation and distribution of diffused light, values for AOPs can be obtained from calculation of IOPs.

The diffuse attenuation coefficient of downward irradiance was described by Kirk (1994) as

$$K_d = \frac{1}{\mu_0}[a^2 + G(\mu_0)ab]^{1/2} \tag{2.4}$$

where μ_0 is the mean cosine of the angles the photons make with the vertical just below the water surface and is dependent on the solar elevation and the proportion of direct and diffuse, respectively. μ_0 is calculated according to the sampling time, latitude, and solar declination. $G(\mu_0)$ is itself a function of solar altitude, described as

$$G(\mu_0) = g_1\mu_0 - g_2 \tag{2.5}$$

where g_1 and g_2 are constants. When K_d is calculated at the whole depth of the euphotic layer, the values of g_1 and g_2 are 0.425 and 0.190, respectively; when K_d is calculated at the half-depth of the euphotic layer, the values of g_1 and g_2 are 0.473 and 0.218, respectively (Spinrad et al., 1994).

The irradiance ratio $R(0)$ is described in relation to backscattering and absorption coefficients (Tassan, 1994):

$$R(0) = f\frac{b_b}{a + b_b} \approx f\frac{b_b}{a} \tag{2.6}$$

where b_b, a, and f are the backscattering coefficient (m^{-1}), absorption coefficient, and the coefficient of a function of solar altitude, respectively (0.33 could be used sometimes). Kirk (1984) obtained the function between f and μ_0 using Monte Carlo modeling:

$$f(\mu_0) = -0.629\mu_0 + 0.975 \tag{2.7}$$

2.1.1 Inherent Optical Properties of Lake Taihu

2.1.1.1 Optical Properties of Pure Water

Pure water appears blue because of slight absorption of blue and green light, and this absorption increases gradually with increases in wavelength of incident light over 550 nm, especially within the red waveband. For example, pure water 0–1 m deep can absorb 35% of incident red light with a wavelength of 680 nm. The absorption coefficients of different wavelength in pure water are given in Fig. 2.1 (Smith & Baker, 1981).

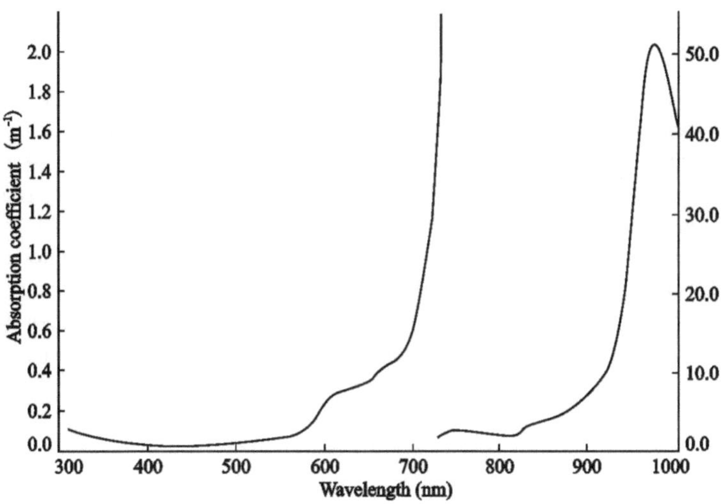

Fig. 2.1 Absorption spectra of pure water (Smith & Baker, 1981)

2.1.1.2 Optical Properties of CDOM

The concentration and formation of CDOM can significantly change the underwater light field because of its light absorption. The strongest absorption, within the ultraviolet waveband, has a marked effect on the aquatic ecosystem by restricting penetration of the harmful UV-B radiation (280–320 nm) and therefore protecting submerged aquatic vegetation (Williamson et al., 1996; Morris & Hargreaves, 1997; Laurion et al., 2000; Huovinen et al., 2003; Zhang et al., 2004a). The absorption of CDOM extends into the blue portion of the spectrum and overlaps with that of chlorophyll *a* (Chl a) in phytoplankton and nonalgal particulate matter. Thus, CDOM affects the primary productivity of water bodies, and affects remote sensing of biomass, primary productivity of phytoplankton, and total suspended matter (TSM) (Carder et al., 1991; Doxaran et al., 2002; Rochelle-Newall & Fisher, 2002a).

Most CDOM is dissolved organic carbon (DOC). Photochemical degradation of CDOM results in reduction of DOC, decrease of average molecular weight, changes in optical properties of water, and production of a complex mixture of reactive oxygen species and carbon photoproducts. Most photoproducts are inorganic compounds, such as carbon monoxide, carbon dioxide, and other dissolved inorganic carbon (DIC), that represent direct photochemical mineralization of the dissolved organic matter (DOM). Other photoproducts are low molecular weight organic molecules, amino acids, carbon disulfide, and carbonyl sulfide, which have a great impact on the biological processes. Photoproducts provide carbon and nitrogen for the growth of microorganisms.

Increasing attention is being paid to the study of CDOM as an integral part of studies on remote sensing of water color, environmental effects of UV-B radiation, microbial food webs, and the global carbon cycle.

Measurement Method

CDOM absorption and its relationship with DOC were investigated at two contrasting locations in Lake Taihu in April 2004, Meiliang Bay, a typical algae zone, and Eastern Taihu Bay, a typical macrophyte zone (sites M0–M11, E1–E14, respectively; Fig. 2.2). In our study, DOM was defined as the organic matter that remained in solution after 0.70-μm pore GF/F filtration (contrasting with CDOM obtained through a 0.22-μm pore filter in many other studies). The CDOM absorption spectra were measured between 240 and 800 nm at 1-nm intervals using a Shimadzu UV-2401PC UV-Vis recording spectrophotometer with a 4-cm quartz cuvette; Milli-Q water was used as the reference. The absorption coefficients were obtained as follows (Kirk, 1994):

$$a_{\mathrm{CDOM}}(\lambda') = 2.303D(\lambda)/r \tag{2.8}$$

where $a_{\mathrm{CDOM}}(\lambda')$ is the uncorrected CDOM absorption coefficient at wavelength λ, $D(\lambda)$ was the optical density at wavelength λ, and r is the cuvette path length in meters (m).

Absorption coefficients were corrected for backscattering of small particles and colloids that pass through filters, using the following equation (Bricaud et al., 1981; Green & Blough, 1994):

$$a_{\mathrm{CDOM}}(\lambda) = a_{\mathrm{CDOM}}(\lambda') - a_{\mathrm{CDOM}}(750') \cdot \lambda/750 \tag{2.9}$$

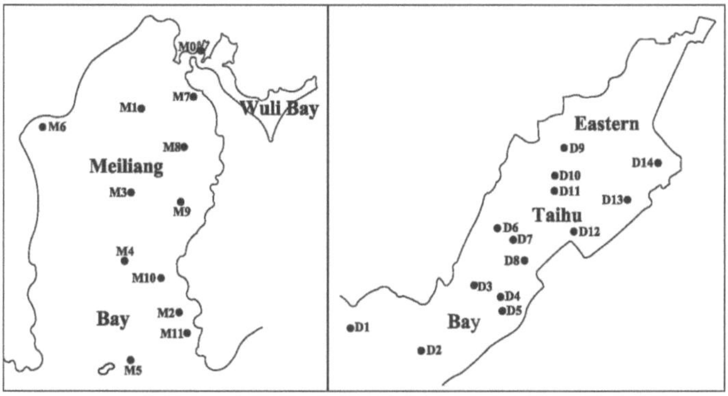

Fig. 2.2 Location of sampling sites in Meiliang Bay and Eastern Taihu Bay

where $a_{CDOM}(\lambda)$ = absorption coefficient at a given wavelength (λ) corrected for scattering, $a_{CDOM}(\lambda')$ = uncorrected absorption coefficient at a given λ, and $a_{CDOM}(750')$ = uncorrected absorption coefficient at 750 nm. The concentration of CDOM is usually described using the absorption coefficient at a wavelength of 355, 375, or 440 nm because it cannot be measured directly (Del Castillo & Coble, 2000; Stedmon et al., 2000; Kirk, 1994); in the present study, 355 nm was used.

A high-temperature catalytic oxidation analyzer (1020 TOC; range, 0.5–500 mg/L; precision, 3% relative standard deviation; I.O. Corporation) was used for the determination of DOC by direct injection of a filtered sample.

The apparent DOC-specific absorption coefficient was calculated by the following equation:

$$a_{CDOM}^{*}(\lambda) = a_{CDOM}(\lambda)/DOC \tag{2.10}$$

where $a_{CDOM}^{*}(\lambda)$, $a_{CDOM}(\lambda)$, and DOC are the DOC-specific absorption coefficient, absorption coefficient of CDOM at the wavelength of λ, and dissolved organic carbon concentration, respectively.

Because the absorption spectrum of CDOM decreases exponentially from the UV to the visible wavelength, the following equation was used to fit the spectral dependence of CDOM absorption (Bricaud et al., 1981):

$$a_{CDOM}(\lambda) = a_{CDOM}(\lambda_0)\exp[S(\lambda_0 - \lambda)] \tag{2.11}$$

where $a_{CDOM}(\lambda)$ is the absorption coefficient at wavelength λ, $a_{CDOM}(\lambda_0)$ is the absorption coefficient at a reference wavelength λ_0, and S is the spectral slope as a measure of absorption decrease with increasing wavelength.

The Spectral Absorption and Spatial Distribution of CDOM

The absorption curves of CDOM at representative stations in Meiliang Bay and Eastern Taihu Bay are shown in Fig. 2.3. The absorption coefficient of CDOM

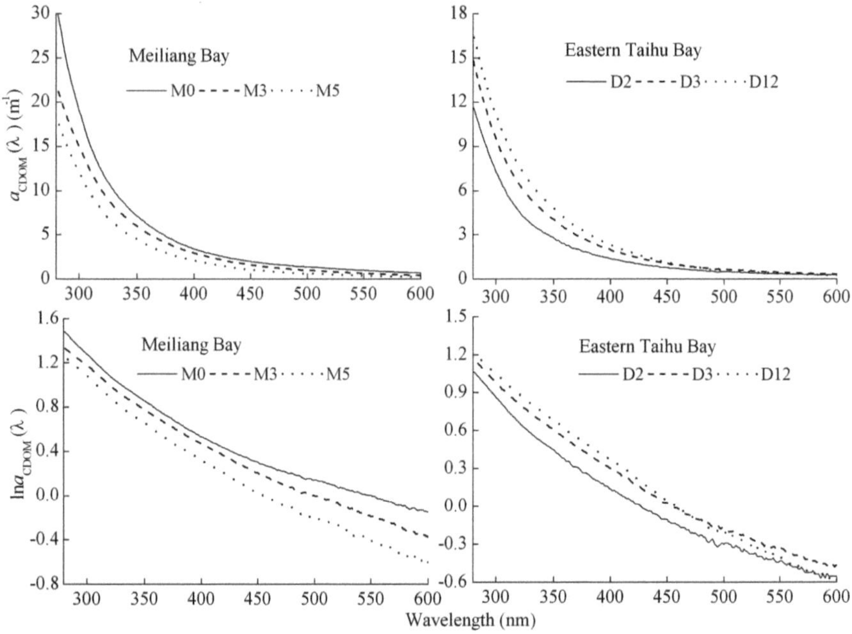

Fig. 2.3 Chromophoric dissolved organic matter (CDOM) absorption spectra of representative stations in Meiliang Bay and Eastern Taihu Bay (the *three lines* represent different sampling sites)

approached 0 near 700 nm and reached the maximum within the ultraviolet range of 280–320 nm. The mean CDOM absorption coefficients $a_{\mathrm{CDOM}}(280)$ and $a_{\mathrm{CDOM}}(355)$ of all 26 sampling stations were 17.46 ± 5.75 and $4.17 \pm 1.47\,\mathrm{m}^{-1}$, respectively.

The CDOM absorption coefficients were significantly different in Meiliang Bay and Eastern Taihu Bay [$n = 26$, $P \leq 0.0001$, one-way analysis of variance (ANOVA)], with the concentrations of CDOM in Eastern Taihu Bay, a macrophyte zone, being significantly lower than those in Meiliang Bay, an algal zone. Maximum DOC concentration, 17.2 mg/L, and maximum CDOM absorption $a_{\mathrm{CDOM}}(355)$, $8.3\,\mathrm{m}^{-1}$, were both recorded in the mouth of the Zhihugang River; minimum DOC concentration of 6.6 mg/L, and $a_{\mathrm{CDOM}}(355)$ of $2.6\,\mathrm{m}^{-1}$ were in the centre of Eastern Taihu Bay. In general, the CDOM absorption coefficients decreased from the outflow of the Liangxi River and the Meiliang Bay inlet to the outlet in Meiliang Bay, indicating that the large amount of CDOM entering the lake by rivers was subsequently diluted by the lake water. The absorption coefficients at the mouths of the Liangxi River ($a_{\mathrm{CDOM}}(355)$ of $6.7\,\mathrm{m}^{-1}$) and the Zhihugang River ($a_{\mathrm{CDOM}}(355)$ of $8.3\,\mathrm{m}^{-1}$) were much greater than those at other sampling stations ($<4\,\mathrm{m}^{-1}$).

In a comparable study of Lake Saint-Pierre, a large shallow lake in Canada, Frenette et al. (2003) also found that CDOM concentrations adjacent to river inflows were higher than those in the main part of the lake. In other bays (the estuarine of the Tyrrhenian Sea, the Orinoco River mouth, and the Huon Estuary), CDOM decreased more sharply from inlet to outlet than was the case in Lake Taihu (Seritti et al., 1998; Del Castillo et al., 1999; Clementson et al., 2004). The values for CDOM absorption coefficients we derived for Lake Taihu were similar to those from Lake Saint-Pierre (Frenette et al., 2003), and from other lakes in the northeastern United States, Colorado, and Alaska (Morris et al., 1995), and were within the range of absorption coefficients reported by Kirk (1994). However, the CDOM absorption coefficients in Lake Taihu were significantly higher than those in coastal zones and the oceans (Kirk, 1994; Ferrari & Dowell, 1998).

Because the strongest absorption of CDOM is in ultraviolet and blue light wavebands, the absorption coefficient of CDOM is often used to describe the attenuation of UV-B radiation in water bodies. Thus, the regression between the CDOM absorption coefficient and the diffuse attenuation coefficient has been determined in many studies (Morris et al., 1995; Laurion et al., 2000). The concentrations of CDOM in Meiliang Bay were significantly higher than those in Eastern Taihu Bay, thus in Meiliang Bay the UV-B-induced damage to submerged aquatic vegetation is much less than that in Eastern Taihu Bay. However, higher CDOM concentration is accompanied by photodegradation, resulting in increased penetration depth of the UV-B radiation.

The analysis of variance of DOC concentration showed that DOC concentration was significantly higher in Meiliang Bay than in Eastern Taihu Bay as CDOM absorption ($n = 26$, $P \leq 0.0001$, one-way ANOVA). Furthermore, a significant correlation was found between DOC concentration and CDOM absorption (Fig. 2.4).

DOC concentration and CDOM absorption coefficient were significantly correlated, and the correlation strengthened with decreasing wavelength within the ultraviolet and blue-green light, and the correlation coefficient decreased from 290 nm (Fig. 2.4), agreeing with measurements made in Meiliang Bay in winter 2003 (Zhang et al., 2004a). A significant positive correlation between DOC concentration and CDOM absorption coefficient and the modeling of the DOC concentration according to a regression have previously been documented (Seritti et al., 1998; Rochelle-Newall & Fisher, 2002b). However, the exact values vary with the difference in optical properties of different water bodies.

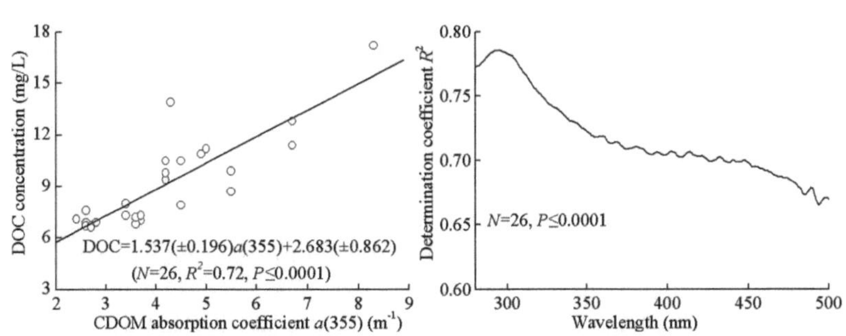

Fig. 2.4 Linear correlation between CDOM absorption coefficients and dissolved organic carbon (DOC) concentrations (*left*) and the determination coefficient of linear correlation between CDOM absorption coefficients and DOC concentrations (*right*)

The DOC-specific absorption coefficient of CDOM can be used to distinguish the sources and types of CDOM, with a lower CDOM absorption coefficient corresponding to a low DOC-specific absorption coefficient. When there is only a small variation in the DOC-specific absorption coefficients, this shows the consistency of CDOM sources. For Lake Taihu, the DOC-specific absorption coefficient of CDOM at 355 nm $a_{\mathrm{CDOM}}{}^{*}(355)$ ranged from 0.31 to 0.64 L/(mg·m) [mean, 0.46 ± 0.08 L/(mg·m)], consistent with that for lakes in the northeastern United States and Colorado and Alaska (Morris et al., 1995). Stedmon et al. (2000) found that there were no significant differences for DOC-specific absorption coefficient in Danish coastal waters, and thus the mean value could be used for the total water area. In contrast, Gallie (1997) found that the DOC-specific absorption coefficient of CDOM was significantly and positively correlated with pH and DOC concentration in lakes.

In Lake Taihu, there were no significant differences in the DOC-specific absorption coefficient of CDOM at 280, 355, and 440 nm between the two lake zones, which may indicate similar CDOM sources. CDOM from the degradation of phytoplankton can be excluded as a major source of total CDOM in April, because phytoplankton were in the growth phase. Furthermore, the significant negative correlation between the absorption coefficient of CDOM and Chl a concentration in both lake zones (Meiliang Bay, $n = 12$, $P \leq 0.05$; Eastern Taihu Bay, $n = 14$, $P \leq 0.01$) also indicated that the contribution of CDOM from phytoplankton degradation was slight.

Estimation of the Spectral Slope, S

The spectral slope, S, indicates the rate at which the CDOM absorption coefficient decreases with increasing wavelength and has been used to distinguish different sources of CDOM and its modification by chemical and biological processes (Carder et al., 1989). The value of S increases with decreasing CDOM absorption coefficient, aromatic content, and molecular weight. Therefore, it is also widely accepted that the spectral slope S can be used as a proxy for the composition of CDOM (Kowalczuk et al., 2003).

The mean S values at the 26 sampling stations in Meiliang Bay and Eastern Taihu Bay (see Fig. 2.2) for the wavelength domains of 280–500 nm, 280–360 nm, and 360–440 nm were 14.37 ± 0.73, 19.17 ± 0.84, and $13.38 \pm 0.82\ \mu\mathrm{m}^{-1}$, respectively. There were slight differences in S values within the same waveband, and significant differences in S values between wavebands. The difference in S values in the same waveband is mainly attributed to the chemical composition of CDOM, that is to say, the ratio of humic acid to fulvic acid. Carder et al. (1989) found that the S value from the turbid flow of the Mississippi River, rich in fulvic acid ($S = 19.4\ \mu\mathrm{m}^{-1}$), was significantly higher than that from the Gulf of Mexico, rich in humic acid ($S = 10.0\ \mu\mathrm{m}^{-1}$). In Lake Taihu, the mean S value in 280–360 nm was 1.43 times higher than in 360–440 nm, showing that CDOM absorption increased more quickly approaching short wavelengths, and that the S value was related to the wavelength selected and the composition of CDOM (the ratio of humic acid to fulvic acid). The

S values obtained from exponential fitting within the three wavebands all fell within the range of 10–25 μm^{-1} reported for shallow lakes by Markager & Vincent (2000).

There were no significant differences in S values between Meiliang Bay and Eastern Taihu Bay; thus, the composition of CDOM was similar in both regions, and we conclude that the CDOM came from rivers because the chlorophyll *a* concentration was low in April.

2.1.1.3 Suspended Matter

To analyze the absorption spectra of total particulates, 37 water samples were collected in Meiliang Bay, the lake centre, and Wuli Bay during an algal bloom in July 2004. The coefficients $a_p(\lambda)$, $a_d(\lambda)$, and $a_{ph}(\lambda)$ were determined by the quantitative filter technique (QFT) of Mitchell & Kiefer (1988). Water samples of 50–200 mL were collected and filtered onto 47-mm Whatman GF/F glass fiber filters under low vacuum. Absorption spectra were recorded every 1 nm from 350 to 800 nm using a Shimadzu UV-2401PC UV-Vis spectrophotometer. A blank filter, wetted with filtered water, was used as a reference. All spectra were set to 0 at 750 nm to minimize differences between sample and reference filters. Measured $OD_f(\lambda)$ were corrected for the increase in pathlength caused by multiple scattering in the glass fiber filter by using the equation from Cleveland & Weidemann (1993):

$$OD_s = 0.378\,OD_f + 0.523\,OD_f^{\,2}\,OD_f \leq 0.4 \tag{2.12}$$

where OD_s and OD_f are the optical densities of the corrected and measured particulate matter, respectively. The absorption coefficients of particulate matter $a_p(\lambda)$ were calculated as follows (Cleveland & Weidemann, 1993):

$$a_p(\lambda) = 2.303\ OD_s(\lambda)S/V \tag{2.13}$$

where 2.303 is the factor used to convert base 10 to a natural logarithm, S is the filter clearance area, and V is the filtered volume. After measurement, the filter was soaked in ethanol to exact phytoplankton pigments for 4 h and rinsed with filtered water. The absorption spectra of the extracted filter were once again measured to obtain the optical densities of nonalgal particulates. The absorption coefficient of nonalgal particulates $a_d(\lambda)$ was determined similarly using equations (2.12) and (2.13). The coefficient $a_{ph}(\lambda)$ was obtained as follows:

$$a_{ph}(\lambda) = a_p(\lambda) - a_d(\lambda) \tag{2.14}$$

Chlorophyll *a*-specific absorption was calculated as follows:

$$a^*{}_{ph}(\lambda) = a_{ph}(\lambda)/\text{Chl a} \tag{2.15}$$

Absorption Spectra of Total Particulates

Absorption spectra of total particulates, nonalgal particulates, phytoplankton, and chlorophyll-specific absorption spectra of phytoplankton are shown in Fig. 2.5. The total particulate absorption coefficient varied significantly between different stations and at different times for the same station. The $a_p(440)$ and $a_p(675)$ ranged from 1.14 to 8.78 m^{-1} and from 0.30 to 3.86 m^{-1}. There were three typical patterns of spectra for total particulate absorption coefficient (see Fig. 2.5). The first was an absorption spectrum of total suspended particulates similar to that of phytoplankton (Fig. 2.6A); the second was similar to that of nonalgal particulates (Fig. 2.6B); and the third was distinct from that of both phytoplankton and nonalgal particulates (Fig. 2.6C). Irrespective of spectrum pattern, the absorption curve of total particulates was very similar to that of phytoplankton at around 675 nm because of the characteristic absorption of phytoplankton at this wavelength.

The first type of spectrum showed that absorption coefficients of phytoplankton were usually higher than those of nonalgal particulates (Fig. 2.6A). There were absorption peaks of total particulates at 440, 675, and 440 nm; however, the former was sometimes not very significant because of nonalgal particulate absorption. Most of the 19 stations with this spectral type were in areas with relatively high Chl a concentrations and relatively low inorganic particulate concentrations in Wuli Bay and Meiliang Bay. This type of particulate matter absorption coefficient appears in

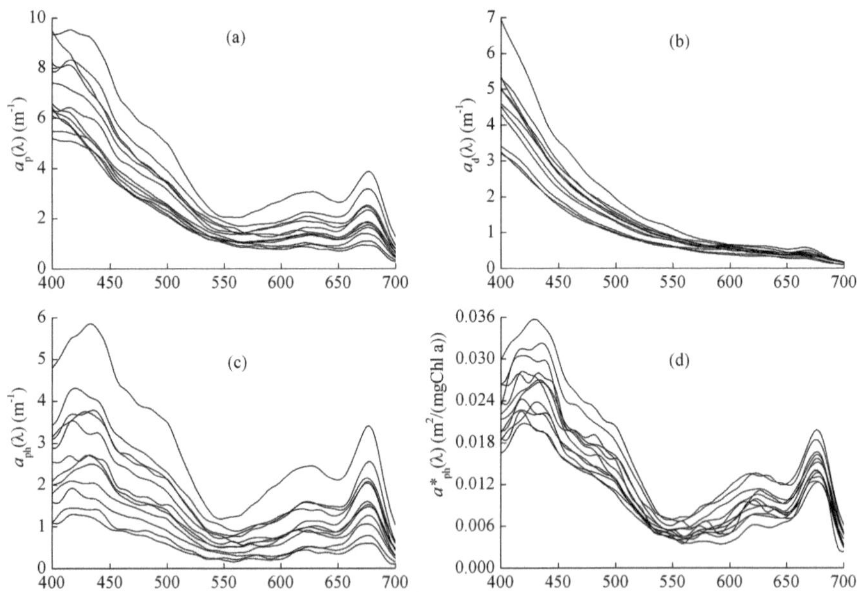

Fig. 2.5 Absorption spectra of total particulates (**a**), nonalgal particulates (**b**), phytoplankton (**c**), and chlorophyll-specific absorption spectra of phytoplankton (**d**) in summer, 2004 (different lines represent different sampling sites)

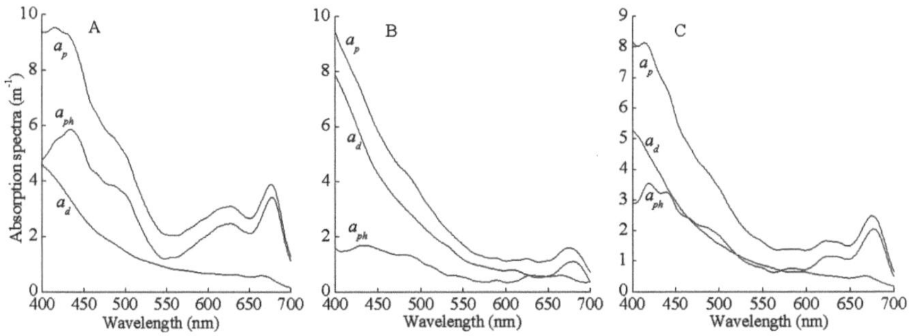

Fig. 2.6 Absorption spectra of total particulates of three typical types (a_p, total particulate absorption; a_d, nonalgal particulate absorption; a_{ph}, phytoplankton absorption)

the summer without wind waves, when algal blooms frequently occurred at stations where the percentage of organic particulates to total particulates exceeded 40%. The absorption spectra of particulate matter in the ocean and in Case I water bodies usually belong to this type (Sosik & Mitchell, 1995; Dupouy et al., 1997; Sasaki et al., 2001).

For the second type of spectrum, there was a single absorption peak of total particulates at 675 nm, at which wavelength the absorption of phytoplankton could exceed that of nonalgal particulates (see Fig. 2.6B). At other wavelengths, the absorption of phytoplankton was dominated by the absorption spectrum of nonalgal particulates. Most of the 12 stations with this spectral type were in the lake centre, where there are relatively low Chl a concentrations and relatively high inorganic particulate matter concentrations. In Lake Taihu, this second spectral type was dominant in Meiliang Bay, when sediment was resuspended by medium to large wind waves. This type of particulate absorption spectrum often occurs in Case II water bodies such as estuaries, littoral belts, and shallow lakes (Cao et al., 2003).

In the third type of spectrum, absorption by phytoplankton and nonalgal particulates each account for about 50% of total particle absorption (see Fig. 2.6C). At wavelengths below 450 nm, the absorption coefficients of nonalgal particulates were higher than those of phytoplankton, whereas above 600 nm the absorption coefficients of nonalgal particulates were lower than those of phytoplankton. This spectral type corresponded to the change in the percentage of organic particulates to total particulates in the range of 30–40% in the stations. The six stations of this type of spectrum were found in Wuli Bay and Meiliang Bay with small wind waves.

Among the 37 samples collected in the summer of 2004, the frequency of the first, second, and third types of spectra were 19, 12, and 6 times, respectively, as already mentioned. Our sampling strategy may have influenced the frequency distribution of different spectral types. For example, the sampling in Wuli Bay was in an enclosed area where there were either no wind waves or only small wind waves. The first type of spectrum only occurred in the absence of wind waves, and in the presence of algal blooms, and lower concentrations of nonalgal particulates in summer. The second type of spectrum occurred with the dominant absorption of nonalgal particulates, or

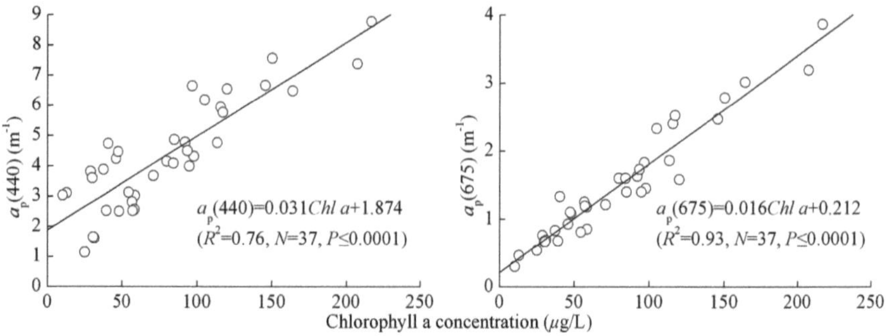

Fig. 2.7 Linear relationships between $a_p(440)$, $a_p(675)$, and Chl a concentration

large wind waves, and was often seen in winter. The third type occurred in Meliang Bay in summer with small wind waves.

There were very significant linear correlations between the particulate absorption coefficients $a_p(440)$, $a_p(675)$, and Chl a concentrations (Fig. 2.7). The samples were taken in summer when Chl a concentrations were at a maximum, and there were only small wind waves at most stations; consequently, there were low concentrations of inorganic particulates. The percentage of organic particulates to total particulates usually exceeded 30%. The determination coefficients between Chl a concentrations and $a_p(675)$ were higher than those between Chl a concentrations and $a_p(440)$, reflecting that the effect of nonalgal particulates absorption on total particulates at 440 nm was more significant than at 675 nm. The correlation was strongest for stations with the first type of spectrum, and a power function fitting was usually better than linear fitting.

The linear regressions between the absorption coefficients of particulates and concentrations of suspended matter are shown in Table 2.1. $a_p(440)$ and $a_p(675)$, the characteristic absorption wavebands of phytoplankton, were more significantly affected by organic particulates than by inorganic particulates during summer algal blooms with very small wind waves or no wind waves. However, at a wavelength of

Table 2.1 Linear regressions between absorption coefficients of particulates $a_p(400)$, $a_p(440)$, $a_p(675)$, and concentrations of suspended matter

Parameters	Regression equations	*P*	R^2
$a_p(\lambda)$, *TSM*	$a_p(400) = 0.087\ (\pm 0.008)\ TSM + 1.327\ (\pm 0.408)$	≤0.0001	0.78
	$a_p(440) = 0.057\ (\pm 0.010)\ TSM + 1.722\ (\pm 0.513)$	≤0.0001	0.48
	$a_p(675) = 0.012\ (\pm 0.006)\ TSM + 0.931\ (\pm 0.313)$	= 0.057	0.10
$a_p(\lambda)$, *OSM*	$a_p(400) = 0.202\ (\pm 0.036)\ OSM + 2.081\ (\pm 0.651)$	≤0.0001	0.47
	$a_p(440) = 0.212\ (\pm 0.020)\ OSM + 0.890\ (\pm 0.363)$	≤0.0001	0.76
	$a_p(675) = 0.102\ (\pm 0.008)\ OSM - 0.185\ (\pm 0.146)$	≤0.001	0.82
$a_p(\lambda)$, *ISM*	$a_p(400) = 0.082\ (\pm 0.013)\ ISM + 2.900\ (\pm 0.460)$	≤0.0001	0.54
	$a_p(440) = 0.042\ (\pm 0.014)\ ISM + 3.096\ (\pm 0.499)$	≤0.005	0.21

TSM, total suspended matter; OSM, suspended organic matter; ISM, suspended inorganic matter.

400 nm (short waveband), which is mainly determined by nonalgal particulates, the effect of inorganic particulates was more significant, and dominated under medium or large wind wave conditions.

Absorption Spectra of Nonalgal Particulates

The absorption coefficient of nonalgal particulates decreased with increase in wavelength (see Fig. 2.5). The $a_d(440)$ ranged from 0.41 to 4.89 m^{-1}, with a mean of $2.24 \pm 1.01\ m^{-1}$.

Many previous studies have shown that nonalgal particulates have a relatively stable characteristic absorption spectrum, which can be described using an exponential power equation (Bricaud et al., 1998; Roesler, 1998; Hojerslev & Aas, 2001). For Lake Taihu, the absorption coefficients of nonalgal particulates in samples in the range of 400–700 nm followed an exponential power, and S_d values (exponential slope) ranged from 7.9 to 12.5 μm^{-1}, with a mean of $10.08 \pm 1.01\ \mu m^{-1}$. These S_d values were lower than those in the estuary of the Zhujiang River (Cao et al., 2003), in eastern coastal inland areas of the United States (Roesler, 1998), and in the tropical Pacific Ocean and eastern Mediterranean Sea (Bricaud et al., 1998), and were higher than those in the estuary of the Lower St. Johns River (United States) (Gallegos, 2005). The S_d values for Lake Taihu were similar to those in the Yellow Sea and East China Sea (Zhu & Li, 2004), where there was a range of 5–16 μm^{-1}, most (95%) of which were within 8–12 μm^{-1}, for the wavelength domain from 380 to 620 nm.

Linear regression relationships between the absorption coefficient of nonalgal particulates and the concentration of total suspended matter (TSM), suspended organic matter (OSM), suspended inorganic matter (ISM), and Chl a are shown in Table 2.2. The absorption coefficient of nonalgal particulates at $a_d(440)$ correlated most closely with the concentration of TSM, followed by ISM; there was only a weak correlation with OSM. The results showed that inorganic particulates dominated nonalgal particulates only with few organic particulates as degradation products of phytoplankton. The absorption coefficient of nonalgal particulates at

Table 2.2 Linear regressions between nonalgal particulate absorption coefficients $a_d(440)$, $a_d(675)$, and suspended matter and Chl a concentrations

Parameter	Regression equations	P	R^2
$a_d(\lambda)$, *TSM*	$a_d(440) = 0.042\ (\pm 0.003)\ TSM + 0.287\ (\pm 0.165)$	≤ 0.0001	0.83
	$a_d(675) = 0.0044\ (\pm 0.001)\ TSM + 0.059\ (\pm 0.038)$	≤ 0.0001	0.50
$a_d(\lambda)$, *OSM*	$a_d(440) = 0.051\ (\pm 0.022)\ OSM + 1.410\ (\pm 0.386)$	<0.05	0.14
	$a_d(675) = 0.0114\ (\pm 0.002)\ OSM + 0.077\ (\pm 0.044)$	≤ 0.001	0.38
$a_d(\lambda)$, *ISM*	$a_d(440) = 0.046\ (\pm 0.003)\ ISM + 0.841\ (\pm 0.144)$	≤ 0.0001	0.79
	$a_d(675) = 0.004\ (\pm 0.001)\ ISM + 0.143\ (\pm 0.035)$	≤ 0.001	0.32
$a_d(\lambda)$, Chl a	$a_d(440) = 0.0075\ (\pm 0.003)\ Chl\ a + 1.635\ (\pm 0.297)$	<0.05	0.14
	$a_d(675) = 0.0019\ (\pm 0.000)\ Chl\ a + 0.111\ (\pm 0.031)$	≤ 0.0001	0.49

TSM, total suspended matter; OSM, suspended organic matter; ISM, suspended inorganic matter.

$a_d(675)$ correlated better with the concentration of organic particulates than inorganic particulates, mainly because of the disturbance of the characteristic absorption of phytoplankton in this waveband.

Studies on Case I waters (the optical properties determined by the concentration of phytoplankton), such as pelagic waters, showed that there was a better correlation between the absorption coefficient of nonalgal particulates and Chl a concentration than pertains in Case II waters (the optical properties determined by the concentration of phytoplankton, suspended matter and CDOM), such as Lake Taihu. In Case I waters, nonalgal particulates mainly come from degradation products of phytoplankton. However, some studies in estuaries in Case II waters showed that there was no significant correlation between the absorption coefficient of nonalgal particulates and Chl a concentration (Cao et al., 2003).

In the present study, we found that although there was only a weak linear correlation between $a_d(440)$ and Chl a concentration ($R^2 = 0.14$, $P < 0.05$), there was a significant linear correlation between $a_d(675)$ and Chl a concentration ($R^2 = 0.49$, $P \leq 0.0001$) (Table 2.2, Fig. 2.8). These results show that the degradation products of phytoplankton, after the algal blooms, are one of the major sources of nonalgal particulates in the summer, in such lake areas as Meiliang Bay and Wuli Bay when there are only small wind waves, if any at all. This result is also shown in the significant linear correlation between the concentration of organic particulates and that of Chl a concentration: $OSM = 0.127Chl\ a + 6.220$ ($R^2 = 0.75$, $n = 37$, $P \leq 0.0001$).

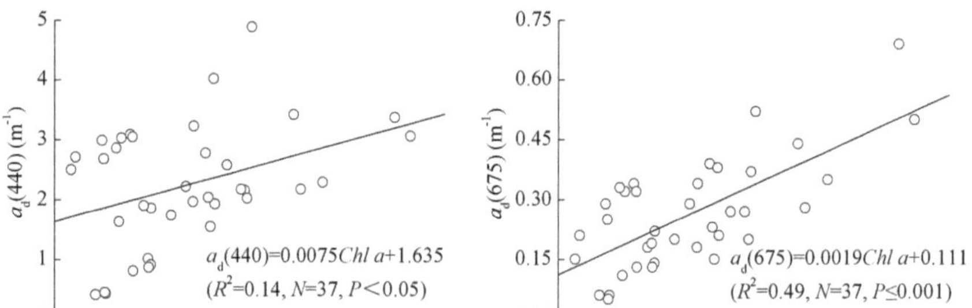

Fig. 2.8 Linear relationships between $a_d(440)$, $a_d(675)$, and Chl a concentration

Absorption Spectra of Phytoplankton

The absorption spectra of phytoplankton showed that pigments had characteristic absorption at 440 and 675 nm (see Fig. 2.5). However, there were large variations in the absorption coefficients of phytoplankton at different stations because of significant differences in pigment concentrations. For example, the $a_{ph}(440)$ values of 0.39–5.68 m^{-1} corresponded with Chl a values of 10.18–216.94 ug/L. The absorption coefficients of phytoplankton were mainly affected by the concentration of phytoplankton and composition of the population, and varied with the change in Chl a concentration (Fig. 2.9).

Studies on Case I water bodies such as deep oceans, and Case II water bodies such as estuaries and littoral belts, showed both linear and nonlinear relationships between the absorption coefficients of phytoplankton and the Chl a concentrations. For example, there was a strong linear relationship in the estuary of the Lower St. Johns River (Gallegos, 2005), a power function relationship in the estuary of the Zhujiang River (Cao et al., 2003), and an exponential function relationship in Case I and II water bodies such as the Sargasso Sea, estuaries and bays, the Mediterranean Sea, and the Atlantic Ocean (Bricaud et al., 1995), similar to that in temperate and tropical seas (Cleveland, 1995). However, a study on Case II water bodies in the northern Atlantic Ocean showed that there was almost no correlation between the absorption coefficients of phytoplankton and the Chl a concentrations (Hoepffner & Sathyendranath, 1993).

The present study in Meiliang Bay in Lake Taihu in the summer of 2004 showed that there was a very good correlation between the absorption coefficients of phytoplankton and Chl a concentration. As shown in Fig. 2.9, the absorption coefficients of phytoplankton increased with increasing Chl a concentrations, which could represented by a linear relation and by a power function, of which the latter was better. The absorption coefficient did not increase linearly, suggesting that specific absorption was not constant, and there was some difference in its specific absorption.

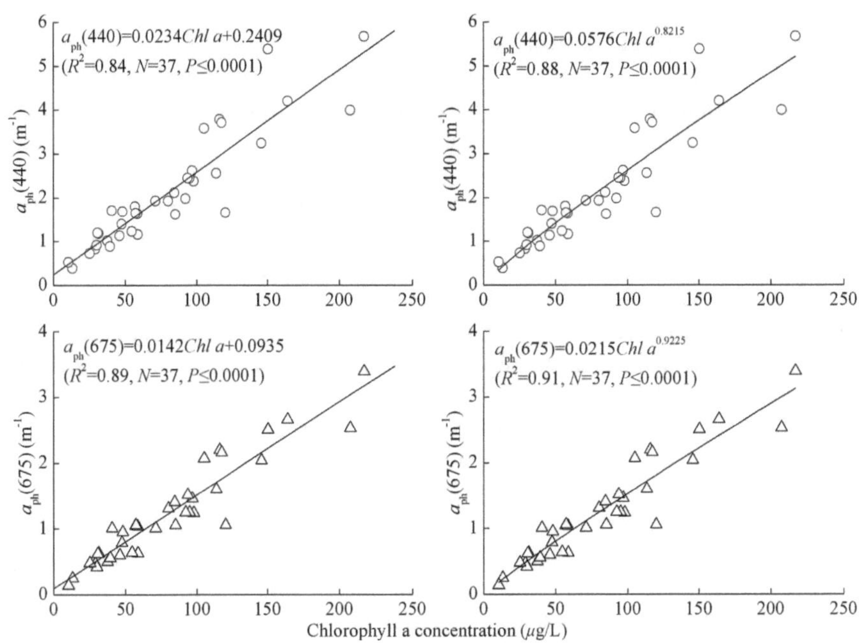

Fig. 2.9 Correlations between $a_{ph}(440)$, $a_{ph}(675)$, and Chl a concentration using linear (*left*) and power function (*right*) fitting

The chlorophyll-specific absorption coefficient is used to describe light absorption capability per unit of Chl a, and is often considered as a relative constant, averaging 0.016 m^2/(mg Chl a), when used in bio-optical modeling, in remote sensing of water color, underwater radiation, and primary production. However, the chlorophyll-specific absorption coefficient actually varies locally and seasonally and is affected by illumination and by the population structure of phytoplankton (Sosik & Mitchell, 1995; Lutz et al., 1996; Stuart et al., 2000). The chlorophyll-specific absorption coefficient lies within a range of a_{ph}*(440) is 0.01–0.18 m^2/(mg Chl a), and is usually higher in oligotrophic waters that in eutrophic waters (Bricaud et al., 1995).

In Lake Taihu, the values for the chlorophyll-specific absorption coefficients were a_{ph}*(440): 0.0135–0.0521 m^2/(mg Chl a) (mean, 0.0282 ± 0.0072 m^2/(mg Chl a)), and a_{ph}*(675) 0.0088–0.0250 m^2/(mg Chl a) (mean, 0.0159 ± 0.0033 m^2/(mg Chl a)). Corresponding values from other studies in Case I water bodies are southern California, a_{ph}*(440) 0.008–0.102 m^2/(mg Chl a), median, 0.041 m^2/(mg Chl a), a_{ph}*(674) 0.001–0.038 m^2/(mg Chl a), median, 0.015 m^2/(mg Chl a) (Millán-Núñez et al., 2004); Western subarctic Gyre, a_{ph}*(440) at the surface, 0.042 ± 0.009, at 25 m deep, 0.031 ± 0.013; Alaskan Gyre in the subarctic North Pacific, a_{ph}*(440) at the surface, 0.064 ± 0.005 m^2/(mg Chl a), at 25 m, 0.064 ± 0.016 m^2/(mg Chl a) (Sasaki et al., 2001); and in the northwest Northern Pacific (off Sanriku), a_{ph}*(440) for depths greater than 40 m, 0.082±0.022 m^2/(mg Chl a), for depths less than 30 m, 0.019 ± 0.009 m^2/(mg Chl a) (Suzuki et al., 1998).

The a_{ph}*(440) recorded in the eutrophic waters of Lake Taihu was similar to, or slightly less than, the median values in oceans, whereas the a_{ph}*(675) was significantly less than the median values in oceans (Churilova & Berseneva, 2004). These lower values in the eutrophic lake than the oligotrophic oceans may reflect that blue-green algae dominate in Lake Taihu, whereas diatoms dominate in the oceans.

Regression analysis showed that there was a significant negative correlation between a_{ph}*(440) and Chl a concentrations (Fig. 2.10), in contrast, there was generally no correlation between a_{ph}*(675) and Chl a concentrations. Millán-Núñez et al. (2004) also found that chlorophyll-specific absorption coefficients significantly and negatively correlated with Chl a concentrations and primary production ($P < 0.05$). The significant and negative correlation at 440 nm might indicate that the absorption of phytoplankton was disturbed by nonalgal particulates. Nonalgal particulates could also play a shielded role in the absorption of pigments except for the mutual shielded role among pigments themselves. Therefore, there was a significant packing effect, and the chlorophyll-specific absorption coefficients decreased with the increase of Chl a concentration. In contrast, at 675 nm, there was only very slight absorption by nonalgal particulates, algae began to dominate particulate absorption, and there was no competition and shielding, so consequently there was no packing effect. A previous study also showed that the packing effect was related to wavelength, and the more obvious the absorption peak, the more significant the packing effect (Suzuki et al., 1998).

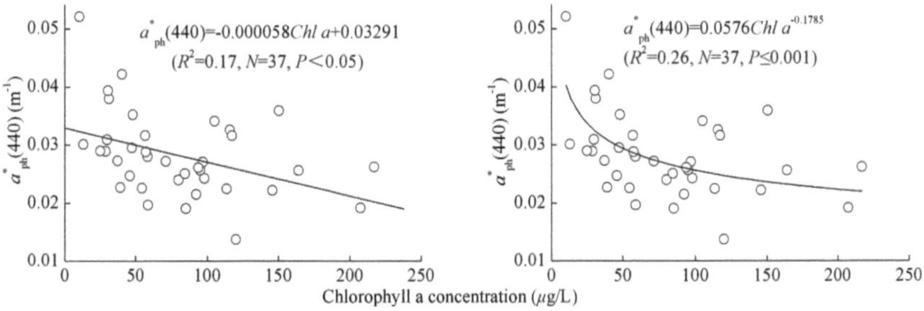

Fig. 2.10 Correlations between $a_{ph}{}^*(440)$ and Chl a concentration (linear and power correlations)

2.1.2 Apparent Optical Properties of Lake Taihu

2.1.2.1 Measurement Method

Underwater irradiance was measured in 2004 in Meiliang Bay, an algal zone (M0, M6, M12: April 5; M0–M11: April 16), and in Eastern Taihu Bay, a macrophyte zone (E4, E7, E9, E11: April 4) (see Fig. 2.2). Incident photosynthetic available radiation (PAR, 400–700 nm) at the water surface, and at different depths (0, 0.2, 0.5, 0.75, 1.0, 1.5 m), was measured using a Li-Cor 192SA sensor connected to a Li-Cor1400 datalogger. The spectrum of underwater irradiance at the different depths was measured with a scanning spectroradiometer (Macam SR9910).

Diffuse attenuation coefficients for downward irradiance $K_d(\lambda)$ were obtained from the nonlinear regression of the underwater irradiance profile (Kirk, 1994). Only $K_{\mathrm{d}}(\lambda)$ values from regression fits of $R^2 \geq 0.95$ were accepted, with at least three depths per fit.

2.1.2.2 Diffuse Attenuation Coefficient

The spectral attenuation coefficients from 400 to 750 nm at stations in typical algal and macrophyte zones are given in Fig. 2.11. Marked differences in spectral attenuation coefficients were found for different stations. At all stations, spectral attenuation coefficients were largest within the blue light band, because of strong absorption by nonalgal particulates and CDOM within the short wavebands. The absorption coefficients of nonalgal particulates and CDOM decreased exponentially with increase in wavelength.

The sampling stations fell into two types, according to the wavelength at which the minimum spectral attenuation coefficients were seen: (i) minimum coefficient in the green light band (about 580 nm), stations E4, E7, E9, E11 in Eastern Taihu Bay, and at stations M2 and M9 in Meiliang Bay without wind waves; and (ii) minimum in the red light waveband (about 700 nm), all other stations.

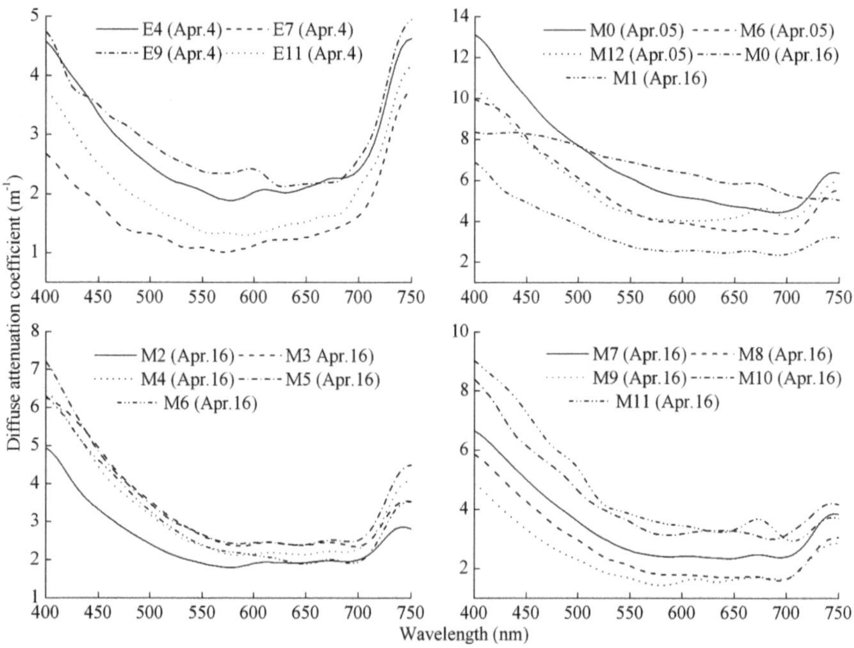

Fig. 2.11 Spectral attenuation coefficients from 400 to 750 nm at different station in Lake Taihu (*E*, sites in Eastern Taihu Bay, a macrophytes zone; *M*, sites in Meiliang bay, an algal zone)

If the minimum spectral attenuation coefficients appeared in the green light band, the corresponding water bodies were usually clear and the concentrations of suspended matter and Chl a were low. In such stations, CDOM and pure water play important roles in light attenuation. The light attenuation of CDOM is concentrated below 550 nm, whereas that of pure water is concentrated above 600 nm (Kirk, 1994). There are very marked difference for pure water absorption at short and long wavelength, for example, the absorption coefficients at wavelengths of 580 and 700 nm were 0.108 and 0.65 m^{-1}, respectively (Kirk, 1994). In these clear water stations, a green shift in the spectrum occurs with increasing depth, that is to say, the ratio of green light in the total light increases with increasing depth.

In the stations where the minimum spectral coefficient was in the red waveband, the water was usually turbid, and concentrations of suspended matter and Chl a were relatively high. The absorption and scattering by nonalgal particulates was concentrated below 600 nm, and phytoplankton had an absorption peak at the red light waveband of 675 nm (Kirk, 1994). Therefore, the absorption and scattering in the green light waveband increased. Furthermore, phytoplankton also had an absorption peak at 440 nm. In this situation, the percent of light attenuation by pure water had decreased significantly, so the minimum spectral attenuation coefficient was not at the long waveband of 700 nm, but at the green light waveband of 580 nm. However, at some stations, attenuation coefficients were relatively low, near 580 nm. With increasing depth, the relative contribution of blue light to PAR decreased, whereas the relative contribution of red light markedly increased. Furthermore, the attenuation coefficient increased significantly because of the increasing absorption of pure water at the waveband above 700 nm. Light at short wavelengths (ultraviolet and blue) is the most attenuated.

Green or red is usually the most penetrating wavelength in shallow lakes. In Lake Taihu, green light is the least attenuated in Eastern Taihu Bay, whereas red is the least attenuated in Meiliang Bay (see Fig. 2.11). An attenuation peak was recorded at 675 nm at station M12 on April 5, and at stations M0 and M11 on April 16 (Fig. 2.11), resulting from phytoplankton absorption because of the relatively higher Chl a concentrations than other stations.

To clearly show the wavebands appearing in the minimum spectral attenuation coefficients, Fig. 2.12 depicts the ratios of spectral attenuation coefficients to attenuation coefficient at the 580- and 700-nm wavelengths for two types of stations. For the first type of station, the attenuation coefficient remained rather constant within the range of 580–700 nm, and increased significantly from 700 nm, with the attenuation coefficient at 750 nm approaching, or exceeding, that at 440 nm. The results show that the pure water, clear water with low concentrations of CDOM and particles, plays an important role in absorption, especially in the longer wavebands. For the second type of stations, the attenuation coefficient was minimal at 700 nm, then increased gradually, but at 750 nm it was still less than at 440 nm, showing that the relative contribution of pure water to the total absorption decreases.

During the measurement period of April 4–5 and April 16 in 2004, the PAR attenuation coefficient ranged from 1.24 to 7.77 m^{-1}. The values in Meiliang Bay were significantly higher than those in Eastern Taihu Bay ($n = 19$, $P < 0.005$). There was a significant spatial difference in PAR attenuation in Meiliang Bay and Eastern Taihu Bay. In Meiliang Bay, of two samplings, K_d(PAR) values on April 5 were much higher than those on April 16. The average K_d(PAR) value on April 5 was 1.9 times of that on April 16.

As a large shallow lake, Lake Taihu has many types of microhabitat, each of which may have a different attenuation coefficient. In general, K_d(PAR) is low in

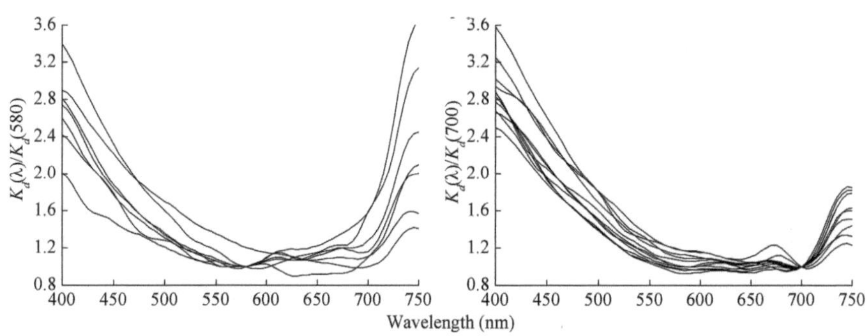

Fig. 2.12 Ratios of spectral attenuation coefficients to attenuation coefficient at 580 (*left*) or 700 nm (*right*)

the macrophyte-dominated regions of the lake because of low concentrations of optical constituents. The PAR attenuation coefficients also vary temporally because Lake Taihu is a shallow lake, and thus physical, chemical, and biological factors are variable, with wind waves being the most significant factor changing TSM concentration. In Meiliang Bay, on April 5, 2004, TSM concentrations reached 86.20 and 33.48 mg/L at stations 0 and 6, respectively, whereas on April 16, 2004, concentrations were only 22.48 and 11.84 mg/L, respectively. The decreased TSM concentration lead to significantly increased PAR attenuation coefficients between the two measurement dates at the same sampling station.

Linear regressions between K_d(PAR) and major environmental factors showed that the most significant factor affecting PAR attenuation coefficients in Lake Taihu was the concentration of ISM, mainly because inorganic particulates dominate the suspended matter in the water bodies without algal blooms (Table 2.3). ISM usually accounted for more than 50% of TSM, and reached 80–90% during large wind waves. The K_d(PAR) values were significantly and positively correlated with the ratios of ISM to OSM and TSM, but were significantly and negatively correlated with the ratios of OSM to the TSM (Table 2.3). The higher the ratio of ISM to TSM, the higher the attenuation coefficient. The results also showed that inorganic

Table 2.3 Correlations between diffuse attenuation coefficients (K_d(PAR)), DOC, absorption coefficients (a_d), Chl a, and suspended matter ($N = 19$) (see separate comments)

Equations	R^2	P
K_d(PAR) =		
$1.25 + 0.078\ TSM$	0.94	≤0.0001
$0.430\ OSM + 0.337$	0.67	≤0.0001
$0.084\ ISM + 1.650$	0.88	≤0.0001
$0.757\ ISM/OSM + 1.190$	0.53	≤0.0001
$0.856\ ISM/TSM - 1.562$	0.26	≤0.05
$-0.856\ OSM/TSM + 5.294$	0.26	≤0.05
$0.34\ DOC - 0.35$	0.34	≤0.01
$0.038\ Chl\ a + 2.43$	0.20	=0.057
$0.800\ a(440) + 0.320$	0.74	≤0.0001
$3.683\ a(580) + 0.108$	0.75	≤0.0001
$2.391 a(675) + 0.730$	0.26	≤0.05
$2.06 + 1.11 a_{CDOM}(380) + 0.037 Chl\ a + 0.073\ TSM$	0.98	≤0.0001
$0.758 + 0.041 DOC + 0.014 Chl\ a + 0.072\ TSM$	0.98	≤0.0001
$K_d(440)$ =		
$1.348 a(440) + 0.855$	0.85	≤0.0001
$1.624 a_p(440) + 2.433$	0.78	≤0.0001
$K_d(580)$ =		
$3.119\ a(580) + 0.237$	0.57	≤0.0001
$5.514\ a_p(580) + 0.964$	0.70	≤0.0001
$K_d(675)$ =		
$2.789\ a(675) + 0.010$	0.54	≤0.0001
$2.465\ a_p(675) + 1.745$	0.46	≤0.001

PAR, photosynthetic available radiation; TSM, total suspended matter; OSM, organic suspended matter; ISM, inorganic suspended matter; DOC, dissolved organic carbon.

particulates play a key role in PAR attenuation, reflecting that suspended matter dominates PAR attenuation.

2.1.2.3 Irradiance Ratio

The ratio of upward irradiance to downward irradiance is the irradiance ratio. The irradiance ratio at the surface water $R(0)$ can be directly related to the composition of matter in water bodies (Kirk, 1994). In remote sensing of water color, this parameter is used to estimate the concentrations of TSM, Chl a, and DOC (or CDOM). The irradiance ratio could be described using the ratio of backscattering coefficient to total absorption coefficient (Tassan, 1994):

$$R(0,\lambda) = \frac{E_u(0,\lambda)}{E_d(0,\lambda)} = f\frac{b_b(\lambda)}{a(\lambda) + b_b(\lambda)} \approx 0.33\frac{b_b(\lambda)}{a(\lambda)} \tag{2.16}$$

where $R(0, \lambda)$, $b_b(\lambda)$, and $a(\lambda)$ are the irradiance ratio at the surface of the water, backscattering coefficient (m^{-1}), and total absorption coefficient (m^{-1}), respectively.

The curve of the irradiance ratio at the surface water, characteristic of inland water bodies, is shown in Fig. 2.13. The irradiance ratio was very low in the range of 400–500 nm because of the high absorption of nonalgal particulates, CDOM, and phytoplankton. The irradiance ratio could be described as the ratio of the backscattering coefficient to the total absorption coefficient (equation (2.16)). Therefore, the irradiance ratio will decrease with the increase of the total absorption coefficient, when back scattering is stable. It is generally thought that CDOM absorbs light, but does not scatter light, and its absorption increases exponentially with decreasing wavelength.

In Meiliang Bay, the lowest irradiance ratio at the 400 nm wavelength was found at site M0 on April 5, 2004, because the concentration of suspended matter reached 86.20 mg/L, significantly higher than that of other stations. At site 12 in Meiliang Bay, the concentration of Chl a reached 45.50 μg/L; a distinct irradiance ratio low

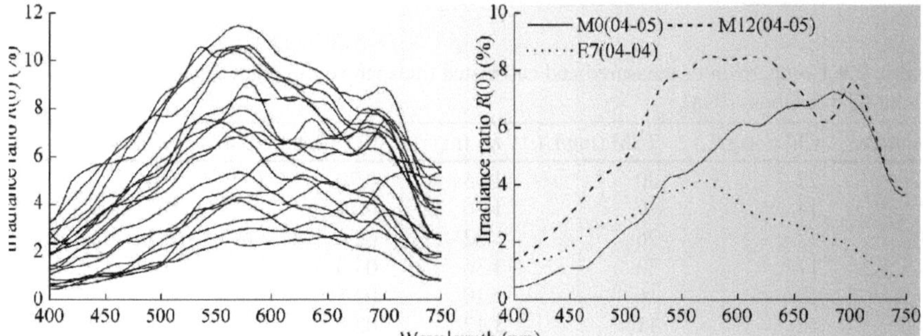

Fig. 2.13 Spectral distribution of irradiance reflectance (all sites at *left* and three typical sites at *right*) in typical lake zones of Lake Taihu

value was found at around 675 nm. The peaks of the irradiance ratio occurred within the range 560–590 nm because of decreasing absorption of photosynthetic pigment and nonalgal particulates and increasing backscattering. The low value at 675 nm was mainly attributed to the characteristic absorption of Chl a in this waveband. A second peak was found around 705 nm.

The values, peaks, and minima of the irradiance ratio in the Lake Taihu field study agreed with the results from lakes outside of China, and from our modeling study (Dekker et al., 1992; Gons et al., 1998; Gons, 1999; Zhang et al., 2004b). For example, Gons (1999) reported that, in the shallow lakes Vecht and Overijssel in the Netherlands, the minimum $R(0, \lambda)$ appeared within the blue and violet waveband, and there were peaks at 584 and 704 nm. Studies by Dekker et al. (1992) in Lake Loosdrecht, the Netherlands, showed that the minimum $R(0, \lambda)$ was within 400–500 nm, there was a peak at 600 nm, then a dip within 675–680 nm from pigment absorption, and within 700–720 nm because of pigment fluorescence.

2.1.3 Modelling of Transparency

2.1.3.1 Calculation of Transparency of Water Bodies in Lake Taihu

Transparency is a fundamental optical property of water bodies and an important indicator of eutrophication. Transparency is usually measured using a Secchi disk and can be calculated from the following equation:

$$ST = \frac{1}{K_{\rm d}}\left[1.946 + \frac{1}{2}\left(\frac{B_{O(d)}}{B_{O(u)} - B_{l(u)}}\right)\right] \tag{2.17}$$

where $K_{\rm d}$, $B_{\rm O(d)}$, $B_{\rm O(u)}$, and $B_{\rm l(u)}$ are the diffuse attenuation coefficient, downward radiance at the surface water, upward radiance at the surface water, and upward radiance at the depth of l meters, respectively. According to the equation above, Cai & Gons (1998) calculated the transparency of different lake zones in Lake Taihu, and these data are given in Table 2.4.

Table 2.4 Comparison of measured and calculated transparency in eight different zones in Lake Taihu (Cai & Gons, 1998).

Stations	Chl a (μg/L)	TSM (mg/L)	$K_{\rm d}$ (m^{-1})	SD measured (m)	SD calculated (m)
1	27	20	1.35	0.70	0.64
2	14	17	1.16	0.70	0.73
3	17	26	1.32	0.50	0.50
4	17	33	1.36	0.40	0.48
5	12	16	1.10	0.85	0.85
6	12	11	1.12	0.90	0.87
7	12	11	1.03	1.00	1.06
8	7	10	1.02	1.00	0.78

SD, Secchi depth.

2.1.3.2 The Relationship Between TSM and Transparency

Transparency of lake water, used to describe the penetration depth of light, varies with differences in absorption and scattering of incident light resulting from lake water, suspended matter, phytoplankton and CDOM. Therefore, the composition and concentration of TSM in lake water significantly affects transparency, with increase of TSM concentration being accompanied by decreased transparency (Fig. 2.13).

For studies in oceans, a logarithmic function was performed on suspended matter and transparency (Qin et al., 1986; Ping, 1993; Su et al., 2001). Therefore, the following equation was obtained, based on numerous measurements in Lake Taihu.

$$TSM^{1/4} = -0.690\,(\pm 0.035)SD + 1.750\,(\pm 0.033)\,(R^2 = 0.74, N = 135, P \leq 0.0001) \quad (2.18)$$

where *TSM* and *SD* are total suspended matter concentration (mg/L) and transparency (m), respectively. Figure 2.14 and equation (2.18) show that TSM was closely related to transparency, and thus transparency could be estimated from the TSM concentration. More than 80% of the suspended matter was inorganic particulates; therefore, it was this component, and not phytoplankton, that determined variation in transparency in Lake Taihu.

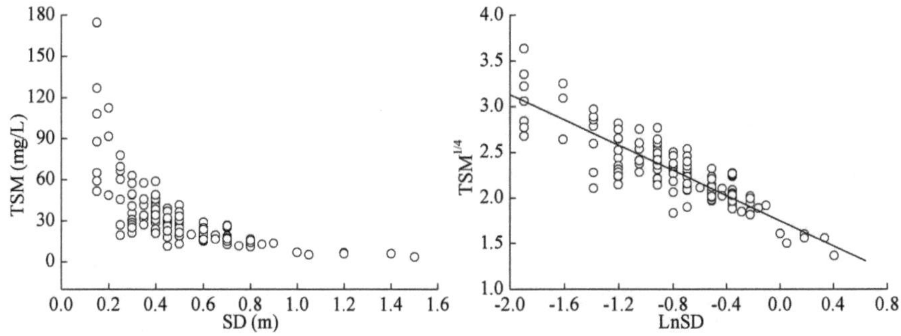

Fig. 2.14 Correlation between total suspended matter (*TSM*) and transparency (*SD*)

2.1.3.3 Relationship Between Chl a Concentration and Transparency

There was no significant relationship between Chl a concentration and transparency (Fig. 2.15). When Chl a concentration was high, transparency was sometimes high and sometimes low, demonstrating that Chl a was not the dominant factor affecting transparency. However, Chl a concentration did have some affect on transparency, because when Chl a concentration was relatively high, transparency was generally relatively low (Fig. 2.15). Our findings showed that in Lake Taihu the effect of Chl a concentration on transparency was less significant than that of TSM, in contrast with the situation in oceans (Prieur & Sathyendranath, 1981; Li et al., 1997).

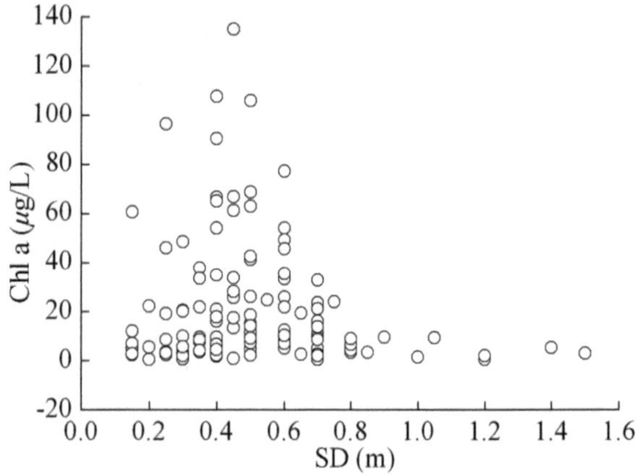

Fig. 2.15 Correlation between Secchi depth (*SD*) and *Chl a* concentration

The linear and nonlinear regression analyses all showed that there was no significant correlation between Chl a and Secchi depth in Lake Taihu. In this large shallow lake, the sediment is easily resuspended by wind waves, resulting in the water becoming turbid, and change of an optical parameter such as transparency is mainly determined by the composition and concentration of TSM. There was only a significant relationship between Chl a concentration and transparency during the algal bloom season.

2.1.4 Ecological Effect of Underwater Radiation

2.1.4.1 The Effects of UV-B Radiation on the Lake Ecosystem

As a result of destruction of the ozonosphere and the appearance of the ozone hole, absorption of UV-B by ozone is becoming weaker, resulting in increased harmful UV-B (280 nm) reaching the earth's surface (Farman et al., 1985; Tevini, 1993; Kerr & Mcelroy, 1993). Although UV-B radiation occupies only a small part of total solar irradiance, it can profoundly influence physiological and ecological entities in lake ecosystems, such as photosynthesis, growth, biomass, population structure, and biodiversity.

Williamson (1995) and Laurion et al. (2000) showed that the effect of UV-B radiation on the lake ecosystem was complex and was not limited to inhibiting the growth of phytoplankton and decreasing primary productivity. Because different individual organisms, species, and populations have different sensitivity to UV-B radiation, the increased radiation will lead to changes in composition and structure of lake ecosystems, including food chains. Organisms living in lakes have a number of physiological mechanisms which, if activated, can serve to reduce UV-B damage. For example, swimming animals could dive deeper to avoid UV-B radiation, and other animals can secrete light-protective and antioxidant compounds, such as mycosporines, scytonemins, and mycosporine-like amino acids (MAAs), to protect their critical tissues from UV-B damage (Hessen, 1993; Karentz et al., 1991). Although these mechanisms may serve to protect individuals, UV-B may still change the existing ecological interactions between organisms and their environments. The response of lake ecosystems to UV-B is mainly embodied in the adaptation of organisms to UV-B and the change of the composition, structure, and function of the lake ecosystems resulting from the effect of UV-B radiation.

Moreover, the increase in the ratio of UV-B to PAR radiation will decrease carbon fixation and increase the release of carbon dioxide in the aquatic ecosystem. The resulting increase of carbon dioxide in the atmosphere will accelerate the global greenhouse effect, which will ultimately affect global climatic change.

Studies of the ecological effects of increased of UV-B radiation on lake ecosystems are focused on (1) the attenuation law of UV-B in lake water bodies; (2) the interaction between UV-B and dissolved organic matter; (3) the harmful effects of UV-B on organisms in the lake; and (4) the response of the lake ecosystem to the increased UV-B radiation.

Transmission of UV-B in water is mainly affected by CDOM, total suspended particulate matter, and phytoplankton. The diverse compositions of these substances differently affect UV-B attenuation, and UV-B attenuation coefficients range from 0.005 to 30 m^{-1}.

In clear sea water, UV-B could be found at depths as great as 30 m (Smith et al., 1992). In 26 mountain lakes in the Alps and Pyrenees, the mean UV-B euphotic depth was 8 m and the maximum was 27 m (Sommaruga, 2001). In eutrophic shallow lakes (or lakes significantly affected by human activity), the UV-B euphotic depth is usually very shallow. For example, in 75% of the lakes in the Midwest of the United State, Florida, Quebec, and the Great Plains of Canada, the UV-B euphotic depth was less than 0.5 m (Williamson et al., 1996; Arts et al., 2000). In Lake Taihu, the depth profiles of 380 nm underwater irradiance have been measured in different regions, and the results are shown in Fig. 2.16. In general, the 1% penetration depth of UV-B was very shallow, and UV-B had been absorbed completely at a depth of 0.5 m.

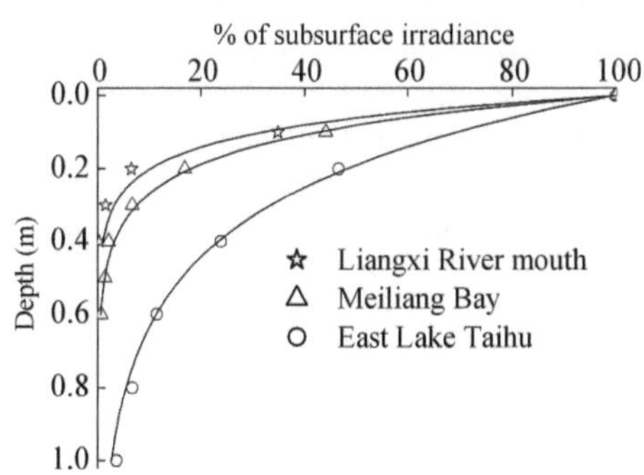

Fig. 2.16 Depth profiles of 380 nm UV irradiance for three regions of Lake Taihu

2.1.4.2 Photochemical Processes

Study of photochemical processes, such as the chemical and biological effects of light on CDOM, is a major topic in aquatic optics. Photochemical degradation can transform dissolved organic matter to inorganic forms (such as inorganic carbon and carbon dioxide) or low molecular weight C-H compounds. The degradation can lead to light absorption by dissolved organic matter, depending on the properties of dissolved organic matter and dissolved inorganic matter. The output of dissolved inorganic C resulting from photochemical degradation is positively correlated with high concentrations of Fe and the high absorption rate of ultraviolet radiation and negatively correlated with pH, conductivity, alkalinity, and Chl a concentration in water.

Therefore, photochemical processes occur more readily in eutrophic lakes, such as Lake Taihu, than in oligotrophic lakes (Bertilsson & Tranvik, 2000). Photochemical process bleach the color of the lake water ("light bleaching"), which influences the optical properties of the water, and consequently changes the penetration depth of radiation in the water, thus affecting photosynthesis and lake primary productivity.

Photochemical processes are of great importance to the carbon cycle in lake water. Photochemical dissolution of organic matter favors bacteria activity and accelerates the decomposition of dissolved organic matter. Many researchers (Vodacek et al., 1997; Gao & Zeep, 1998; Del Castillo et al., 1999; Moran et al., 2000) have indicated that this photochemical degradation leads to a slight change of CDOM spectral slope (S); decreased S values were reported by Gao & Zeep (1998) and Del Castillo et al. (1999), whereas increased S values were reported by Vodacek et al. (1997) and Moran et al. (2000). Some quantitative models studied the effect of light on the cycle of dissolved organic matter (Miller & Zepp, 1995; Amon & Benner, 1996; Moran & Zeep, 1997), but these studies were limited by the use of short-term data to predict long-term effects.

2.1.4.3 Bio-optical Model and Estimation of Primary Production

Study of the relationship between light and primary productivity is the first task of lake optics, hydrology, and biology, and will deepen our understanding of the distribution of PAR of lake water and lake primary productivity. Such knowledge can support lake management, particularly of eutrophication. Many bio-optical models are successfully used to calculate primary productivity (Kiefer & Mitchell, 1983; Smith et al., 1989; Sathyendranath et al., 1989; Kirk, 1994). Among them, the model of Smith et al. (1989) is the most representative. The primary productivity of phytoplankton at depth Z can be expressed as

$$P(Z) = \varphi(Z) \times 12\,000 Q_{\text{phar}}(Z) \tag{2.19}$$

where $P(Z)$ is the rate of phytoplankton photosynthesis, $\varphi(Z)$ is the quantum yield, 12,000 is a constant for converting moles of carbon to milligrams of carbon, and $Q_{\text{phar}}(Z)$ is the photosynthetically absorbed radiation, defined as

$$Q_{\text{phar}}(Z) = \int_{400nm}^{700nm} Q(\lambda, Z) \left[\sum_{i=1}^{n} a_i(\lambda) C_i(Z) \right] d\lambda \tag{2.20}$$

in which $Q_{\text{phar}}(\lambda, Z)$ is the spectral quantum irradiance at depth Z, $a_{\text{i}}(\lambda)$ is the pigment-specific absorption coefficient for the major absorbing pigment groups presented in the sample, and $C_{\text{i}}(Z)$ is the pigment concentration at depth Z.

Yu & Cai (1998) used the following Photosynthesis-Irradiance (P-I) model to compute primary productivity in Lake Taihu. The model gave good estimates for the influence of suspended matter, CDOM, and Chl a on primary productivity:

$$P_v(z) = B(Z)\left[P_{\max} f(I_b, \beta, I) \tanh(I/I_k) + r_b\right] \tag{2.21}$$

$$P_A = \int_0^{Z_{\max}} P_v(Z)\, dZ \tag{2.22}$$

where $P_v(\text{Z})$ is the rate of phytoplankton photosynthesis (mgC/(m^3·h)), $B(\text{Z})$ is Chl a concentration at depth $Z(\text{mg/m}^3)$, $P_{\max}$ is maximum rate of phytoplankton photosynthesis, I is light intensity, I_{k} is P-I saturation parameter, I_{b} is light inhibition parameter, β is light inhibition index, h is intercept, r_{b} is amelioration value, P_{A} is primary productivity of the lake water body (mgC/(m^3·h)), and $Z_{\max}$ is the maximum depth.

Cai & Yang (1993a) also used a bio-optical model and showed that the primary productivity of Lake Taihu was concentrated in the first 0.8 m of the surface water; primary productivity below a depth of 1 m was very low (Fig. 2.17). The results from field samples, using the dark and light bottle dissolved oxygen method (the Winkler method), are also shown in Fig. 2.17. The modelled value for primary productivity agreed closely with measured values. In autumn (October), the measured value was between the modeled ones for the winter and the summer.

The day-to-day variation in the P_{v} value was very significant, because primary production of the water column was affected by light, temperature, and nutrients. Therefore, the monthly mean primary production in a season could not be equal to the primary production in a day. In the model, the parameters such as Chl a concentration and suspended matter concentration were not measured on the same day; therefore, the comparison of modeled and measured values above only showed that the modeled value was reasonable on the whole. More studies are required for a detailed comparison between modeled and measured primary productivity.

There was significant daily variation in lake surface radiation in both summer and winter. In summer, there were many hours of sunshine, solar radiation was strong, and the radiant energy that entered water within 1 day was more than that on 1 day in winter. Thus, primary production in the water columns in summer was much higher than that in winter (Fig. 2.18). Moreover, in summer, primary production varied significantly between sunrise and sunset, and was saturated between 1000 and 1500. In winter, primary production was much less than in summer and was saturated for only a short time.

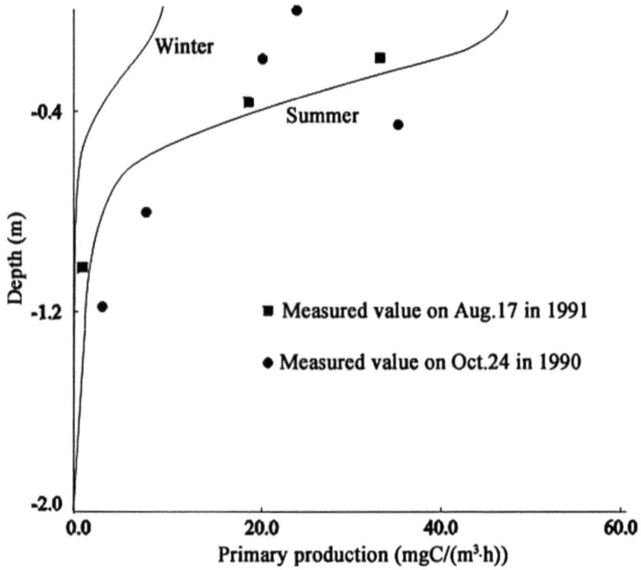

Fig. 2.17 Comparison of modeled and measured values for average primary production in Lake Taihu as a function of water depth (Cai & Yang, 1993a)

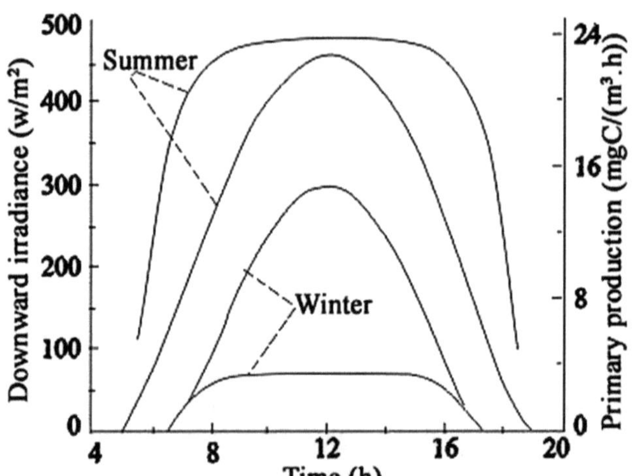

Fig. 2.18 Diurnal variation of downward irradiance (*dashed lines*) on the surface of Lake Taihu and primary productivity of the water column per unit area (*solid lines*) for summer and winter (Cai & Yang, 1993a)

2.2 Simulation of Radiative Transfer Between the Atmosphere and the Water Body

Qiming Cai and Yunlin Zhang

Transmission of solar radiation between the atmosphere and the water body is a complex and interrelated process. The solar radiation is absorbed and scattered by various components in the atmosphere, then enters the water body, where its intensity and spectral composition change as a result of the influence of pure water, CDOM, phytoplankton, and nonalgal suspended particulates. Part of the radiation in the water body will be reflected back to the atmosphere (this is the radiant energy detected by remote sensing), and a part of the radiation returning to the atmosphere will return to the water body again because of backscattering by the atmosphere.

To determine the properties of solar radiation in the water body in Lake Taihu, we studied the atmosphere and the water body as a united system. Yang et al. (1992), using the long-term monthly average weather data for Wuxi City, and a clear sky, radiative transfer spectrum model, presented the distribution of PAR at the water surface in summer and winter. Cai & Yang (1993b) then presented a numerical model of solar radiation transfer in the atmosphere–water body system to determine the distribution of the underwater light field in Lake Taihu.

2.2.1 The Radiative Transfer Spectrum Model in the Atmosphere

In the model, the principal factors affecting radiative transfer in the atmosphere are (1) ozone absorption in the upper atmosphere, (2) Rayleigh scattering of the molecules in the atmosphere, (3) water vapor absorption in the atmosphere, (4) aerosol scattering and absorption, and (5) absorption by carbon dioxide and other gas mixtures. Based on the previous studies of Brine & Iqbal (1983), Justus & Paris (1985), and Bird & Riordan (1986), the following model is proposed:

2.2.1.1 The Clear Sky Radiative Transfer Parameter Model

The Direct Transmittance of the Atmosphere, $T_{A\lambda}$

The direct transmittance of the atmosphere can be computed according to the Lambert-Beer law:

$$T_{A\lambda} = T_{r\lambda} T_{o\lambda} T_{w\lambda} T_{ps\lambda} T_{Pa\lambda} T_{u\lambda} \tag{2.23}$$

Cai, Q.M.
State Key Laboratory of Lake Science and Environment, Nanjing Institute of Geography and Limnology, Chinese Academy of Sciences, 73 East Beijing Road, Nanjing 210008, P. R. China

where λ is the wavelength of radiation, $T_{r\lambda}$, $T_{o\lambda}$, $T_{w\lambda}$, $T_{ps\lambda}$, $T_{pa\lambda}$, and $T_{u\lambda}$ are the transmittance caused by Rayleigh scattering, ozone absorption, water vapor absorption, aerosol scattering, aerosol absorption, and the atmosphere mixture absorption, respectively.

The Diffuse Transmittance of the Atmosphere, $t_{A\lambda}$

The diffuse transmittance of the atmosphere is influenced by scattering in the atmosphere. The diffuse transmittance $t_{A\lambda}$ can be calculated by the following equation:

$$t_{A\lambda} = T_{o\lambda} T_{\mu\lambda} T_{w\lambda} T_{pa\lambda} [0.5 \left(1 - T_{r\lambda}^{0.95}\right) + f_\lambda T_{r\lambda}^{1.5} (1 - T_{ps\lambda}) \qquad (2.24)$$

where f_λ is the percentage of the aerosol scattering vertically toward the lower space.

The Reversal Reflectivity of the Atmosphere to the Direct Radiation and Diffuse Radiation, $R_{A\lambda}$ and $r_{A\lambda}$

After being reflected by the water surface, radiation travels up to the atmosphere, where it will be reversely reflected. After reflection by the water surface, direct radiation travels vertically, because the average reflection angle approaches the Snell reflection angle, which is formed on the smooth water surface. This part of the reflected radiation can be statistically managed as the direct radiation, and the reversal reflectivity of the atmosphere is now:

$$R_{A\lambda} = T_{u\lambda} T_{w\lambda} T_{pa\lambda} [0.5(1 - T_{r\lambda}) + (1 - f_\lambda) T_{r\lambda} (1 - T_{ps\lambda})] \qquad (2.25)$$

The diffuse radiation remains diffuse after reflection on the water surface, and in this case, the reversal reflectivity of the atmosphere is

$$r_{A\lambda} == T'_{u\lambda} T'_{w\lambda} T'_{pa\lambda} [0.5(1 - T'_{r\lambda}) + (1 - f'_\lambda) T_{r\lambda} (1 - T'_{ps\lambda})] \qquad (2.26)$$

In the formula, the value with " ′ " is the corresponding value of formula (2.25) taken from the diffuse value under the average atmosphere conditions.

2.2.1.2 The Reflection on the Water Surface

The light reflection on the smooth water surface belongs to Fresnel reflection because of the unbiased direct light, and its reflectivity can be calculated by the Fresnel equation:

$$\rho_s = \frac{1}{2} \left| \frac{\sin^2(i-j)}{\sin^2(i+j)} + \frac{\tan^2(i-j)}{\tan^2(i+j)} \right| \qquad (2.27)$$

where ρ_s is the direct radiation reflectivity, and i and j are the incident angle and the refraction angle, respectively. The relationship between the incident angle and the refraction angle can be described using the cosine law:

$$\frac{\sin i}{\sin j} = n \tag{2.28}$$

where n is the refractive index of water, which is approximately equal to 1.33.

From the foregoing equation, the direct radiation reflectivity of different incident angles can be obtained directly (Table 2.5).

Because diffuse radiation and the factors influencing it are extremely complicated, quantification of its reflectivity is difficult. Currently, the quantification is generally replaced by the first-order approximation, that is to say considering diffuse light with equal radiance as the atmospheric radiance coming from all directions, of which the radiance is L; thus, the computation expression of the diffuse light reflectivity can be described as follows (Jerlov, 1976):

$$\rho_d = \frac{2\pi \int_0^{\frac{\pi}{2}} \rho(i) L \sin i \cos i di}{2\pi \int_0^{\frac{\pi}{2}} L \sin i \cos i di} = \int_0^{\frac{\pi}{2}} \rho(i) \sin 2i di \tag{2.29}$$

where ρ_d is the reflectivity of diffuse radiation, $\rho(i)$ is the direct radiation reflectivity of the incident angle, and i is the incident angle, respectively. According to Jerlov (1976), the value of ρ_d is equal to 6.6%, derived through the integrated equation (2.29).

According to the reflectivity of the direct and diffuse radiation, the total reflectivity of the lake surface can be written as follows (Jerlov, 1976):

$$\rho = \frac{E_r}{E} = \rho_s (1 - m) + \rho_d m \tag{2.30}$$

Table 2.5 The direct radiation reflectivity of different incident angles from 0° to 90° on the smooth water surface

Incident angle (°)	0	1	2	3	4	5	6	7	8	9
	(°)									
0	2.1	2.1	2.1	2.1	2.1	2.1	2.1	2.1	2.1	2.1
10	2.1	2.1	2.1	2.1	2.1	2.1	2.1	2.1	2.1	2.1
20	2.1	2.1	2.1	2.1	2.1	2.2	2.2	2.2	2.2	2.2
30	2.2	2.3	2.3	2.3	2.3	2.4	2.4	2.4	2.4	2.5
40	2.5	2.5	2.6	2.7	2.8	2.9	3.0	3.1	3.2	3.4
50	3.5	3.6	3.8	4.0	4.2	4.4	4.7	5.0	5.4	5.8
60	6.2	6.6	7.0	7.5	8.2	8.8	9.6	10.4	11.4	12.4
70	13.6	14.8	16.2	17.8	19.6	21.5	23.8	26.0	28.8	31.4
80	35.0	38.6	42.8	47.6	52.9	58.6	65.0	72.0	80.6	89.6
90	100.0									

where E and E_r are the irradiance of the incidence and reflection, respectively, and m is the ratio of the diffuse radiation to the total radiation.

If there are no wind waves, the reflectivity of the water surface can be obtained by the Fresnel equation. However, the water surface fluctuates in response to wind and waves, and reflectivity on the wavy surface differs from that on the smooth surface, especially in the morning and evening.

Reflectivity of the wavy water surface can be obtained using the following random method. Considering a small surface element of the wavy surface, we can omit the curvature of the surface element and perform probability statistics of the various states occurring on the surface element, and the probability statistic expectation of the reflectivity on the surface element is the reflectivity of the wavy surface. According to Cox and Munk's probability distribution function of wave orientation (Cox & Munk, 1954) and the sampling principle of the random variable, we show that the equation obtained by random sampling from all possible wave orientations of the normal vector is

$$\begin{cases} \vec{n}(\mu \cdot \varphi) = \left[-(1-\mu^2)^{1/2}\cos\varphi, (1-\mu^2)^{1/2}\sin\varphi, \mu\right] \\ \mu = 1/(1-\sigma^2 \ln \eta_1)^{1/2} \\ \varphi = 2\pi\eta_z \end{cases} \tag{2.31}$$

where σ is the parameter related to the wind speed on the water surface, $\vec{n}$ is the unit normal vector of the surface element, μ is the cosine of the surface element obliquity, φ is the azimuth of the normal vector of the surface element, and η_1, η_2, and η_3 mentioned below are the random numbers distributing equally between 0 and 1.

Because the variation of the azimuth in the wave orientation has been already considered, the azimuth of the incident radiation is not considered here. The unit vector in different directions of the incident radiation on the surface element can be calculated by the following formula:

$$\vec{e} = (0, \sin\theta, -\cos\theta) \text{ and } \cos\theta = \begin{cases} \cos Z, \text{ for the direct radiation} \\ (Z \text{ is the solar zenith angle}) \\ 1-\eta_3, \text{ for the diffusion radiation} \end{cases} \tag{2.32}$$

After being given $\vec{e}$ and $\vec{n}$, the Fresnel reflectivity on this surface element can be obtained:

$$\begin{cases} r(\mu, \varphi, \theta) = \frac{1}{2}\left[\dfrac{\sin^2(\xi_i - \xi_t)}{\sin^2(\xi_i + \xi_t)} + \dfrac{\mathrm{tg}^2(\xi_i - \xi_t)}{\mathrm{tg}^2(\xi_i + \xi_t)}\right] \\ \cos\xi_i = -(\vec{e} \cdot \vec{n}) \\ \sin\xi_t = \sin\xi_i / m (m \text{ is the reflective index of water}) \end{cases} \tag{2.33}$$

In addition, the quantity of the incident radiation received by the wave surface is determined by the geometry of the area of the projection from the incident direction to the surface, so we introduce $\cos \xi_i / \mu$ to determine the statistical average of the reflectivity on the wave surface element and to determine the reflectivity, r, of the wavy surface:

$$r = \left[\sum_{k=1}^{N} \cos \xi_i^{(k)} r(\mu^{(k)}, \varphi^{(k)}, \theta^{(k)}) / \mu^{(k)} \right] \Bigg/ \left(\sum_{k=1}^{N} \cos \xi_i^{(k)} / \mu^{(k)} \right) \tag{2.34}$$

Solving the foregoing formula under direct and diffuse conditions, we can calculate the direct reflectivity R_f and the diffuse reflectivity r_f on the wavy surface.

2.2.1.3 The Downward Radiation on the Lake Surface

The downward radiation on the lake surface is formed by direct solar radiation through the atmosphere, diffuse atmospheric radiation, and the numerous reflections between the atmosphere and the water body. According to the radial tracer method from the accumulation principle of the radiation, the downward radiation on the lake surface is as follows:

$$F_{D\lambda} = \left[T_{A\lambda} + \frac{T_{A\lambda} R_f R_{A\lambda}}{1 - r_{A\lambda} r_f} C_\lambda \right] D F_\lambda \cos Z \tag{2.35}$$

where F_λ is the solar radiation flux of the upper atmosphere from the average sun-earth distance, D is the correction coefficient for F_λ considering the annual variation of the sun–earth distance, and C_λ is the parameter of the clear sky radiative transfer model.

The total downward radiation on the lake surface F_D can be obtained by integrating all the solar radiation spectrum wavebands. Similarly, the PAR downward radiation on the lake surface P_D can be obtained by integrating PAR. The atmospheric parameters used in the calculation of equation (2.35) are given in Table 2.6.

The measured and calculated values of the diurnal changes in the downward radiation on the lake surface on September 27 and 28, using long-term monthly average meteorological data, are shown in Fig. 2.19. During the measurements, there was some cloudy weather, so the flux of the radiation on the lake surface varied greatly with time. The measured value was lower than that calculated by the numerical model. Because the value of the radiation in a clear sky is the upper

Table 2.6 Atmospheric parameters used in calculation of equation (2.35) for downward radiation on the lake surface (Yang et al., 1992)

Parameters	February	June	September
Air pressure (hPa)	1023.9	1005.9	1012.7
Wind velocity (m/s)	3.6	3.3	2.7
Vapor content of the per section in the air column (g/cm^2)	0.85	2.93	1.9
Ozone content (atm-cm)	0.4	0.32	0.32
Optical thickness of aerosol at 550 nm	0.4	0.05	0.2

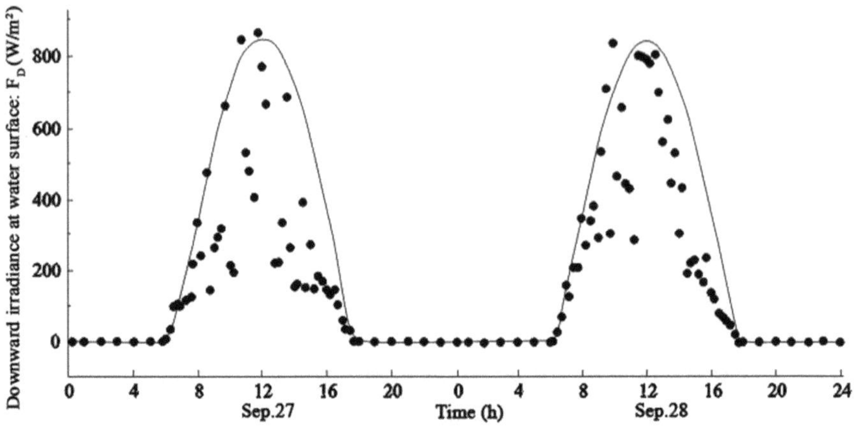

Fig. 2.19 Comparison of the values for diurnal changes in total solar radiation arriving at the lake surface, derived from measurement (*dots*), and calculated from the model (*solid line*) (Yang et al., 1992)

limit of the radiation on the water surface, so the envelope of the observed/measured radiation values under minimal cloud conditions is the radiation flux corresponding with clear sky conditions. Many researchers determine the daily variation curve of the atmosphere–earth radiation in a clear sky using the envelope method. The calculated radiation value almost enveloped the observed data points, which meant the model reflected the condition of the radiation on the water surface in the clear sky.

Calculated values for the downward radiation spectrum at the lake surface at midday and late afternoon on February 15 and June 15 in are shown in Fig. 2.20. The minimal value for radiation in the spectrum distribution curve was determined by the characteristics of the absorption components in the atmosphere; among these, ozone mainly absorbs ultraviolet radiation and visible light and water vapor and carbon dioxide absorb infrared radiation.

At noon, the peak radiation in the 500–700 nm waveband dominated the whole lake surface radiation (see Fig. 2.20). In contrast, in the afternoon, when the solar altitude angle is lower, this maximum peak area is less dominant, and contains many peaks and troughs. The atmospheric quality affects the radiative transfer and the radiation flux reaching the water surface. The characteristic absorption of components in the atmosphere is more obvious in the afternoon than at noon because of the increases of atmospheric quality. Meanwhile, the increase of the atmospheric quality increases the attenuation of Rayleigh scattering; because Rayleigh scattering is inversely proportional to the biquadrate of the incident wavelength, when atmospheric quality increases, attenuation of short-wave radiation is much greater than that of long-wave radiation.

The calculated PAR spectra at the water surface (corresponding to the downward radiation in Fig. 2.20) are shown in Fig. 2.21. The proportion of the PAR that was long-wave radiation, beyond 0.6 μm, was relatively low at noon, and increased near sunset at 1600 and 1800 (Fig. 2.21).

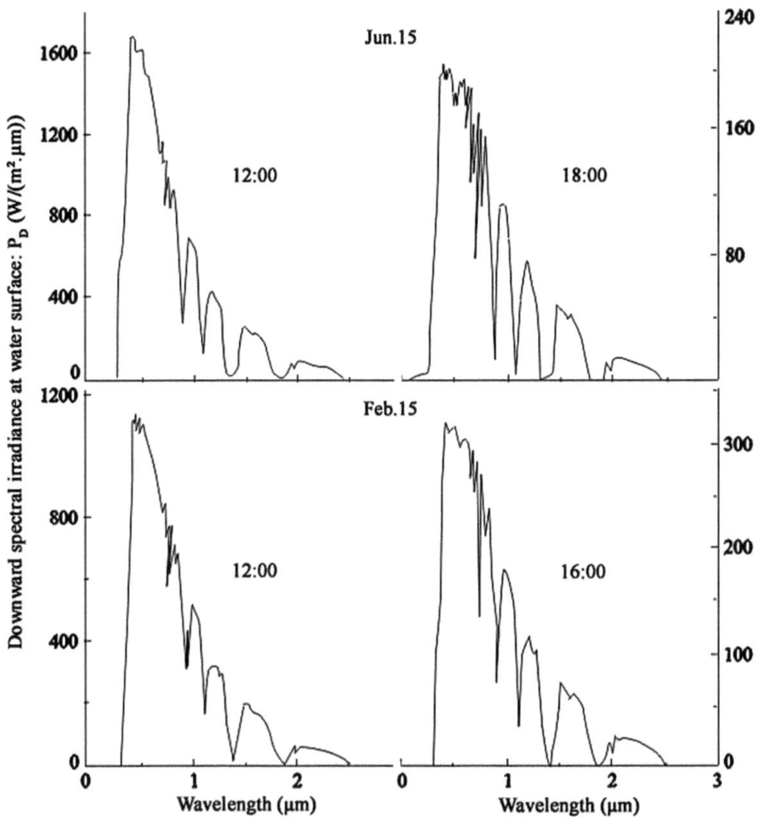

Fig. 2.20 The calculated value of the downward radiation reaching the lake surface at midday and late afternoon in February and June (Yang et al., 1992)

In water ecology research, the radiation value in the broad band is often converted to the PAR value. Daily variation of the proportion of PAR in the total radiation was calculated for February 15 and June 15, according to long-term average monthly climatological data (Fig. 2.22). In June, the percentage of PAR accounting for the total radiation was higher than in February. For example, on June 15, PAR accounted for approximately 46% of the radiation from 8:00 to 16:00; however on February 15, the maximal value was lower, and lasted for a shorter time, 43% at noon (Fig. 2.22). The percentage of PAR in the total radiation was minimal at sunrise and sunset, which indicated that the attenuation of PAR by the atmosphere was much stronger near sunrise and sunset than at other times.

2.2.2 Model of Light Transmission in Water

Suspended particles and water molecules resulted in multi-scattering of light. Direct light and diffuse light were separated using the processing method of atmosphere

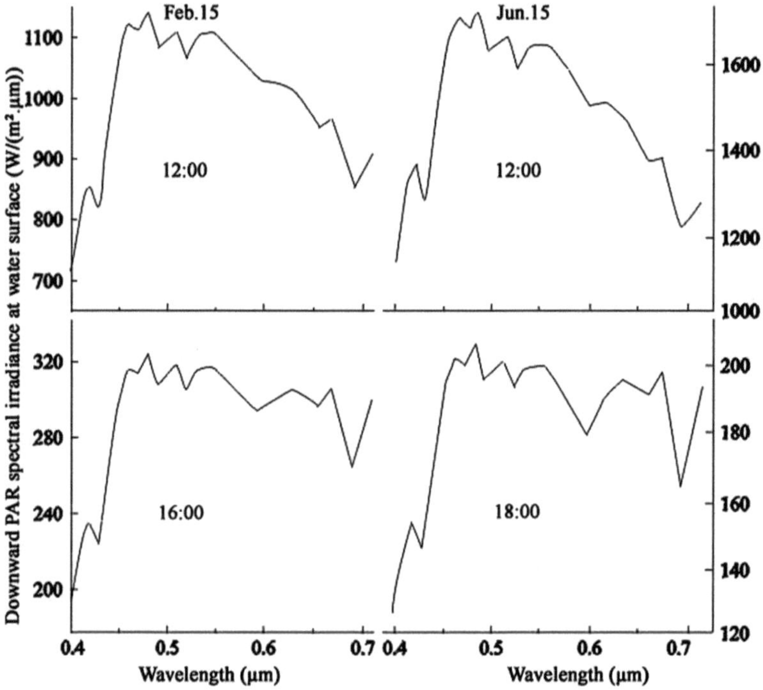

Fig. 2.21 The calculated value of photosynthetic available radiation (PAR) reaching the lake surface at mid day and late afternoon in February and June (Yang et al., 1992)

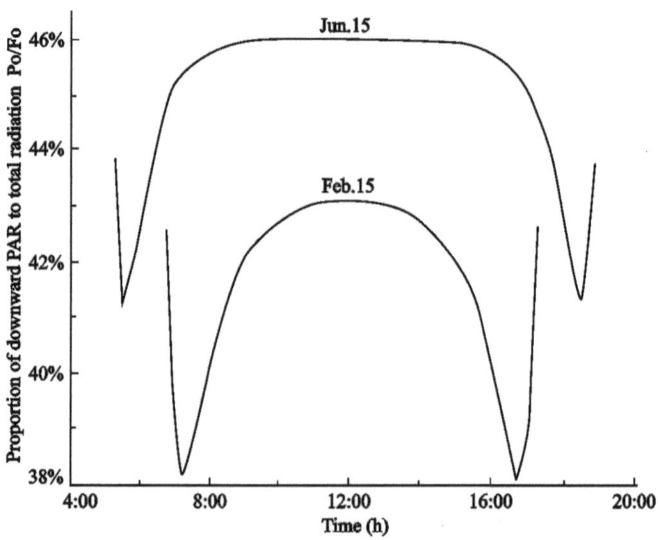

Fig. 2.22 Daily variation curve of the proportion of PAR in the total downward radiation for February 15 and June 15 (Yang et al., 1992)

radiative transmission. The underwater irradiance was calculated using the accumulation principle of thin-layer water. The water body was divided into many thin vertical layers, and the optical characteristic of all thin layers was regarded as being equal to each other. The first-model approximation could be solved by the radiative transmission equation of the second approximation.

$$\frac{dF^+}{d\tau} = r_1 F^+ - r_2 F^- - \pi S_0 \omega_0 \beta(\mu_0) e^{-\tau/\mu_0} \tag{2.36}$$

$$\frac{dF^-}{d\tau} = r_2 F^- - r_1 F^- + \pi S_0 \omega_0 \left[1 - \beta(\mu_0)\right] e^{-\tau/\mu_0} \tag{2.37}$$

where $F^{\pm} = \int_0^1 \mu I(\tau, \pm\mu) d\mu$ is the underwater irradiance; the positive sign and minus sign represent the upward and downward irradiance, respectively; ω_0 is the single reflectivity of the water body; and $\beta(\mu_0)$ is the integral backscattering coefficient.

$$\beta(\mu_0) = \frac{1}{4\pi\omega_0} \int_0^1 \int_0^{2\pi} P(\mu_0, \varphi_0, \mu, \varphi) d\mu d\varphi \tag{2.38}$$

The coefficients in the formula (2.36) and (2.37) are as follows:

$$r_1 = \frac{7 - 3g^2 - \omega_0(4 + 3g) + \omega_0 g^2(4\beta_0 + 3g)}{4[1 - g^2(1 - \omega_0)]} \tag{2.39}$$

$$r_2 = -\frac{1 - g^2 - \omega_0(4 - 3g) - \omega_0 g^2(4\beta_0 + 3g - 4)}{4[1 - g^2(1 - \mu_0)]} \tag{2.40}$$

where g is the dissymmetry factor before and after the particle scattering of the energy in water.

$$g = \frac{1}{2\omega_0} \int_{-1}^1 P(\cos\theta) \cos\theta d \cos\theta \tag{2.41}$$

where θ is the scattering angle and P is the scattering function.

To solve equations (2.36) and (2.37), the direct light and diffuse light are considered separately. For direct light, the boundary condition is $F^-(0) = F^+(\tau) = 0(\Delta\tau$ is the optical thickness of the thin layer), and the reflectivity and transmissivity of the thin layer are as follows:

$$\begin{aligned} R =& F^+(0)/\pi\mu_0 S_0 \\ =& \left\{\omega_0 \left[(1 - K\mu_0)(\alpha_2 + K\beta_0) e^{K\Delta\tau} - (1 + K\mu_0)(\alpha_2 - K\beta_0) e^{-K\Delta\tau} - \right.\right. \\ & \left.\left. 2K(\beta_0 - \alpha_2\mu_0) e^{-\Delta\tau/\mu_0}\right]\right\} / \left\{\left(1 - K^2\mu_0^2\right)\left[(K + r_1) e^{K\Delta\tau} + (K - r_1) e^{-K\Delta\tau}\right]\right\} \end{aligned} \tag{2.42}$$

$$
\begin{aligned}
T &= F^{-}(\Delta\tau)/\pi\mu_0 S_0 \\
&= \Big\{-\omega_0 \mathrm{e}^{-\Delta\tau/\mu_0}\Big[(1+K\mu_0)(\alpha_1+K-K\beta_0)\mathrm{e}^{K\Delta\tau} \\
&\quad -(1-K\mu_0)(\alpha_1-K+K\beta_0)\mathrm{e}^{-K\Delta\tau}+2K(1-\beta_0+\alpha_1\mu_0)\mathrm{e}^{\Delta\tau/\mu_0}\Big]\Big\}/ \\
&\quad \Big\{(1-K^2\mu_0^2)\Big[(K+r_1)\mathrm{e}^{K\Delta\tau}+(K-r_1)\mathrm{e}^{-K\Delta\tau}\Big]\Big\}
\end{aligned}
\tag{2.43}
$$

where $\alpha_1 = r_1(1-\beta_0)+r_2\beta_0$, $\alpha_2 = r_2(1-\beta_0)+r_1\beta_0$, $K = (r_1^2-r_2^2)^{1/2}$. For diffuse light, the boundary condition is $F^{-}(0) \neq 0$, $F^{+}(\Delta\tau) = 0$, and the reflectivity and transmissivity are as follows:

$$
\tilde{R} = F^{+}(0)/F^{-}(0) = \frac{r_2(1-e^{-2K\Delta\tau})}{K+r_1+(K-r_1)\mathrm{e}^{-2K\Delta\tau}} \tag{2.44}
$$

$$
\tilde{T} = F^{-}(\Delta\tau)/F^{-}(0) = \frac{2Ke^{-K\Delta\tau}}{K+r_1+(K-r_1)\mathrm{e}^{-2K\Delta\tau}} \tag{2.45}
$$

According to the results of the two kinds of thin layers above, we can use the accumulation principle of the heterosphere in radiative transmission to solve the irradiance distribution of the whole water body. It can be proved that the irradiance in any depth (thin layer) of the water body follows the recurrence relation below:

$$
\begin{aligned}
& d_1 = 0 \\
& d_{2l} = D < t_l >< R_l > \\
& d_{2l+1} = D < t_l >< T_l >, l = 1, 2, \cdots, n-1 \\
& d_{2n} = D < t_n > R_g \\
& F_1 = d_2 + < \tilde{R}_1 > F_2 + < \tilde{T}_1 > F_3 \\
& F_2 = d_1 + r^{\bullet} F_1 \\
& F_{2l-1} = d_{2l} + < \tilde{R}_l > F_{2l} + < \tilde{T}_l > F_{2l+1} \\
& F_{2l} = d_{2l-1} + < \tilde{R}_{l-1} > F_{2l-1} + < \tilde{T}_{l-1} > F_{2l-2}, i = 2, 3, \cdots, n-1 \\
& F_{2n-1} = d_{2n} + R_g F_{2n} \\
& F_{2n} = d_{2n-1} + < \tilde{R}_{n-1} > F_{2n-1} + < \tilde{T}_{n-1} > F_{2n-2}
\end{aligned}
\tag{2.46}
$$

where d_{2l-1} and d_{2l} are the direct upward and downward irradiance in the l layer, respectively; F_{2l-1} and F_{2l} are the scattering upward and downward irradiance in the l layer, respectively; $<>$ is the average angular distribution caused by the effect of wind waves on the reflectance and transmission in the l layer; D is the direct irradiance after transmitting through the water surface; and t_l is the direct irradiance in the l layer, which can be expressed as follows:

$$
< t_l > = \int\int e^{\tau_l/\mu_0} f(\mu_0, \varphi_0) \mathrm{d}\mu_0 \mathrm{d}\varphi_0 \tag{2.47}
$$

where $f(\mu_0, \varphi_0)$ is the divergence function of the transmissive quantity on the wavy surface. In addition, the 0 layer and the n layer represent the water surface and the bottom of the lake, respectively; R_g is the reflectivity of the bottom of the lake; and r^* is the reflectivity of the irradiance from the water to the surface layer.

According to this model, we calculated the depth profiles of underwater irradiance under different wind waves and different suspended matter concentrations (Fig. 2.23), and at different times of day (Fig. 2.24). With the increase in suspended matter, the percentage of downward irradiance at different depth accounting for surface irradiance significantly decreased. For pure water ($C_0 = 0$), the percentage of downward irradiance at depth 2.0 m below the surface was approximately 90% of surface irradiance. However, the percentage was lower than 1% at 0.8 m for the water with 46.5 mg/L suspended matter ($C_0 = 1.5$). Similarly, the percentage of different depth reflective irradiance accounting for surface irradiance decreased with the increase of suspended matter.

The depth profiles of underwater irradiance at 550 and 800 nm for different wavelength and incident angles are shown in Fig. 2.25 when the wind velocity was 2 m/s. It is easy to find that the transmission of light with different wavelengths in the water was different, and varied with the incident angle. Because scattering of short-wave light exceeded that of long-wave light, the relative contribution of diffuse light to total downward light was higher for short-wave light than for longer wavelength light. Moreover, a significant increase in absorption of pure water was found from 700 nm. Therefore, the attenuation of the short-wave light (550 nm) decreased with depth more slowly than that of long-wave light (800 nm). With increasing depth, the percentage of green light in the water body increased more than that of red light. However, considering only the absorption of the suspended particles, the depth profiles of different underwater irradiance were more complicated in actual water bodies because of the absorption of CDOM and phytoplankton. In addition, Fig. 2.25 also shows that the changes of the direct irradiance with water depth were not linear in the single logarithmic coordinate because of scattering. The shorter the wavelength, the greater the curve.

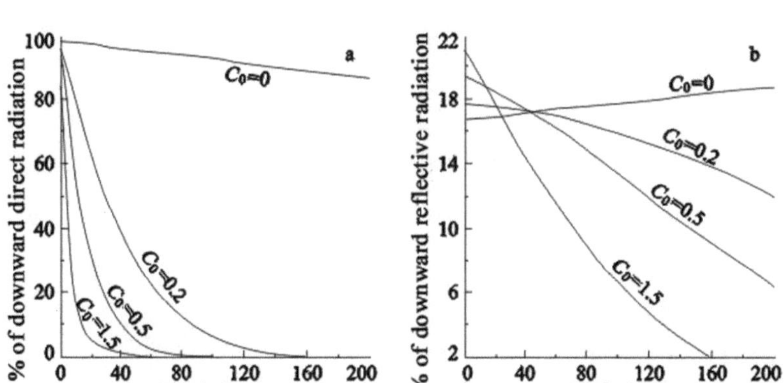

Fig. 2.23 Depth profiles of underwater irradiance under different wind waves and suspended matter concentrations: **a**, direct irradiance; **b**, reflective irradiance ($C_0 = 0$, pure water; $C_0 = 1$, 31 mg/L suspended matter concentration) (Cai et al., 1993b)

Fig. 2.24 Depth profiles of underwater irradiance at midday and early evening in summer (F_T, downward irradiance; F_N, net irradiance; F_{TD}, downward direct irradiance) (Cai & Yang, 1993b)

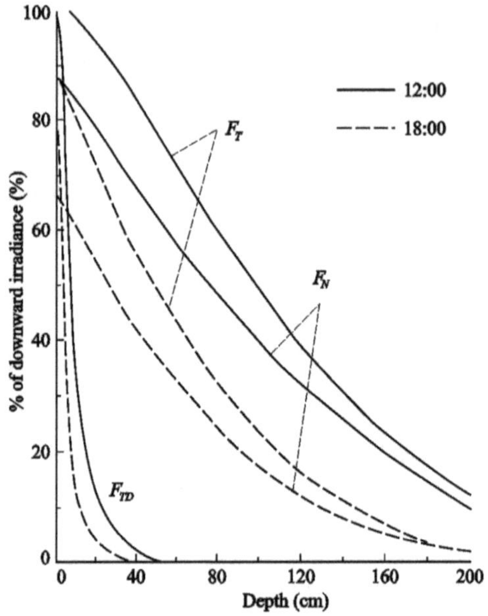

The numerical model described above allowed us to calculate light transmission in the vertical and nonuniform water body of Lake Taihu. The model was suitable for study of the distribution of the downward irradiance, upward irradiance, direct irradiance, and diffuse irradiance in the different water depths, and of the

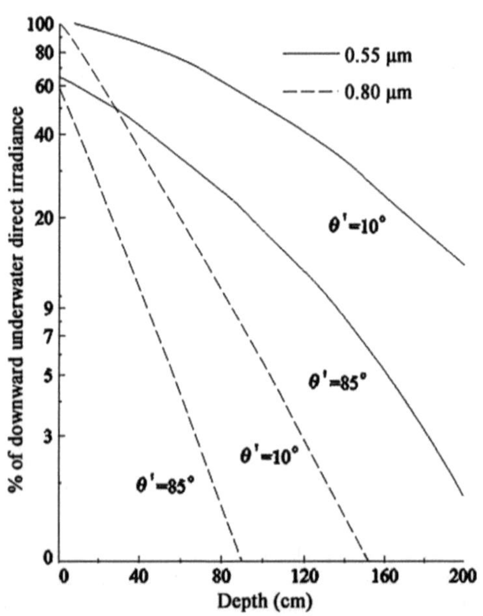

Fig. 2.25 Depth profiles of underwater irradiance of 0.55 and 0.80 μm wavelength under four different incident angles; winds 2 m/s (Cai & Yang, 1993b)

relationships between these parameters and the concentration and particle size of suspended matter, waves, solar altitude, and other factors.

References

Amon, R. M. W. & R. Benner, 1996. Photochemical and microbial consumption of dissolved organic carbon and dissolved oxygen in the Amazon River system. Geochimica et Cosmochimica Acta 60: 1783–1792.

Arts, M. T., R. D. Robarts, F. Kasai, M. J. Walser, V. P. Tumber, A. J. Plante, H. Rai & H. J. Delange, 2000. The attenuation of ultraviolet radiation in high dissolved organic carbon waters of wetlands and lakes on the northern Great Plains. Limnology and Oceanography 45: 292–299.

Bertilsson, S. & L. J. Tranvik, 2000. Photochemical transformation of dissolved organic matter in lakes. Limnology and Oceanography 45: 753–762.

Bird, R. E. & C. Riordan, 1986. Simple solar spectral model for direct and diffuse irradiance on horizontal and titled planes at the earth's surface for cloudless atmosphere. Journal of Climate and Applied Meteorology 25: 87–97.

Bricaud, A., A. Morel & L. Prieur, 1981. Absorption by dissolved organic matter of sea (yellow substance) in the UV and visible domains. Limnology and Oceanography 26: 43–53.

Bricaud, A., M. Babin, A. Morel & H. Claustre, 1995. Variability in the chlorophyll-specific absorption coefficients of natural phytoplankton: analysis and parameterization. Journal of Geophysical Research 100: 13321–13332.

Bricaud, A., A. Morel, M. Babin, K. Allali & H. Claustre, 1998. Variations of light absorption by suspended particles with chlorophyll a concentration in oceanic (case 1) waters: analysis and implications for bio-optical models. Journal of Geophysical Research 103: 31033–31044.

Brine, D. T. & M. Iqbal, 1983. Diffuse and global solar spectral irradiance under cloudless skies. Solar Energy 30: 447–453.

Cai, Q. M. & H. Gons, 1998. Numerical model and application of spectral radiation in atmosphere-water system for Taihu Lake. In: Cai, Q. M. (ed), Ecological and environmental studies of Taihu Lake. Beijing: Meteorology Press, 5–18 (In Chinese with English abstract).

Cai, Q. M. & P. Yang, 1993a. A model of PAR transfer in atmosphere-water system and primary production of lake. Advances in Water Science 4: 171–178 (In Chinese with English abstract).

Cai, Q. M. & P. Yang, 1993b. Model of irradiance transfer in wavy water of Lake Taihu. Oceanologia et Limnologia Sinica 14: 286–293 (In Chinese with English abstract).

Cao, W. X., Y. Z. Yang, X. Q. Xu, L. M. Huang & J. L. Zhang, 2003. Regional patterns of particulate spectral absorption in the Pearl River estuary. Chinese Science Bulletin 48: 2344–2351.

Carder, K. L., R. G. Steward, G. R. Harvey & P. B. Ortner, 1989. Marine humic and fulvic acids: their effects on remote sensing of ocean chlorophyll. Limnology and Oceanography 34: 68–81.

Carder, K. L., S. K. Hawes, K. A. Baker, R. C. Smith & R. G. Steward, 1991. Reflectance model for quantifying chlorophyll a in the presence of productivity degradation products. Journal of Geophysical Research 96: 20599–20611.

Churilova, T. Y. & G. P. Berseneva, 2004. Absorption of light by phytoplankton, detritus, and dissolved organic substances in the coastal region of the Black Sea (July–August 2002). Physical Oceanography 14: 221–233.

Clementson, L. A., J. S. Parslow, A. R. Turnbull & P. I. Bonham, 2004. Properties of light absorption in a highly coloured estuarine system in south-east Australia which is prone to blooms of the toxic dinoflagellate *Gymnodinium catenatum*. Estuarine, Coastal and Shelf Science 60: 101–112.

Cleveland, J. S., 1995. Regional models for phytoplankton absorption as a function of chlorophyll a concentration. Journal of Geophysical Research 100: 13333–13344.

Cleveland, J. S. & A. D. Weidemann, 1993. Quantifying absorption by aquatic particles: a multiple scattering correction for glass-fiber filter. Limnology and Oceanography 38: 1321–1327.

Cox, C. & W. Munk, 1954. Statistics of the sea surface derived from sun glitter. Journal of Marine Research 13: 198–227.

Dekker, A. G., T. J. Malthus, M. M. Wijnen & E. Seyhan, 1992. Remote sensing as a tool for assessing water quality in Loosdrecht lakes. Hydrobiologia 233: 137–189.

Del Castillo, C. E. & P. G. Coble, 2000. Seasonal variability of the colored dissolved organic matter during the 1994–95 NE and SW monsoons in the Arabian Sea. Deep-Sea Research II 47: 1563–1579.

Del Castillo, C. E., P. G. Coble, J. M. Morell, J. M. Lopez & J. E. Corredor, 1999. Analysis of the optical properties of the Orinoco River plume by absorption and fluorescence spectroscopy. Marine Chemistry 66: 35–51.

Doxaran, D., J. M. Froidefond, S. Lavender & P. Castaing, 2002. Spectral signature of highly turbid waters application with SPOT data to quantify suspended particulate matter concentrations. Remote Sensing of Environment 81: 149–161.

Dupouy, C., J. Neveux & M. J. André, 1997. Spectral absorption coefficient of phytosynthetically active pigments in the equatorial Pacific Ocean (165 E–150 W). Deep-Sea Research II 44: 1881–1906.

Farman, J. C., B. G. Gardiner & J. D. Shanklin, 1985. Large losses of total ozone in Antarctica reveal seasonal Cox/Nox interaction. Nature (London) 315: 207–210.

Ferrari, G. M. & M. D. Dowell, 1998. CDOM absorption characteristics with relation to fluorescence and salinity in coastal areas of the southern Baltic Sea. Estuarine, Coastal and Shelf Science 47: 91–105.

Frenette, J. J., M. T. Arts & J. Morin, 2003. Spectral gradients of downwelling light in a fluvial lake (Lake Saint-Pierre, St. Lawrence River). Aquatic Ecology 37: 77–85.

Gallegos, C. L., 2005. Optical water quality of a blackwater river estuary: the Lower St. Johns River, Florida, USA. Estuarine, Coastal and Shelf Science 63: 57–72.

Gallie, E. A., 1997. Variation in the specific absorption of dissolved organic carbon in northern Ontario lakes. In: Steven, G. A. & F. Robert (eds), Ocean Optics XIII. Proceedings of the International Society for Optical Engineering 2963: 417–422.

Gao, H. & R. G. Zeep, 1998. Factors influencing photoreactions of dissolved organic matter in a coastal river of the southeastern United States. Environmental Science & Technology 32: 2940–2946.

Gons, H. J., 1999. Optical teledetection of chlorophyll *a* in turbid inland waters. Environmental Science & Technology 33: 1127–1133.

Gons, H. J., J. Ebert & J. Kromkamp, 1998. Optical teledetection of the vertical attenuation coefficient for downward quantum irradiance of photosynthetically available radiation in turbid inland waters. Aquatic Ecology 31: 299–311.

Green, S. & N. Blough, 1994. Optical absorption and fluorescence properties of chomophoric dissolved organic matter in natural waters. Limnology and Oceanography 39: 1903–1916.

Hessen, D., 1993. DNA-damage and pigmentation in alpine and arctic zooplankton as bioindicators of UV radiation. Internationale Vereinigung für Theoretische und Angewandte Limnologie 25: 482–486.

Hoepffner, N. & S. Sathyendranath, 1993. Determination of the major groups of phytoplankton pigments from absorption spectra of total particulate matter. Journal of Geophysical Research 98: 22789–22803.

Hojerslev, N. K. & E. Aas, 2001. Spectral light absorption by yellow substance in the Kattegat-Skagerrak area. Oceanologia 43: 39–60.

Huovinen, P. S., H. Penttol & M. R. Soimasuo, 2003. Spectral attenuation of solar ultraviolet radiation in humic lakes in central Finland. Chemosphere 51: 205–214.

Jerlov, N. G., 1976. Marine optics. Amsterdam: Elsevier Scientific, 231.

Justus, C. G. & Paris, M. V., 1985. A model for solar spectral irradiance and radiation at the bottom and top of a cloudless atmosphere. Journal of Climate and Applied Meteorology 24: 193–205.

Karentz, D., F. S. McEuen, M. C. Land & W. C. Dunlap, 1991. Survey of mycosporine-like amino acid compounds in Antarctic marine organisms: potential protection from ultraviolet exposure. Marine Biology 108: 157–166.

Kerr, J. B. & C. T. McElroy, 1993. Evidence for large upward trends of ultraviolet-B radiation linked to ozone depletion. Science 262: 1032–1034.

Kiefer, D. A. & B. G. Mitchell, 1983. A simple, steady state description of phytoplankton growth based on absorption cross section and quantum efficiency. Limnology and Oceanography 28: 770–776.

Kirk, J. T. O., 1984. Dependence of relationship between inherent and apparent optical properties of water on solar altitude. Limnology and Oceanography 29: 350–356.

Kirk, J. T. O., 1994. Light and photosynthesis in aquatic ecosystem. Cambridge, UK: Cambridge University Press 1–431.

Kowalczuk, P., W. J. Cooper, R. F. Whitehead, M. J. Durako & W. Sheldon, 2003. Characterization of CDOM in an organic-rich river and surrounding coastal ocean in the South Atlantic Bight. Aquatic Sciences 65: 384–401.

Laurion, I., M. Ventura, J. Catalan, R. Psenner & R. Sommaruga, 2000. Attenuation of ultraviolet radiation in mountain lakes: factors controlling the among- and within-lake variability. Limnology and Oceanography 45: 1274–1288.

Li, W., S. K. Zhang, S. C. Wu & H. Y. Xin, 1997. Preliminary study on variations in vertical distribution of irradiance in the North Yellow Sea. Journal of Oceanography of Huanghai & Bohai Seas 15: 16–24 (In Chinese with English abstract).

Lutz, V. A., S. Sathyendranath & E. J. H. Head, 1996. Absorption coefficient of phytoplankton: regional variations in the North Atlantic. Marine Ecology Progress Series 135: 197–213.

Markager, W. & W. F. Vincent, 2000. Spectral light attenuation and absorption of UV and blue light in natural waters. Limnology and Oceanography 45: 642–650.

Millán-Núñez, E., M. E. Sierackib, R. Millán-Núñez, J. R. Lara-Lara, G. Gaxiola-Castro & C. C. Trees, 2004. Specific absorption coefficient and phytoplankton biomass in the southern region of the California current. Deep-Sea Research II 51: 817–826.

Miller, W. L. & R. G. Zepp, 1995. Photochemical production of dissolved inorganic carbon from terrestrial organic matter: significance to the oceanic organic carbon cycle. Geophysical Research Letters 22: 417–420.

Mitchell, B. G. & D. A. Keifer, 1988. Chlorophyll a specfic absorption and fluorescence excitation spectra for light-limited phytoplankton. Deep-Sea Research, 35(5): 639–663.

Moran, M. A. & R. G. Zeep. 1997. Role of photoreactions in the formation of biologically labile compounds from dissolved organic matter. Limnology and Oceanography 42: 1307–1316.

Moran, M. A., W. M. Sheldon, Jr. & R. G. Zepp, 2000. Carbon loss and optical property changes during long-term photochemical and biological degradation of estuarine dissolved organic matter. Limnology and Oceanography 45: 1254–1264.

Morris, D. P. & B. P. Hargreaves, 1997. The role of photochemical degradation of dissolved organic carbon in regulating the UV transparency of three lakes on the Pocono Plateau. Limnology and Oceanography 42: 239–249.

Morris, D. P., H. Zagarese, C. E. Williamson, E. G. Balseiro & B. R. Hargreaves, 1995. The attenuation of solar UV radiation in lakes and the role of dissolved organic carbon. Limnology and Oceanography 40: 1381–1391.

Ping, Z. L., 1993. Calculating the suspended matter content in the Yellow Sea with the sea water transparency data and NOAA satellite data. Oceanologia et Limnologia Sinica 24: 24–29 (In Chinese with English abstract).

Prieur, L. & S. Sathyendranath, 1981. An optical classification of costal and oceanic waters based on the specific spectral absorption curves of phytoplankton pigments, dissolved organic matter, and other particulate materials. Limnology and Oceanography 26: 671–689.

Qin, Y. S., F. Li, T. M. Zheng & S. M. Xu, 1986. A study on total suspended matter in winter in the South Yellow Sea. Marine Sciences 10: 1–6 (In Chinese with English abstract).

Rochelle-Newall, E. J. & T. R. Fisher, 2002a. Production of chromophoric dissolved organic matter fluorescence in marine and estuarine environments: an investigation into the role of phytoplankton. Marine Chemistry 77: 7–21.

Rochelle-Newall, E. J. & T. R. Fisher, 2002b. Chromophoric dissolved organic matter and dissolved organic carbon in Chesapeake Bay. Marine Chemistry 77: 23–41.

Roesler, C. S., 1998. Theoretical and experimental approaches to improve the accuracy of particulate absorption coefficients derived from the quantitative filter technique. Limnology and Oceanography 43: 1649–1660.

Sasaki, H., S. I. Saitoh & M. Kishino, 2001. Bio-optical properties of seawater in the Western Subarctic gyre and Alaskan Gyre in the subarctic north Pacific and the southern Bering Sea. Journal of Oceanography 57: 275–284.

Sathyendranath, S., T. Platt, C. Caverhill, Warnock & M. Lewis, 1989. Remote sensing of oceanic primary production: computations using a spectral model. Deep Sea Research 36: 431–453.

Seritti, A., D. Russo, L. Nannicini & R. Del Vecchio, 1998. DOC, absorption and fluorescence properties of estuarine and coastal waters of the Northern Tyrrhenian Sea. Chemical Speciation and Bioavailability 10: 95–106.

Smith, R. C. & K. S. Baker, 1981. Optical properties of the clearest natural water (200–800 nm). Applied Optics 20: 177–184.

Smith, R. C., B. B. Prezelin, R. R. Bidigare & K. S. Baker, 1989. Bio-optical modeling of photosynthetic production in coastal waters. Limnology and Oceanography 34: 1524–1544.

Smith, R. C., B. B. Prezelin, K. S. Baker, R. R. Bidigare, N. P. Boucher & T. Ciley, 1992. Ozone depletion: ultraviolet radiation and phytoplankton biology in Antarctic water. Science 225: 952–959.

Sommaruga, R., 2001. The role of solar UV radiation in the ecosystem of alpine lakes. Journal of Photochemistry and Photobiology Series B: Biology 62: 35–42.

Sosik, H. M. & B. G. Mitchell, 1995. Light absorption by phytoplankton, photosynthetic pigments and detritus in the California Current System. Deep Sea Research Part I: Oceanographic Research Papers 42: 1717–1748.

Spinrad, R. W., K. L. Carder & M. J. Perry (eds). 1994. Ocean Optics. Oxford University Press, New York, 165–201.

Stedmon, C. A., S. Markager & H. Kaas, 2000. Optical properties and signatures of chromophoric dissolved organic matter (CDOM) in Danish coastal waters. Estuarine, Coastal and Shelf Science 51: 267–278.

Stuart, V., S. Sathyendranath, E. J. H. Head, T. Platt & B. Irwin, 2000. Bio-optical characteristics of diatom and prymnesiophyte populations in the Labrador Sea. Marine Ecology Progress Series 201: 91–106.

Su, J., W. S. Jiang & W. X. Sun, 2001. Analysis of SPM data obtained in ocean investigation in the Bohai Sea. Journal of Ocean University of Qingdao 31: 647–652 (In Chinese with English abstract).

Suzuki, K., M. Kishino, K. Sasaoka, S. Saitoh & T. Saino, 1998. Chlorophyll-specific absorption coefficients and pigments of phytoplankton off Sanriku, Northwestern North Pacific. Journal of Oceanography 54: 517–526.

Tassan, S., 1994. Local algorithms using SeaWiFS data for the retrieval of phytoplankton, pigments, suspended sediment, and yellow substance in coastal waters. Applied Optics 33: 2369–2378.

Tevini, M., 1993. UV-B radiation and ozone depletion. Boca Raton: Lewis Publishers.

Vodacek, A., N. V. Blough, M. D. Degrandpre, E. T. Peltzer & R. Nelson, 1997. Seasonal variation of CDOM and DOC in the Middle Atlantic Bight: terrestrial inputs and photooxidation. Limnology and Oceanography 42: 674–686.

Williamson, C. E., 1995. What role does UV-B radiation play in freshwater ecosystem? Limnology and Oceanography 40: 386–392.

Williamson, C. E., R. S. Stemberger, D. P. Morris, T. M. Frost & S. G. Paulsen, 1996. Ultraviolet radiation in North American lakes: attenuation estimates from DOC measurements and implication for plankton communities. Limnology and Oceanography 41: 1024–1034.

Yang, P., Y. J. Xu & Q. M. Cai, 1992. PAR in the solar radiation at the water surface of Taihu Lake. Journal of Lake Sciences 4: 10–16 (In Chinese with English abstract).
Yu, H. & Q. M. Cai, 1998. The simulation of underwater light field and primary productivity in Taihu Lake. In: Cai Q M (ed) Ecological and environmental studies of Taihu Lake. Beijing: Meteorology Press, 178–187 (In Chinese with English abstract).
Zhang, Y. L., S. C. Wu, B. Q. Qin, W. M. Chen & G. W. Zhu, 2004a. Absorption of light by chromophoric dissolved organic matter (CDOM) in Meiliang Bay of Taihu Lake. China Environmental Science 24: 405–409 (In Chinese with English abstract).
Zhang, Y. L., B. Q. Qin & W. M. Chen, 2004b. Experimental study of underwater light field and affect mechanism. Progress in Natural Science 14: 792–798 (In Chinese).
Zhu, J. H. & T. J. Li, 2004. Spectral model research about absorption coefficient of the de-pigment particulate and yellow substance in Yellow Sea and East China Sea. Ocean Technology 23: 7–13 (In Chinese with English abstract).

Chapter 3
Hydrodynamics and Its Effects on the Aquatic Ecosystem

Hydrodynamics, integral to the study of lake ecosystems, includes currents, waves, and related processes such as water level changes, secchi depth, and seasonal stratification in deep areas. The hydrodynamics of a water body affect the transport of materials and energy, especially exchange at the water–air and water–sediment interfaces.

In shallow lakes, such as Lake Taihu, hydrodynamics is of particular significance because wind-induced sediment resuspension may cause nutrient release from the lake bottom, and currents may subsequently redistribute these nutrients. The sediment resuspension can finally lead to changes of water color, transparency, underwater light climate, and primary production. Following such strong storm-induced disturbances in Lake Taihu, algal blooms always occur.

Technical limitations make it difficult to collect long-term, high-resolution data for waves and currents in shallow lakes, especially when the latter are affected by unstable winds or current speeds are too small to be measured. A further complication in shallow lakes is that the water column may be well mixed by surface disturbance. For these reasons, hydrodynamics in Lake Taihu is now studied by a combination of field observations and numerical simulations. Our research mainly involves wind-driven currents, wave-forming mechanisms, distribution characteristics, and sediment resuspension and their influences on the aquatic ecosystem based on in situ measurements and numerical models.

Qin, B. Q. (ed.), *Lake Taihu, China*,

3.1 Wave Characteristics

Liancong Luo and Boqiang Qin

The waves in Lake Taihu are caused by surface wind disturbance, and their characteristics mainly depend on wind speed, wind fetch, and water depth (Wang, 1987; Qiao, 1989; Zhang, 1992; Pang et al., 1994, Pang & Pu, 1995).

3.1.1 Wind

The dominant wind direction in Lake Taihu is ESE (Qin et al., 2004). However, the winds change seasonally. The dominant wind direction is ENE in January and February, ENE and ESE in March, ESE in April, SSW in May, ESE in June, July, and August, ENE in September and October, E in November, and ENE, WNW, and NNW in December.

3.1.2 Waves

Many field observations have been conducted, especially during storms, and two main investigations are now discussed. (1) Studies undertaken by the Nanjing Hydraulic Research Institute and Nanjing Institute of Geography and Limnology, at Mashan in Meiliang Bay (water depth 2.15–3.40 m), during August–October 1992 and August–November 1993. Continuous wave data were collected using a capacitive wave meter. The highest wind speed ranged 10–15 m/s, and the dominant wind direction was E. (2) Studies by Nanjing Institute of Geography and Limnology, at the center of Meiliang Bay in September 2001, Wuguishan Island in July 2002, and in the southwest of the lake near Changdougang in December 2002 (water depths 2.53–2.75 m). The studies investigated the effects of wind waves on sediment resuspension, internal nutrient release, and cyanobacterial blooms. Wave data were collected by a DJ800 wave sensors. Six wave sensors were set up in September 2001 (wind SE), three in July 2002 (wind SE), and three in December 2002 (wind NW).

3.1.3 Wave Characteristics

For calculating the wave parameters, 12,000 wave records $\zeta_i (i = 1, 2, 3, \ldots, 12{,}000)$ were collected in 20 min, and their mean value ($\bar{\zeta}$) was calculated and used as the "baseline" for further computation (Luo, 2004; Luo et al., 2004a). The wave

Luo, L.C.
State Key Laboratory of Lake Science and Environment, Nanjing Institute of Geography and Limnology, Chinese Academy of Sciences, 73 East Beijing Road, Nanjing 210008, P.R. China
e-mail: lcluo@niglas.ac.cn

period was defined as the duration of two adjacent records at the "baseline." To calculate the mean wave period, the wave numbers M in the 12,000 wave records, and the wave records $T_i(i = 1, \ldots, M)$ in each wave, were predetermined according to the data obtained. The mean wave period could be determined from the following equation (Qin et al., 2004):

$$\bar{\mathrm{T}} = 0.1 \cdot \sum_{i=1}^{M} T_i / M \tag{3.1}$$

The 12,000 records were evenly divided into 10 sections. At the beginning and end of each section, 20 waves were read according to the mean period obtained from the above equation, and the mean value of these 40 waves was calculated and used as the baseline for the 11th wave from the beginning and end of each section. Based on this baseline, all the wave record values were recalculated and new wave records were produced. In the new wave records, the wave period $T_i(i = 1, \ldots, N - 1)$ and the wave height $H_i = \zeta_{\max_i} - \zeta_{\min_i}$ were calculated. The mean wave height and period at 20-min intervals were as follows (Qin et al., 2004):

$$\overline{H} = \frac{1}{N-1} \sum_{i=1}^{N-1} H_i, \ \overline{T} = \frac{1}{N-1} \sum_{i=1}^{N-1} t_i \tag{3.2}$$

The N waves were ordered in decreasing order of height. For 1/10 wave, the wave height and period could be computed from the following equation (Qin et al., 2004):

$$\overline{H}_{1/10} = \frac{1}{(N-1)/10} \sum_{i=1}^{(N-1)/10} H_i, \overline{T}_{1/10} = \frac{1}{(N-1)/10} \sum_{i=1}^{(N-1)/10} t_i \tag{3.3}$$

For 1/3 (significant wave), the values could be computed from the following equation:

$$\overline{H}_{1/3} = \frac{1}{(N-1)/3} \sum_{i=1}^{(N-1)/3} H_i, \overline{T}_{1/3} = \frac{1}{(N-1)/3} \sum_{i=1}^{(N-1)/3} t_i \tag{3.4}$$

The wavelength was acquired from the following equation:

$$\left(\frac{2\pi}{T}\right)^2 = g\frac{2\pi}{L}\mathrm{th}\left(\frac{2\pi}{L}h\right) \tag{3.5}$$

The data for wave height, length, and period, for mean wave, 1/3 wave (significant wave), and 1/10 wave, based on data observed from September 8 to 11, 2001, are shown in Table 3.1.

There were only small variations in wave height, wavelength, period, and speed over the 6-day observation period (see Table 3.1). The mean wave height ranged

Table 3.1 Diurnal changes in wave height, length, and period in Meiliang Bay, July 23–28, 2002 (mean wind speed 3.3 m/s, wind direction SE on July 23, mean wind speed 3.7 m/s, wind direction SE on July 24; mean wind speed 2.9 m/s; wind direction NE on July 25; mean wind speed 3.6 m/s; wind direction NE on July 26; mean wind speed 2.5 m/s; wind direction N on July 27; winds were not recorded due to a technical problem on July 28th)

Time M:dd:hh:mm	Mean wave			Significant wave			1/10 wave		
	Height (cm)	Period (s)	Length (m)	Height (cm)	Period (s)	Length (m)	Height (cm)	Period (s)	Length (m)
7231130	12.32	1.63	4.10	21.17	1.99	5.99	33.75	1.93	5.68
7231224	9.61	1.54	3.67	15.13	1.89	5.46	18.99	1.92	5.59
7231304	9.31	1.48	3.39	14.45	1.83	5.12	17.68	1.85	5.22
7231556	6.41	1.30	2.62	9.89	1.62	4.05	12.09	1.61	4.00
7231636	4.62	1.32	2.72	7.28	1.62	4.05	9.14	1.63	4.10
7231828	6.34	1.15	2.05	9.91	1.37	2.91	12.73	1.41	3.10
7231925	12.59	1.71	4.50	19.77	2.02	6.13	24.34	2.02	6.15
7240925	13.17	1.70	4.45	20.61	2.06	6.37	25.40	2.07	6.42
7241026	12.65	1.62	4.05	19.95	1.96	5.83	25.36	1.97	5.89
7241122	11.16	1.55	3.74	17.75	1.93	5.65	22.04	1.97	5.89
7241242	9.13	1.42	3.12	14.28	1.72	4.55	17.81	1.77	4.83
7241340	7.78	1.34	2.80	11.98	1.64	4.18	14.69	1.67	4.30
7241436	7.21	1.29	2.58	11.00	1.57	3.81	13.48	1.60	3.98
7241536	8.42	1.40	3.06	12.57	1.69	4.40	15.19	1.68	4.35
7241632	8.15	1.42	3.12	12.49	1.71	4.53	15.58	1.70	4.45
7241742	11.14	1.56	3.76	17.51	1.89	5.46	22.39	1.93	5.65
7241836	14.96	1.78	4.86	23.33	2.12	6.69	29.01	2.12	6.69
7241933	14.51	1.77	4.81	22.84	2.13	6.74	28.52	2.17	6.96
7242030	13.65	1.73	4.60	21.92	2.08	6.48	28.20	2.07	6.42
7250954	5.21	1.15	2.06	8.08	1.39	3.01	10.04	1.40	3.03
7251052	4.29	1.15	2.06	6.92	1.47	3.34	9.37	1.52	3.57
7251158	4.28	1.07	1.79	6.59	1.27	2.50	8.11	1.30	2.64
7251244	5.13	1.21	2.27	7.83	1.42	3.14	9.53	1.41	3.10
7251344	6.67	1.26	2.46	10.20	1.48	3.39	12.42	1.47	3.34
7251425	6.41	1.21	2.26	9.79	1.44	3.23	11.95	1.46	3.30

Table 3.1 (continued)

Time M:dd:hh:mm	Mean wave			Significant wave			1/10 wave		
	Height (cm)	Period (s)	Length (m)	Height (cm)	Period (s)	Length (m)	Height (cm)	Period (s)	Length (m)
7260937	5.35	1.09	1.85	8.33	1.30	2.62	10.40	1.32	2.70
7261033	5.65	1.13	1.99	8.79	1.41	3.08	10.87	1.42	3.14
7261139	5.65	1.16	2.08	8.90	1.44	3.23	10.86	1.47	3.34
7261242	3.84	1.00	1.56	5.96	1.28	2.56	7.44	1.33	2.74
7261337	4.56	1.00	1.54	6.99	1.21	2.26	8.67	1.20	2.25
7261437	6.69	1.18	2.15	10.35	1.43	3.17	12.67	1.48	3.39
7261534	7.12	1.27	2.50	11.11	1.56	3.76	13.80	1.58	3.86
7261635	8.72	1.39	2.99	13.53	1.65	4.22	16.78	1.65	4.22
7261731	10.35	1.50	3.48	15.94	1.76	4.78	19.55	1.75	4.73
7270952	5.85	1.27	2.52	9.00	1.49	3.44	11.24	1.48	3.41
7271047	4.73	1.25	2.44	7.40	1.49	3.46	9.18	1.51	3.53
7271155	4.79	1.13	1.99	7.42	1.38	2.95	9.19	1.38	2.97
7271253	5.15	1.07	1.77	7.88	1.21	2.28	9.64	1.21	2.28
7271350	2.85	1.06	1.74	4.38	1.21	2.28	5.49	1.23	2.34
7271446	1.55	1.25	2.42	2.20	1.54	3.67	2.62	1.53	3.62
7271543	1.59	1.68	4.37	2.51	1.67	4.32	3.17	1.68	4.37
7271650	2.80	1.80	4.99	4.30	1.77	4.83	5.37	1.75	4.70
7271747	4.24	1.49	3.46	5.73	1.63	4.13	6.55	1.63	4.10
7271859	4.54	1.16	2.08	6.60	1.59	3.91	7.60	1.63	4.13
7271947	5.18	0.94	1.38	7.60	1.09	1.85	9.08	1.10	1.87
7280940	2.52	0.87	1.18	3.82	1.11	1.92	4.71	1.21	2.28
7281034	1.87	0.78	0.94	2.81	0.98	1.48	3.46	1.01	1.58
7281114	3.65	0.88	1.19	5.53	0.99	1.51	6.83	1.00	1.56
7281235	3.69	1.13	1.99	5.73	1.41	3.08	6.93	1.43	3.19
7281343	3.08	0.99	1.53	4.69	1.24	2.38	5.67	1.26	2.46
7281436	4.07	1.22	2.30	7.20	1.70	4.45	9.96	1.82	5.09
7281524	5.70	1.50	3.48	8.73	1.75	4.73	10.65	1.76	4.76
7281639	3.54	1.36	2.86	5.47	1.58	3.88	6.69	1.59	3.93

from 1.55 to 14.96 cm, period ranged from 0.78 to 1.78 s, and length ranged from 0.94 to 4.99 m. Water depth ranged from 2.7 to 2.85 m, wind speed ranged from 1 to 7 m/s, and the dominant wind direction was east. To show the relationships among mean wave, 1/3 wave (significant wave), and 1/10 wave, correlation analysis was conducted, with the following results:

$$\frac{H_3}{\overline{H}} = 1.58925 - \frac{0.00126}{H^*}\ (R = 0.99897);\ \frac{T_3}{\overline{T}} = 1.0951 + \frac{0.1346}{\overline{T}}\ (R = 0.96405) \quad (3.6)$$

$$\frac{H_{10}}{\overline{H}} = 2.00998 - \frac{0.01}{H^*}\ (R = 0.99208);\ \frac{T_{10}}{\overline{T}} = 1.04995 + \frac{0.21128}{\overline{T}}\ (R = 0.94688) \quad (3.7)$$

$$\frac{H_{10}}{H^3} = 1.2686 - \frac{0.006343}{H_3^*}\ (R = 0.99613);\ \frac{T_{10}}{T_3} = 0.97072 + \frac{0.06341}{T_3}\ (R = 0.99443) \quad (3.8)$$

where $H^* = \overline{H}/d$(d is water depth), and $H_3^* = H_3/d,\ R$ is correlation coefficient.

These observed relationships among different wave heights and periods agree well with the theoretical results. Thus, these observed results are in good agreement with Raley distribution theory (Wen & Yu, 1984).

3.1.4 Wave Spectrum

The wave spectrum is the wave energy distribution along the wave frequency, and this is commonly analyzed using self-correlation, Fast Fourier Transform (FFT), and maximal entropy methods. In our study, self-correlation and FFT were used to calculate wave energy and then wave period at which peak value of wave energy appears based on the wave data collected on July 24, 2002, when wind speed ranged from 3 to 5 m/s and wind direction was SE. The results from the two methods are compared in Table 3.2.

The wave periods calculated from the two methods are in good agreement with each other and have the same trend with wind variation. This result suggests that both methods can be used for wave spectrum analysis in Lake Taihu.

To test the accuracy of wave periods calculated from the self-correlation method, we compared them (*Ts* in Fig. 3.1) with the significant wave periods (*T3* in Fig. 3.1) calculated from the previously used wave analysis method (equation 3.4); the comparison is shown in Fig. 3.1.

Table 3.2 Comparison between wave periods calculated from self-correlation and FFT methods based on the waves observed at Wuguishan Island at 1340, 1756, and 1853 on July 24, 2002

Time 24 July	Period calculated from the maximum spectrum value – self-correlation (s)	Period calculated from the maximum spectrum value – FFT (s)	Wind speed (m/s)	Wind direction
1340	1.923	2.020	3.1	SE
1756	2.222	2.110	3.4	SE
1853	2.381	2.381	4.7	SE

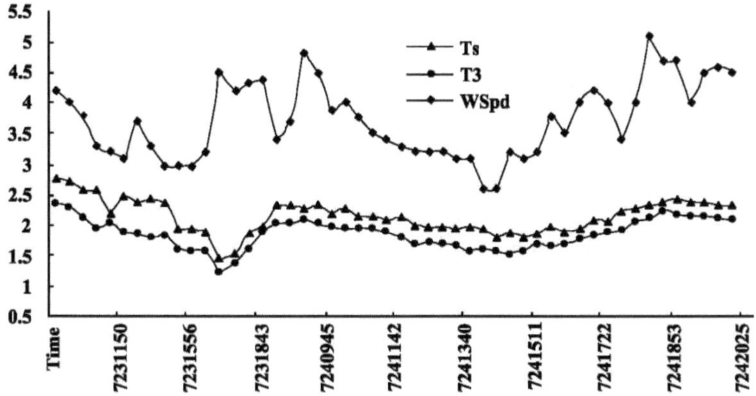

Fig. 3.1 Comparison between significant wave periods (*T3*, unit: s) and calculated periods (*Ts*, unit: s) from self-correlation method. *WSpd* represents wind speed (m/s). The time is given in the format m:dd:hh:mm

The difference between T3 and Ts varies from 0.1 to 0.5 s, and T3 is always slightly smaller than Ts. The linear relationship between them is

$$T_3 = 0.77T_s + 0.2 \tag{3.9}$$

and the following equation shows the relationship between significant wave height (H3) and the maximum spectrum value ($Spec_M$) from the self-correlation method:

$$H_3 = 0.87\sqrt{Spec_M} - 0.04 \tag{3.10}$$

Based on the equations (3.6), (3.7), and (3.8), all the other wave heights and periods can be computed.

3.2 Field Investigations and Analysis of Currents

Boqiang Qin and Liancong Luo

Although the currents in Lake Taihu have not been studied comprehensively, intermittent observations were made over a 30-year period beginning in 1960, and these are now presented. Basic knowledge has been acquired about the character of the current fields, especially in the north part of the lake; however, more investigation and numerical simulation are needed.

Investigation of lake currents in 1960, and between 1986 and 1987, 1991, and 1995, although made at different sites and under different wind conditions, showed that the dominant south–east or south–west winds of summer caused an anticlockwise surface circulation in the centre of Lake Taihu (Fig. 3.2) (Nanjing Institute of Geography, Chinese Academy of Sciences, 1965). The mean current speed was about 10 cm/s, and the maximum was 30 cm/s. In addition, there was a stable southerly current along the west coast of the lake, a stable easterly current along the south of the lake, and an unstable, clockwise circulation in Meiliang Bay. The

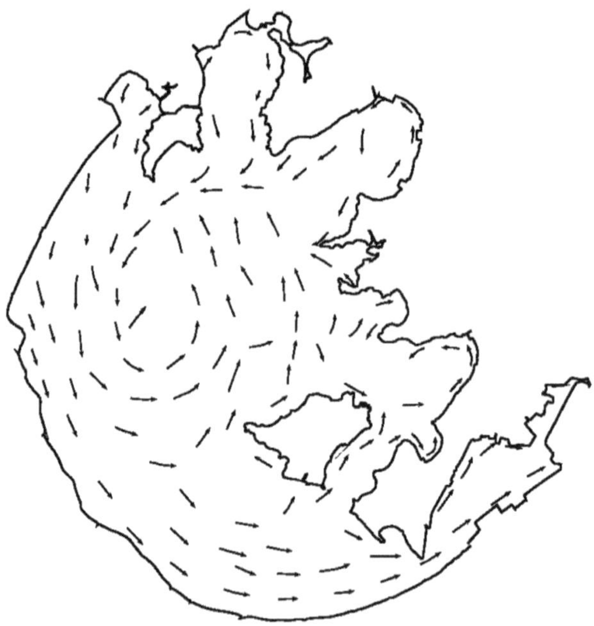

Fig. 3.2 Prevailing currents in Lake Taihu in summer

Qin, B.Q.
State Key Laboratory of Lake Science and Environment, Nanjing Institute of Geography and Limnology, Chinese Academy of Sciences, 73 East Beijing Road, Nanjing 210008, P.R. China
e-mail: qinbq@niglas.ac.cn

east coast of Meiliang Bay, especially the mouth of the Liangxi River, was the zone in which floating material accumulated, caused by the local small-scale circulation.

In August 1997, observations using buoys showed that although the surface and bottom currents were usually the same, the direction of the bottom current was occasionally opposite to that of the surface current. In the same month, a large field investigation using a three-dimensional (3-D) ultrasonic current meter and ANDRAA sea current meter, set 1.0–1.5 m below the water surface at the mouth of Meiliang Bay (120.188° E, 31.41° N), showed that currents were unstable and greatly influenced by waves (see Fig. 3.2). The direction of the lake current changed markedly, and to a greater degree than did the wind field (Fig. 3.3). This finding may have reflected the location of the observation site, which was near the coast, which would have influenced the currents. However, the water always flowed from Meiliang Bay to the lake centre with SE or SW winds, suggesting that water exchange was mainly caused by the water level difference between these two regions. Currents were greatly influenced by waves in shallow Lake Taihu, and their vertical and horizontal speeds were of the same order of magnitude.

On July 28, 1998, meteorological data, surface current (measured by a 3-D ultrasonic current meter), and nutrients (TP, TN, and Chl a) were recorded at 14 sites in

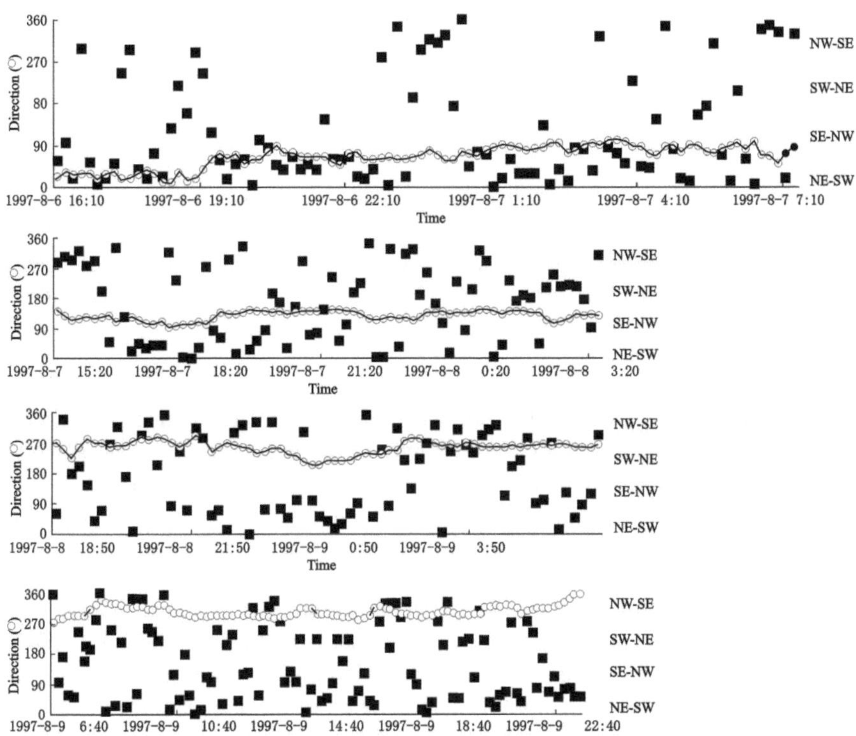

Fig. 3.3 Temporal changes in direction of wind and lake currents at the mouth of Meiliang Bay, at 3- to 4-h intervals between August 6 and 9, 1997 (○, wind; ■, lake current)

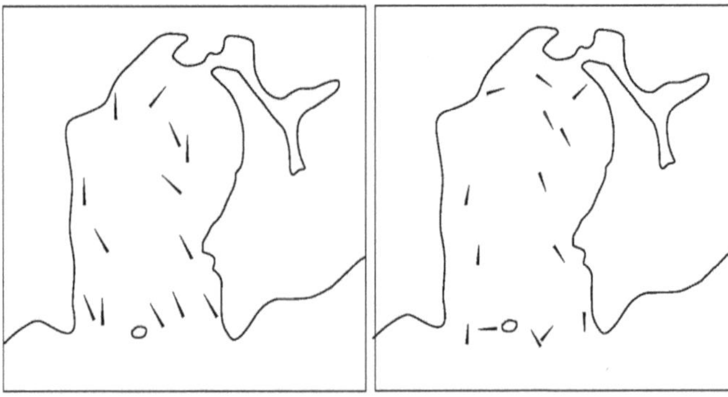

Fig. 3.4 Directions of wind (*left*) and surface current (*right*) in Meiliang Bay on July 28, 1998

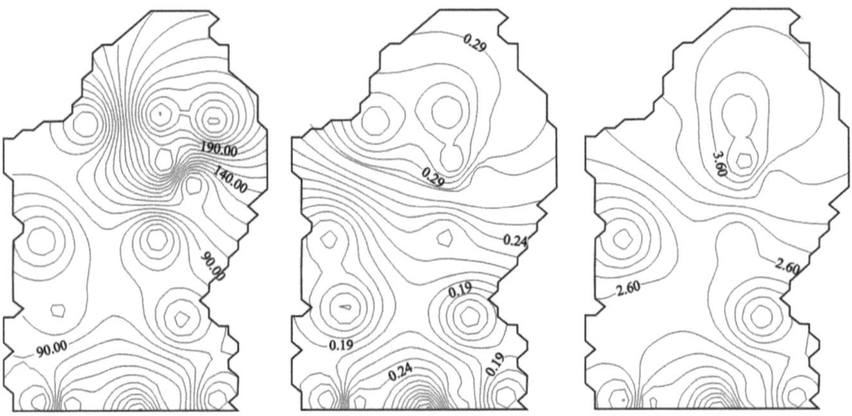

Fig. 3.5 Distribution patterns of cholorophyll *a* (Chl a: ug/L, *left*), total phosphorus (TP: mg/L, *middle*), and total nitrogen (TN: mg/L, *right*) in Meiliang Bay on July 28, 1998 (contour value represents concentration)

Meiliang Bay (Figs. 3.4, 3.5). The dominant wind was southeast, and wind speed was 1.5–2.5 m/s. There was a clockwise gyre in the whole bay but a small anti-clockwise gyre in the north area of the bay resulting from the SE wind, with water moving northwards in the west of Meiliang Bay and south in the east of the bay. The concentrations of TN, TP, and Chl a reflected the current patterns (Fig. 3.4).

On July 28, 1998, water exchange between Meiliang Bay and the open lake area was studied at five sites and three depths per site (Table 3.3); this investigation was undertaken to provide information about water quality in the Bay, which is affected in part by the water exchange. The dominant wind direction was SE. The surface current (0.5 m) was from south to north in the west of the area, and from north to south in the east of the area; in contrast, the bottom currents (2.5, 2.6 m) were from north to south at all five sites. In summer, water mainly moved from the bay to the open area of Lake Taihu in the east, with an exchange rate of 271 m^3/s, and in

Table 3.3 Currents at the mouth of Meiliang Bay on July 28, 1998

Site	Conditions			Currents		
	Wind velocity (m/s)	Wind direction	Depth (m below surface)	Horizontal velocity (cm/s)	Direction (degree)	Vertical velocity (cm/s)
F1	1.6	SE	0.5	4.0	181	1.5
			1.4	4.6	178	1.4
			2.5	2.0	193	1.5
J2	1.4	SE	0.5	3.5	46	1.8
			1.5	4.0	195	1.7
			2.6	6.9	149	1.1
F3	1.7	SE	0.5	4.2	298	1.7
			1.5	2.4	195	1.5
			2.6	6.5	158	1.1
J4	2.4	S	0.5	2.5	271	1.7
			1.4	2.8	58	0.9
			2.5	3.5	155	1.2
F5	2.2	SE	0.5	3.5	5	−0.2
			1.4	2.1	175	1.8
			2.6	0.3	233	1.1

Locations of the sites were F1, 120.205 °E, 31.409 °N; J2, 120.188 °E, 31.410 °N; F3, 120.176 °E, 31.304 °N; J4, 120.150 °E, 31.407 °N; F5, 120.140 °E, 31.408 °N.

the opposite direction in the west, with exchange rate of only 9 m^3/s. However, the mechanism of this exchange is not yet known.

On January 19, 2003, the current pattern in winter was measured by Acoustic Doppler Current Profile (ADCP; SonTek) at the same sites as in summer 1998, and at five depths per site. The dominant wind direction was NNW. The current pattern (Fig. 3.6) was similar to that of summer, with the surface current following the wind direction, especially in the southern bay. In the center of the bay, there was a clockwise circulation with the current flowing into the bay along the western bank and out along the eastern bank, whereas there was an anticlockwise circulation in the northeastern bay.

The current profile in Lake Taihu was further investigated near Wuguishan Island from July 13 to 17, 2003, where the water was 2.4 m deep; an ADCP (SonTek) was again used. Current direction was recorded at 11 depths simultaneously with interval of 10 min (depth from 65, 80, 95, 110, 125, 140, 155, 170, 185, 200, and 215 cm below surface). At a low wind speed of 3.3 m/s from ESE, (10.27–13.47 on July 14, 2003), the currents down to a depth of 185 cm flowed in the same direction as the wind; however, below 185 cm, the currents were weak and their direction was reversed (Fig. 3.7). Thus, at this wind speed, the wind-induced water movement and energy exchange could be transferred only down to 185 cm (Fig. 3.7).

At a higher wind speed of 5–6 m/s (1142–1502, July 13, 2003), all the currents from the surface to the bottom followed the wind direction; thus, the wind-induced water movement and energy dissipation could transfer downwards right to the bottom of the lake (Fig. 3.8).

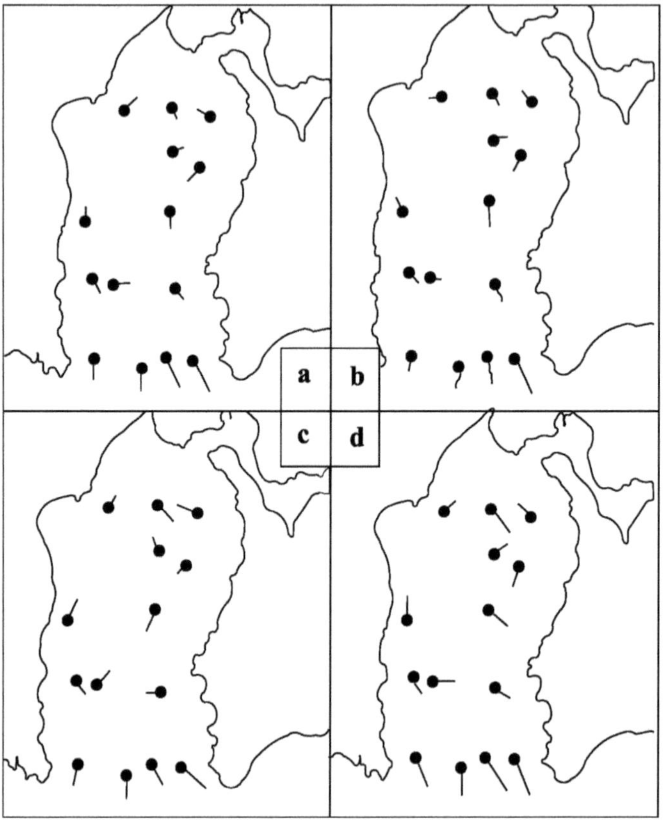

Fig. 3.6 Lake currents in different depths of Meiliang Bay on January 19, 2003 (Luo et al., 2004b): (**a**) 60 cm below surface; (**b**) 135 cm below surface; (**c**) 210 cm below surface; (**d**) vertical average

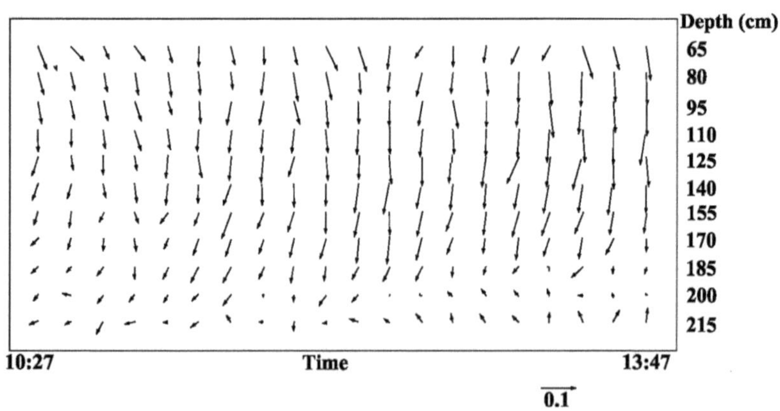

Fig. 3.7 Current vectors at 65, 80, 95, 110, 125, 140, 155, 170, 185, 200, and 215 cm below surface with a wind speed of 3.3 m/s from ESE during 1027–1347, July 14, 2003 (time at X-axis and depth at Y-axis; the *bottom* represents the current vector scale in unit of m/s)

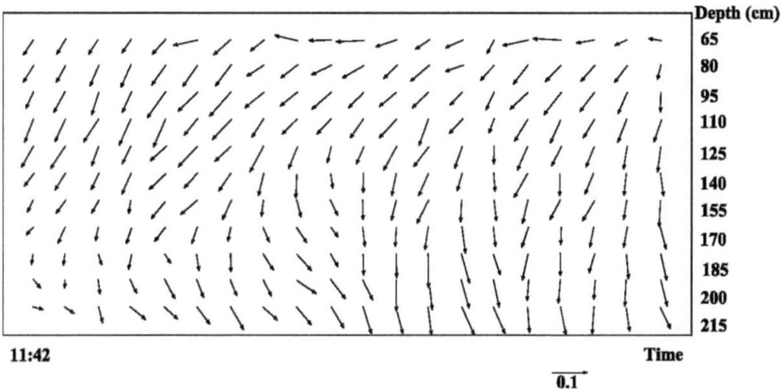

Fig. 3.8 Current vectors at 65, 80, 95, 110, 125, 140, 155, 170, 185, 200, and 215 cm below surface during 1142–1502, July 13, 2003 (time at X-axis and depth at Y-axis; the *bottom* represents the current vector scale in units of m/s)

3.3 Modelling of Lake Hydrodynamics

Liancong Luo and Boqiang Qin

One of the best ways to understand hydrodynamics in lakes and reservoirs is numerical simulation, and this method has been widely used in the world. Numerical model results can provide substantial information about horizontal and vertical current and wave distribution with different prevailing winds, which data are difficult to obtain from field measurements.

Because of the large size and diurnal stratification of Lake Taihu, a 3-D hydrodynamic model is necessary for studying the vertical current distribution and vertical transport of materials in the water column, especially the vertical migration of cyanobacterial cells induced by currents.

3.3.1 Review of Modelling in Lake Taihu

Study of lake hydrodynamics can include field observation, physical modelling, and numerical simulation. The former two methods have not been effective because of technical limitations, and the focus is now given to numerical modelling, particularly because of recent advances in computational fluid dynamics. Development of a universal numerical model suitable for simulation of lake hydrodynamics has become necessary.

Both two-dimensional (2-D) and 3-D hydrodynamic mathematical models are used for simulation of lake currents. The depth-integrated model (2-D) is usually used for shallow lakes without seasonal stratification [for example, the 2-D model for Lake Constance used by Hutter (1983)]. The 3D model is usually employed for deep lakes, although it can be used to obtain the current profile for shallow lakes (Pang & Pu, 1996; Zhu & Cai, 1998a,b).

In the numerical simulations, the differential equations are usually solved by the finite difference method (FDM) or the finite element method (FEM). The former can be used for solving equations simply and quickly, but has poor adaptability to complex terrain. The latter is more complicated, but is more flexible, and is suitable for complex lake coast terrain. Other, less commonly used, methods for solving the differential equations include split operator, irregular grid FDM, finite analysis, and the nested grid.

In numerical simulation of hydrodynamics in Lake Taihu, Wang (1987) used a 2-D differential model to simulate the wind-driven currents under constant wind field conditions. Three current patterns were identified: (i) anticlockwise circulation (S, SW, and W winds), (ii) clockwise circulation (N, NE, and E winds), and (iii) episodic (SE and NW winds). Subsequently, Wu & Pu (1989) designed a

---

Luo, L.C.
State Key Laboratory of Lake Science and Environment, Nanjing Institute of Geography and Limnology, Chinese Academy of Sciences, 73 East Beijing Road, Nanjing 210008, P.R. China
e-mail: lcluo@niglas.ac.cn

depth-mean model (2-D). The irregular FDM model was adopted, with an exact fit function with the lake shore and bottom topography. Wu & Pu (1989) found that (i) the wind-induced current was unstable, changing quickly with changing wind force, and (ii) wind could significantly increase or decrease the water level, with the magnitude of the change mainly depending on wind speed and fetch. Thereafter, Wang et al. (1992) used a nested 2-D hydrodynamic grid model to simulate wind-driven currents in Lake Taihu and accurately predicted flow direction. Liu (1993), using the 2-D hydrodynamic model and FEM, simulated the currents in Meiliang Bay.

The aforementioned studies were based on 2-D models, which excluded the vertical variation of currents, and results reflected the integrated water column. In Lake Taihu, however, there are distinct vertical gradients in many factors such as dissolved oxygen, chlorophyll, light intensity. and turbidity. Thus, a 3-D hydrodynamic model is needed to provide their vertical variations.

Pang et al. (1994) used a 3-D hydrodynamic model to study the influence of terrain on the lake currents. Their results suggested that the narrow pass between Xishan Island and Dongshan Peninsula was an important cause of the large-scale circulation in Lake Taihu.

Subsequently, Zhu & Cai (1998a) established a simple 3-D hydrodynamic model for current simulation in Meiliang Bay. The water column was divided into five layers, and the differential equation was solved by the frog leap scheme (Zhu & Cai, 1998a). The results showed the horizontal current distribution in different layers, with the surface current being consistent with the wind direction, and the bottom current opposite to that at the surface; thus, there was an apparent compensation current. The current speed gradually decreased from the surface downwards.

Although this 3-D model yielded further knowledge of the layered currents, the vertical structure of the lake currents was poorly reflected, because the layer currents in deep areas cannot be computed once the maximum depth for the model has been decided. To compensate for this deficiency, a σ coordinate from ocean studies (Durance & Hughes, 1983) and Zhu & Fang (1994) was applied to the hydrodynamic studies of Lake Taihu. Liang & Zhong (1994) applied this to compute wind-driven currents in Lake Taihu, Hu et al. (1998a) applied it to simulate the effects of Typhoon 9711 on the water level and current fields, and Hu et al. (1998b) used this model to investigate the influence of different wind fields on currents. Hu et al. (1998b) also analyzed the difference between the depth-mean current field and each individual depth current field, and compared the 2-D and 3-D models.

Wang et al. (2001) developed a 3-D numerical model to simulate the effects of stable and unstable wind fields on currents in Lake Taihu. The vertical eddy viscosity coefficients were distinct for different depths, and a C-grid net and an improved alternating direction implicit method (ADI) were adopted. The results showed that the changes in vertical eddy viscosity coefficient with depth had a larger influence on current speed and bottom shear stress (BSS) than did water level. Sun (2001) modeled wind-driven currents in Lake Taihu and showed that wind direction largely determined the circulation and current field pattern in Lake Taihu, and that wind speed determined the current speed.

The results from either the 2-D model (Wang, 1987) or the 3-D model (Liang & Zhong, 1994) showed that as a consequence of the southerly winds, the current in the west of Meiliang Bay flowed south to north whereas that in the east flowed from north to south. In the north of Meiliang Bay, near the Liangxi River, there was a small anticlockwise circulation, which may account for the higher chlorophyll, total phosphorus, and nitrogen in this area compared to other areas of the bay.

3.3.2 Model Description

The basic equations describing the meso-scale water motion in Lake Taihu consist of a continuity equation (3.11), momentum equations (3.12) and (3.13), and a static equilibrium equation (3.14). For simplicity, a constant water density was employed and all the following assumptions were made:

1. Water is incompressible
2. Pressure distribution is vertically hydrostatic
3. β-Plane approximation and Boussinesq approximation are valid
4. Curvature effects and tide-generating forces were ignored
5. Air pressure above the water surface is constant
6. Constant eddy viscosity diffusivities are used to describe turbulence

The resulting equations are as follows (Qin & Stevens, 2001):

$$\frac{\partial u}{\partial x}+\frac{\partial v}{\partial y}+\frac{\partial w}{\partial z}=0 \tag{3.11}$$

$$\frac{\partial u}{\partial t}+u\frac{\partial u}{\partial x}+v\frac{\partial u}{\partial y}+w\frac{\partial u}{\partial z}=fv-\frac{1}{\rho_0}\frac{\partial P}{\partial x}+\frac{\partial}{\partial x}\left(A_h\frac{\partial u}{\partial x}\right)+\frac{\partial}{\partial y}\left(A_h\frac{\partial u}{\partial y}\right)+\frac{\partial}{\partial z}\left(A_v\frac{\partial u}{\partial z}\right) \tag{3.12}$$

$$\frac{\partial v}{\partial t}+u\frac{\partial v}{\partial x}+v\frac{\partial v}{\partial y}+w\frac{\partial v}{\partial z}=-fu-\frac{1}{\rho_0}\frac{\partial P}{\partial y}+\frac{\partial}{\partial x}\left(A_h\frac{\partial v}{\partial x}\right)+\frac{\partial}{\partial y}\left(A_h\frac{\partial v}{\partial y}\right)+\frac{\partial}{\partial z}\left(A_v\frac{\partial v}{\partial z}\right) \tag{3.13}$$

$$\frac{\partial p}{\partial z}=-\rho g \tag{3.14}$$

where u, v, and w are velocity components in x, y, and z directions, t is time, ρ is water density, f the Coriolis force coefficient by $2\omega\sin\varphi$, ω is rotation angular velocity of the earth, φ is latitude, g is the gravitational acceleration and P the water pressure, and A_h, A_v represent turbulent viscosities in horizontal and vertical directions.

A vertical stretched coordinate system (sigma coordinate system) was employed because it gives a smooth representation of the bottom topography and the same order of vertical resolution for shallow and deeper parts of the water. A derivation of the sigma coordinate equations is based on the following transformation:

$$x = x', y = y', \sigma = \frac{z - h(x, y)}{\zeta(x, y, t) - h(x, y)} = \frac{z - h(x, y)}{H(x, y, t)} \tag{3.15}$$

where $\xi(x, y, t)$ is the surface elevation, $h(x, y)$ is the bottom topography, $H(x, y, t)$ is the water depth, and *x*, *y*, *z* are the conventional Cartesian coordinates.

After conversion to sigma coordinates, the original equations (3.11), (3.12), and (3.13) may be written:

$$\frac{\partial(Hu)}{\partial x} + \frac{\partial(Hv)}{\partial y} + \frac{\partial(Hw)}{\partial \sigma} + \frac{\partial \zeta}{\partial t} = 0 \tag{3.16}$$

$$\begin{aligned} &\frac{1}{H}\left(\frac{\partial(Hu)}{\partial t} + \frac{\partial(Huu)}{\partial x} + \frac{\partial(Huv)}{\partial y} + \frac{\partial(Huw)}{\partial \sigma}\right) \\ &= \frac{\partial}{\partial x}\left(A_h\frac{\partial u}{\partial x}\right) + \frac{\partial}{\partial y}\left(A_h\frac{\partial u}{\partial y}\right) + \frac{1}{H^2}\frac{\partial}{\partial \sigma}\left(A_v\frac{\partial u}{\partial \sigma}\right) + fv - g\frac{\partial \zeta}{\partial x} + \varepsilon_1 \end{aligned} \tag{3.17}$$

$$\begin{aligned} &\frac{1}{H}\left(\frac{\partial(Hv)}{\partial t} + \frac{\partial(Hvu)}{\partial x} + \frac{\partial(Hvv)}{\partial y} + \frac{\partial(Hvw)}{\partial \sigma}\right) \\ &= \frac{\partial}{\partial x}\left(A_h\frac{\partial v}{\partial x}\right) + \frac{\partial}{\partial y}\left(A_h\frac{\partial v}{\partial y}\right) + \frac{1}{H^2}\frac{\partial}{\partial \sigma}\left(A_v\frac{\partial v}{\partial \sigma}\right) - fu - g\frac{\partial \zeta}{\partial y} + \varepsilon_2 \end{aligned} \tag{3.18}$$

The last terms in equations (3.17) and (3.18) represent high-order minor terms that can be ignored. Then, the vertical velocity *W* in the (*x*, *y*, *z*) coordinates is related to w in the sigma coordinates as

$$\begin{aligned} W &= Hw + \sigma\left(\frac{\partial H}{\partial t} + u\frac{\partial H}{\partial x} + v\frac{\partial H}{\partial y}\right) \\ &= Hw + \sigma\left(\frac{\partial H}{\partial t} + u\frac{\partial H}{\partial x} + v\frac{\partial H}{\partial y}\right) + \left(u\frac{\partial h}{\partial x} + v\frac{\partial h}{\partial y}\right) \end{aligned} \tag{3.19}$$

In the transformed coordinates, the kinetic and kinematic boundary conditions at the free surface and the bottom can be separately expressed as

$$\frac{\rho A_v}{H}\left(\frac{\partial u}{\partial \sigma}, \frac{\partial v}{\partial \sigma}\right) = \left(\tau_{sx}, \tau_{sy}\right) = \rho_a C_{da}\left(u_w^2 + v_w^2\right)^{1/2}(u_w, v_w)\ w = 0$$
$$\sigma = 1 \text{ (surface)} \tag{3.20}$$

$$\frac{\rho A_v}{H}\left(\frac{\partial u}{\partial \sigma}, \frac{\partial v}{\partial \sigma}\right) = \left(\tau_{bx} \tau_{by}\right) = \rho C_{db}\left(u^2 + v^2\right)^{1/2}(u, v)\ w = 0$$
$$\sigma = 0 \text{ (bottom)} \tag{3.21}$$

τ_{sx}, τ_{sy} are the wind stresses in x and y directions, and τ_{bx}, τ_{by} the bottom current stresses in these two directions, respectively; C_{da} and C_{db} are the wind and bottom drag coefficients, respectively; $(u_w,\ v_w)$ are wind velocities at X and Y directions at 10 m above the water surface, and $(u,\ v)$ the bottom current velocities at X and Y directions; and (ρ, ρa) are the densities of water and air. For lateral close boundaries the conditions are as follows:

$$\frac{\partial V_s}{\partial n} = 0;\ V_n = 0 \tag{3.22}$$

where V_s and V_n are velocities tangential to and normal to the boundary.

The vertical velocities at different levels in the sigma coordinate system can be calculated by vertically integrating the continuity equation:

$$w(\sigma) = -\frac{1}{H}\left(\sigma w'(1) - w'(\sigma)\right) \tag{3.23}$$

where $w'(\sigma)$ is given as

$$w'(\sigma) = -\frac{\partial}{\partial x}\left(H\int_{\sigma}^{1} u d\sigma\right) - \frac{\partial}{\partial y}\left(H\int_{\sigma}^{1} v d\sigma\right) \tag{3.24}$$

The equations in the sigma-coordinate system contain fast-moving external gravity waves at the water surface and slowly moving internal gravity waves in the water body. For both computer economy and accuracy, the vertically integrated equations (external model) are separated from the vertical structure equations (internal model). The so-called model splitting technique (Madale & Piacsek, 1977), allows for computation of the three-dimensional flow structures with minimal additional cost over computation of the two-dimensional flow structures with a vertically integrated model (Sheng, 1982), and also permits calculation of the free surface elevation with little sacrifice in computational time (Mellor, 1998). Significant improvement in computational efficiency results from the use of this technique.

The external mode, as described by the water level (ζ) and the vertically integrated mass fluxes (U and V), is governed by the following equations, which are obtained by vertically integrating equations (3.16)–(3.18) from $\sigma = 0$ to $\sigma = 1$ and using the corresponding boundary conditions:

$$\frac{\partial \zeta}{\partial t} + \frac{\partial U}{\partial x} + \frac{\partial V}{\partial y} = 0 \tag{3.25}$$

$$\begin{aligned}\frac{\partial U}{\partial t} &= -\frac{\partial}{\partial x}\left(\frac{\overline{UU}}{H}\right) - \frac{\partial}{\partial y}\left(\frac{\overline{UV}}{H}\right) + \frac{\tau_{sx}}{\rho} - \frac{\tau_{bx}}{\rho} + fV - gH\frac{\partial \zeta}{\partial x} + (HH_x)\\ &= -gH\frac{\partial \zeta}{\partial x} + D_x\end{aligned} \tag{3.26}$$

$$\frac{\partial V}{\partial t} = -\frac{\partial}{\partial x}\left(\frac{\overline{VU}}{H}\right) - \frac{\partial}{\partial y}\left(\frac{\overline{VV}}{H}\right) + \frac{\tau_{sy}}{\rho} - \frac{\tau_{by}}{\rho} - fU - gH\frac{\partial \zeta}{\partial y} + (HH_y)$$
$$= -gH\frac{\partial \zeta}{\partial y} + D_y \tag{3.27}$$

$$(U, V) = \int_0^1 H(u, v)d\sigma \tag{3.28}$$

where HH_x, HH_y denote the horizontal diffusion terms.

The internal mode of the flow is described by the vertical flow structures. Defining perturbation velocities as $u' = u\text{-}U/H$ and $v' = v\text{-}V/H$, the equations for the internal mode are obtained by subtracting the vertically averaged momentum equations from the three-dimensional equations:

$$\frac{1}{H}\frac{\partial\left(Hu'\right)}{\partial t} = B_x - \frac{D_x}{H} + \frac{1}{H^2}\frac{\partial}{\partial \sigma}\left(A_v\frac{\partial}{\partial \sigma}\left(u' + \frac{U}{H}\right)\right) \tag{3.29}$$

$$\frac{1}{H}\frac{\partial\left(Hv'\right)}{\partial t} = B_y - \frac{D_y}{H} + \frac{1}{H^2}\frac{\partial}{\partial \sigma}\left(A_v\frac{\partial}{\partial \sigma}\left(v' + \frac{V}{H}\right)\right) \tag{3.30}$$

where B_x and B_y represent all terms in the transformed 3-D momentum equations, except the surface slopes and vertical diffusion terms and $D_x,\ D_y$ as defined in equations (3.26) and (3.27). The surface slope terms are not included in these equations, so that a larger time step may be used in the numerical computation.

A nonuniform grid is generally needed in shallow water models to better resolve the complex shoreline geometries and bottom features (Sheng, 1982). A space-staggered C-grid was employed in both the horizontal and vertical directions of the computational domain. The horizontal velocities, u and v, are defined at the center of each layer, and the vertical velocity, w, is defined at the layer interfaces (Drago & Iovenitti, 2000; Zhang & Gin, 2000). Along the shoreline, the normal velocity is set at zero. In the vertical direction, the free surface and the bottom both fall on the full grid points where the vertical velocities and turbulent fluxes are computed or specified. The horizontal velocities are computed at the half-grid points in the vertical direction.

In solving the time-dependent partial differential equations, all the terms in equation (3.25) are treated implicitly, but only the time derivatives and the surface slopes in equations (3.26) and (3.27) are treated implicitly. Then, the following finite difference equations can be obtained:

$$\left(I + (1-\varphi)\lambda_x + (1-\varphi)\lambda_y\right)\Phi^{n+1} = \left(I - \varphi\lambda_x - \varphi\lambda_y\right)\Phi^n + \Delta t D^n \tag{3.31}$$

where $A,\ B, \varphi,\ D,\ \lambda_x$ and λ_y are defined as

$$A = \begin{bmatrix} 0 & 1 & 0 \\ gH & 0 & 0 \\ 0 & 0 & 0 \end{bmatrix}; \ B = \begin{bmatrix} 0 & 0 & 1 \\ 0 & 0 & 0 \\ gH & 0 & 0 \end{bmatrix}; \ \Phi = \begin{bmatrix} \varsigma \\ U \\ V \end{bmatrix}; \ D = \begin{bmatrix} 0 \\ D_x \\ D_y \end{bmatrix}$$

$$\lambda_x = \frac{A\Delta t}{\Delta x}\delta_x; \lambda_y = \frac{A\Delta t}{\Delta y}\delta_y \tag{3.32}$$

and I is the identity matrix, (δ_x, δ_y) are the horizontal grid spacings, Δt is the time step, D_x and D_y are terms defined in equations (3.26) and (3.27), superscripts $n+1$ and n indicate the present and previous step of integration, Δx, Δy are central difference spatial operators, and φ is a weighting factor in the range of 0–1. If $\varphi = 1$, equation (3.31) reduces to a two-step explicit scheme; if $\varphi < 1$, the resulting schemes are implicit, and $\varphi = 0$ corresponds to the fully implicit scheme. The following equations can be obtained by using the ADI scheme (Qin & Stevens, 2001; Wang & Hutter, 2001):

$$(I + (1-\varphi)\lambda_x)\Phi' = \left(I - \varphi\lambda_x - \lambda_y\right)\Phi^n + \Delta t D^n \text{ x-sweep} \tag{3.33}$$

$$\left(I + (1-\varphi)\lambda_y\right)\Phi^{n+1} = \Phi' + (I-\varphi)\lambda_y\Phi^n \text{ y-sweep:} \tag{3.34}$$

By using the Gauss–Seidel method, the two equations can be solved and a large time step can be employed, with the maximum time step decided by the following equation generated from the Courant–Friedrichs–Lewy (CFL) condition:

$$\min\left(\frac{\Delta x, \Delta y}{(gH_{\max})^{0.5}}\right) << \Delta t \leq \min\left(\frac{U}{H\Delta x} + \frac{V}{H\Delta y}\right)^{-1} \tag{3.35}$$

The calculated (U, V) from the external mode are immediately input into the internal mode to solve the vertical structure of velocities.

Treating equations (3.29) and (3.30) with a two-level scheme and vertically implicit scheme, one can obtain the following finite difference equations:

$$\frac{H^{n+1}}{H^n}u'^{n+1} = u'^n + \Delta t\left(B_x - \frac{D_x}{H}\right) + \frac{\Delta t\partial}{H^2\partial\sigma}\left(A_v\frac{\partial}{\partial\sigma}\left(\alpha\left(u'^n + \frac{U^n}{H^n}\right)\right.\right.$$
$$\left.\left.\left. + (1-\alpha)\left(u'^{n+1} + \frac{U^{N+1}}{H^{n+1}}\right)\right)\right)\right) \tag{3.36}$$

$$\frac{H^{n+1}}{H^n}v'^{n+1} = v'^n + \Delta t\left(B_y - \frac{D_y}{H}\right) + \frac{\Delta t\partial}{H^2\partial\sigma}\left(A_v\frac{\partial}{\partial\sigma}\left(\alpha\left(v'^n + \frac{U^n}{H^n}\right)\right.\right.$$
$$\left.\left.\left. + (1-\alpha)\left(v'^{n+1} + \frac{U^{N+1}}{H^{n+1}}\right)\right)\right)\right) \tag{3.37}$$

where α is a weighting factor in the range of 0–1, as φ in equation (3.31). The bottom friction terms in both equations are also treated implicitly to ensure unconditional numerical stability. In addition, care must be taken to ensure that the vertically integrated perturbation velocities at each horizontal location are always equal to zero.

The time step in the internal model can be larger than or equal to that of the external model within the permission of the CFL condition.

3.3.3 Model Verification

The model has been used to simulate the water level and current circulation in Lake Taihu. The horizontal grid net was composed of 44 grid points in the x-direction and 39 in the y-direction, with a horizontal resolution of $\Delta x = 1.55\,\text{km}$ and $\Delta y = 1.974\,\text{km}$. The water column was divided into five layers. The same time step ($\Delta t = 240s$) was used in the internal and external models. Horizontal and vertical diffusion coefficients were kept constant at $50\,\text{m}^2/\text{s}$ and $2.5 \times 10^{-4}\,\text{m}^2/\text{s}$, respectively.

During typhoon 9711, August 18–21, 1997, the model was applied to simulate the water levels at the six hydrometric stations (Dapukou, Xiaomeikou, Taipu, Xiahsn, Xukou, and Wangting; H1–H6 in Fig. 3.9) and the current circulation at the mouth of Meiliang Bay (C1 in Fig. 3.9) where observations were made. The wind data were simultaneously recorded at the six weather stations (Taihu Laboratory for Lake Ecosystem Research: TLLER, Yixing, Changxing, Dongshan, Wujiang, and Huzhou, M1–M6 in Fig. 3.9), and used to calculate the mean values, as there was little difference among the weather stations.

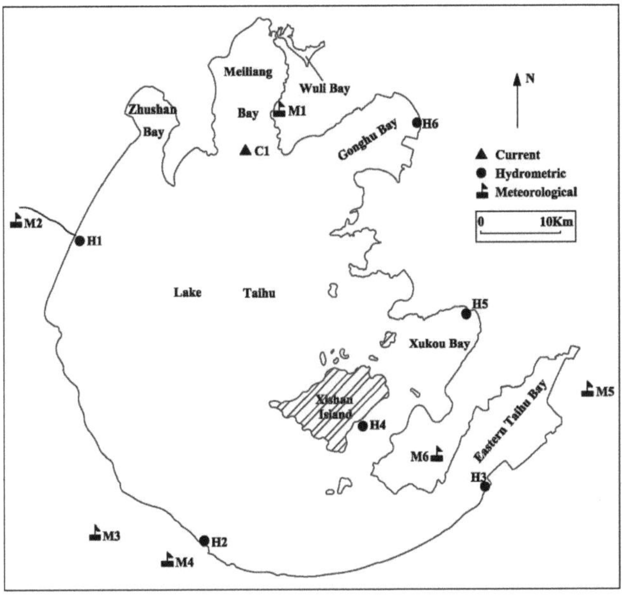

Fig. 3.9 Locations of the meteorological, hydrometric, and current observation sites in Lake Taihu: *C1*, mooring site for current observation; H1–H6, hydrometric stations (*H1*, Dapukou; *H2*, Xiaomeikou; *H3*, Taipu; *H4*, Xishan; *H5*, Xukou; *H6*, Wangting); M1–M6, meteorological stations (*M1*, TLLER; *M2*, Yixing; *M3*, Changxing; *M4*, Huzhou; *M5*, Wujiang; *M6*, Dongsha)

The wind speed oscillated frequently, with a mean of 7.5 m/s and a maximum of 11.7 m/s. From midnight of August 18 to midday of August 19, wind speed decreased dramatically from the maximum of 11.7 m/s down to 6.0 m/s, and then gradually increased to 9.6 m/s. During the typhoon, there was significant variation in the wind direction; the dominant wind was NE on August 18 and SE on August 20. The wind shear caused significant water level changes.

The trend of the modeled water levels was in reasonable agreement with the observed results (Fig. 3.10). However, observed water levels were somewhat higher than the predicted levels, most probably because the model omitted inflow and outflow data. At Dapukou and Xiaomeikou (Fig. 3.10), the main inflow rivers, the observed water levels were higher than the modeled results, probably because a large volume of water flowed into the lake during the typhoon. At Taipu, in the southeast of Lake Taihu (Fig. 3.10), observed water levels were much higher than the predicted levels after the wind shear on August 19. At this site, water levels can drop quickly with SE wind forcing; there are many complications affecting the accuracy of the measured data. At the other three stations, Xishan, Xukou, and Wangting (Fig. 3.10),

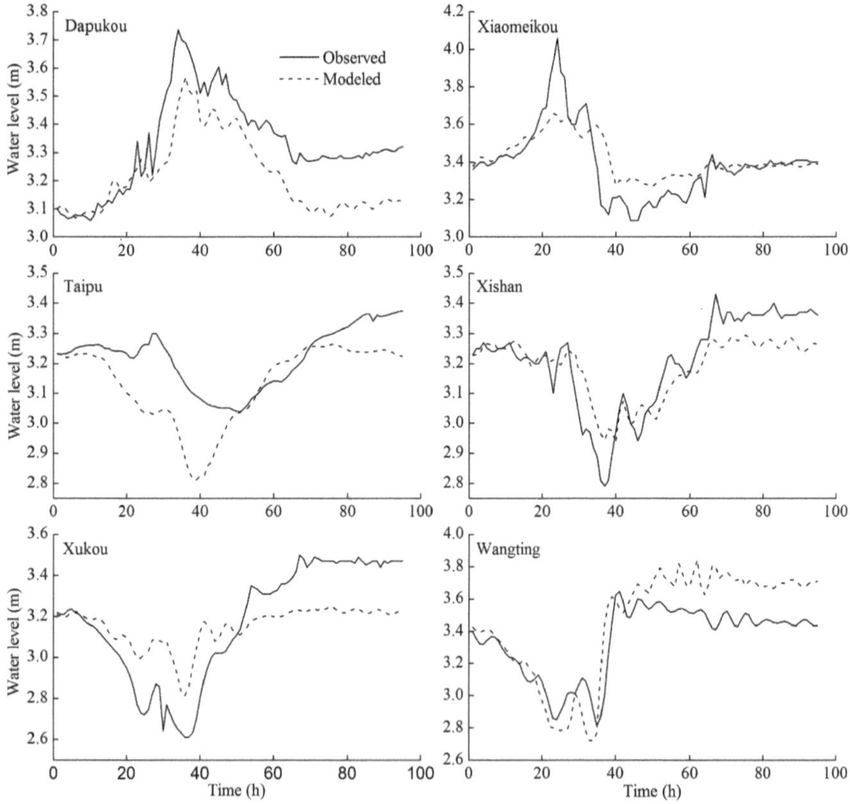

Fig. 3.10 Comparison of water levels between modeled (*dotted lines*) and observed (*solid lines*) at the six hydrometric stations

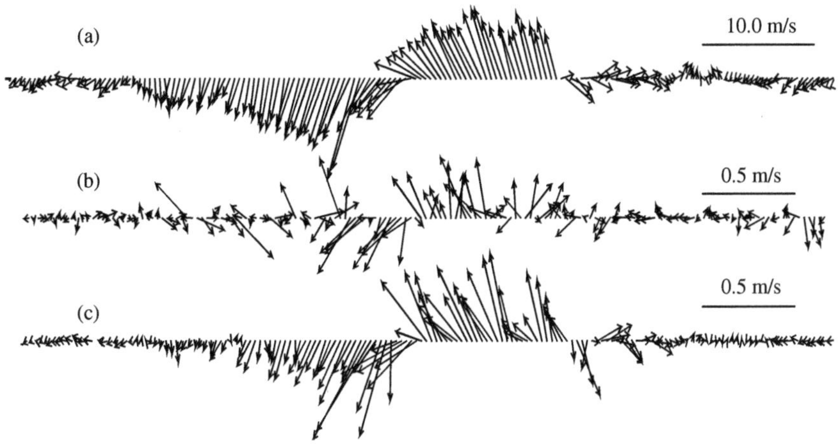

Fig. 3.11 The measured winds (**a**), the measured surface currents (**b**), and the simulated surface currents (**c**) at the mouth of Meiliang Bay during typhoon 9711 in month year

observed water levels were slighter higher than the predicted levels, probably from the lake discharges.

The observed winds and surface currents, and the simulated currents, at the mouth of Meiliang Bay during the typhoon are shown in Fig. 3.11. Both the observed currents and the simulated currents are in good accordance with the wind vectors. However, the simulated currents are more regular and efficient in reflecting the wind shear than the observed currents. The weaker agreement of the observed currents with the measured wind vectors is likely caused by small-scale or meso-scale wave disturbances.

3.3.4 Current Field and Dominant Wind

The monsoon climate of the Lake Taihu area is characterized by prevailing south-easterly winds in summer and northwesterly winds in winter. Numerical modeling has been conducted for both these stable wind forcings to evaluate the impact of wind speed on the water currents in the lake.

A closed-boundary condition was used in the simulation to reflect the negligible influence of the inflow and outflow rivers on the lake current circulation. The free surface is acted on by a constant 10.0 m/s SE wind (summer) and or a constant 10.0 m/s NW wind (winter), with a constant wind-induced stress ($\tau_x = \tau_y = 0.1825\,\text{N/m}^2$). The results of the simulations are shown in Figs. 3.12 and 3.13, respectively.

With 10.0 m/s SE wind forcing (summer), the surface elevation is positive at NW and negative at SE, which means the water mass has been transported from SE to NW of the lake (Fig. 3.12a). There are two clockwise subcirculations in the west and north and two anticlockwise subcirculations in the south and east (Fig. 3.12b).

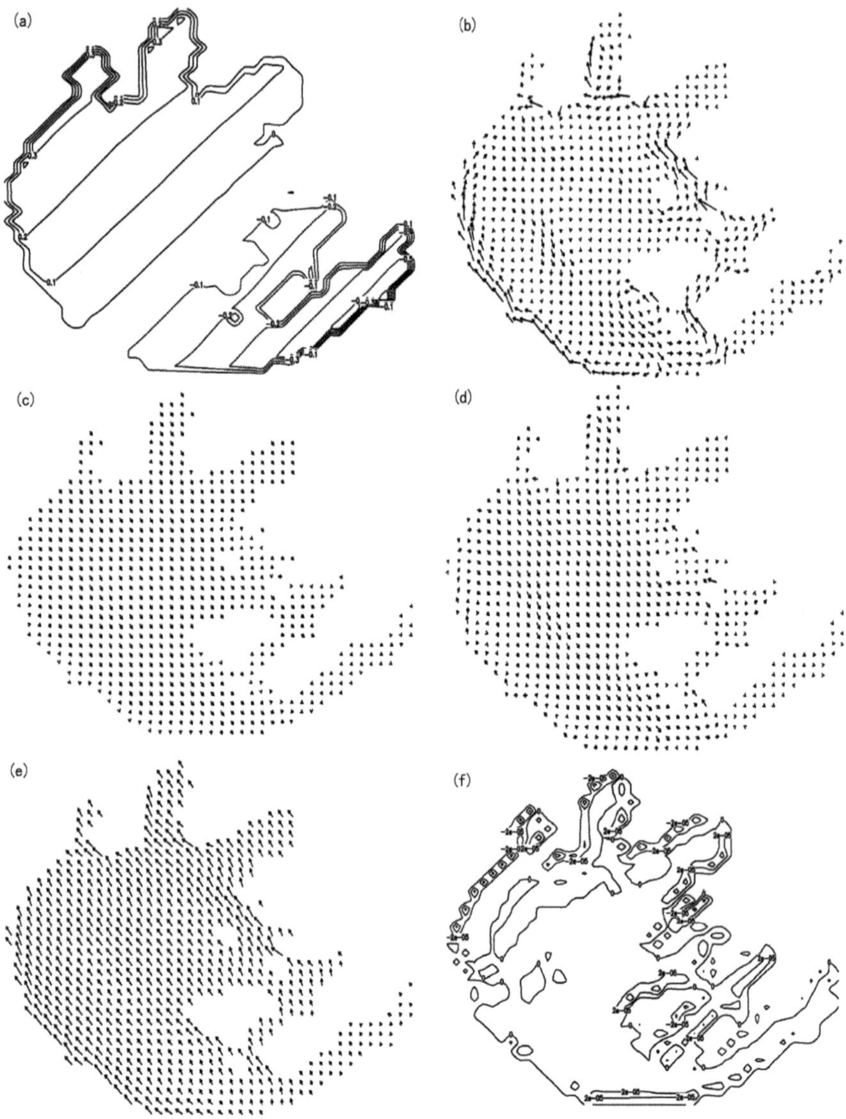

Fig. 3.12 The hydrodynamic characteristics of (with 10.0 m/s SE wind forcing, in summer) (**a**) free surface elevation; (**b**) depth-mean current circulation; (**c–e**) the current circulations of layer 1 (*top*), layer 3 (*middle*), and layer 5 (*bottom*); (**f**) the vertical velocity field of the third layer

The current flows from SE to NW at the surface of the lake, and from NW to SE at the bottom of the lake current (i.e., counter to the wind direction), and the transition area is at the middle layer (Fig. 3.12c–e). As a result, the vertical movement of water is very intense at the boundaries of the SE upward flow and the NW downward flow (Fig. 3.12f).

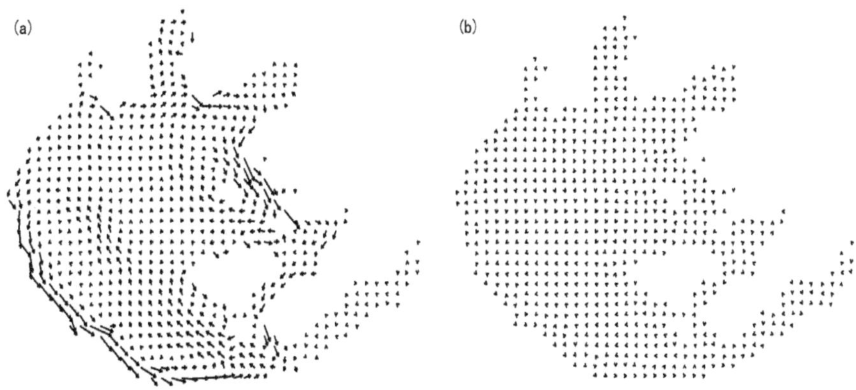

Fig. 3.13 (**a**) The depth-mean current circulations with 10.0 m/s NW wind forcing. (**b**) The depth-mean current circulations with smooth bottom and 10.0 m/s SE wind forcing

For the opposite wind field, 10.0 m/s NW wind forcing (winter), the depth-mean current circulations are shown in Fig. 3.13a.

To examine the effects of bottom topography on eddy vorticity, all the depth-mean current vectors were computed with the same depth at all the grids, and SE 10.0 m/s wind forcing (Fig. 3.13b). The simulation suggests that vorticity would develop even if the wind forcing at all the grids was uniform.

Vorticity can be expressed as

$$\varsigma = \frac{\partial v}{\partial x} - \frac{\partial u}{\partial y} \tag{3.38}$$

Here ς denotes vorticity, and u, v are current speed components at x and y directions. The time derivation of vorticity becomes as follows:

$$\frac{\partial \varsigma}{\partial t} = \frac{1}{\rho}\left[\frac{1}{h}\left(\frac{\partial \tau_{sy}}{\partial x} - \frac{\partial \tau_{sx}}{\partial y}\right) + \left(\tau_{sy}\frac{\partial \left|\frac{1}{h}\right|}{\partial x} - \tau_{sx}\frac{\partial \left|\frac{1}{h}\right|}{\partial y}\right)\right] \tag{3.39}$$

where h is bottom elevation, τ_{sx}, τ_{sy} are wind stresses in x and y directions, and ρ is water density.

In the right-hand part of equation (3.39), the first term will be zero due to the uniform wind. Hence, equation (3.39) becomes (Wang, 1987):

$$\frac{\partial \varsigma}{\partial t} = k\frac{|W|}{h^2} W \Lambda \nabla h \tag{3.40}$$

where k is constant and W is the unit vector of wind. From equation (3.40), vorticity change results from the bottom elevation gradient. The time derivation of vorticity increases with the increased angle between wind direction and bottom elevation gradient, and the maximum value is reached when the two vectors are orthogonal

where ρ is water density, f_w the wave friction coefficient, and U_m the maximum horizontal bottom speed caused by wave disturbance. U_m could be obtained using equation (3.42) (Madsen, 1976):

$$U_m = \pi H_s/(T_S \sin(2\pi d/L_d)) \tag{3.42}$$

where L_d represents the wavelength in water depth d, obtained using the following equation:

$$L_d = gT_S^2 \tan(2\pi d/L_d)/2\pi \tag{3.43}$$

where f_w is related to lake bottom roughness and wave Reynold number. For a deep-water lake, the value of f_w was generally 0.004 (Sheng & Lick, 1979), and for a shallow lake, 0.015. Here we used the following equation (Jiang et al., 2000):

$$f_w = \exp[-6 + 5.2(A_\delta/K_s)]^{-0.19} \tag{3.44}$$

where, $A_\delta = H_s/[2\sinh(\frac{2\pi}{L_s}h)]$; K_s is lake bottom roughness, normally set at 0.2 mm (Nielsen et al., 2001).

Shear stress produced by the lake current could be computed with the following formula (Sheng & Lick, 1979; Hawley, 2000):

$$T_c^b = \rho u_b^* \cdot u_b^* \tag{3.45}$$

where ρ is water density; u_b^* was computed using the following equation:

$$u_b^* = \frac{k_0 u_z}{\ln \frac{z}{Z_0}} \tag{3.46}$$

where k_0 is Karman constant (0.4); u_z was lake current speed at z meters (generally taken as 0.5 m) above the lake bottom; and Z_0 was bottom roughness set at 0.2 mm (Hawley, 2000; Nielsen et al., 2001).

Using the wave data and bottom current speed recorded by wave meter and acoustic Doppler profile (ADP) in the center of Lake Taihu in July 2002, we computer the shear stresses at the water–sediment interface resulting from currents and waves (Fig. 3.14).

With southeast winds of 1–6 m/s, significant wave height was 10–30 cm, wave length was 3–8 m, and the wave period was 1–2 s, their variations corresponding closely with wind speed (Fig. 3.14). However, variations in bottom current speed of 1–20 cm/s did not correspond closely with wind speed. The homochronous wave shear stress was 0.001–0.400 N/m^2 (Zhang & Xu, 2003), and the lake current shear stress was only 0.001–0.010 N/m^2, which was much smaller than the wave shear stress. At several times in the afternoons of July 23 and 24, when wind speeds were lower than 4–5 m/s, the shear stress of the lake currents is of the same order of that

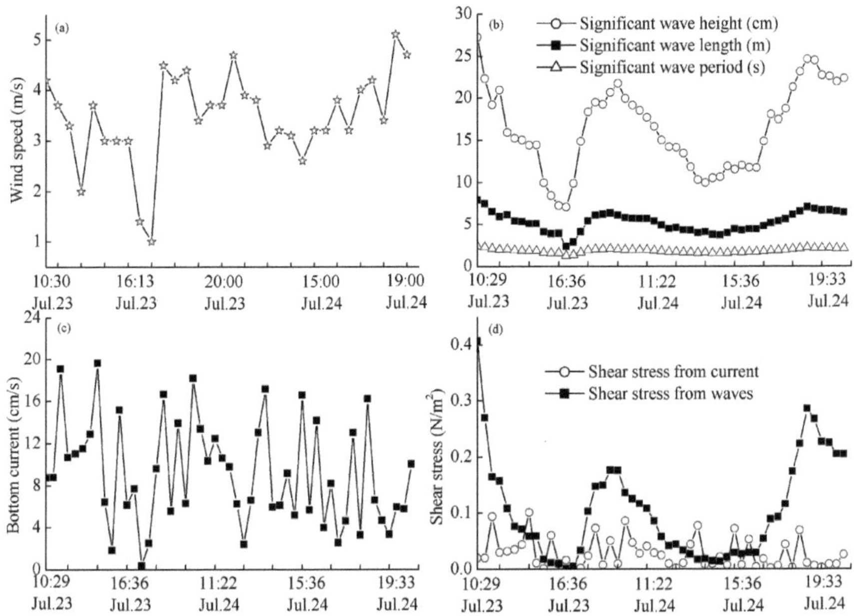

Fig. 3.14 Measured wind speed (**a**), wave height, period, and length (**b**), bottom current (**c**), and their resulting shear stresses (**d**) at the center of Lake Taihu during July 23–24, 2002

of waves. Therefore, with small wind speed, the shear stress at the water–sediment interface from lake currents and waves was almost the same; however, in high winds, the shear stress from waves was dominant.

3.4.2 Critical Stress for Sediment Resuspension

The critical stress of lake sediment resuspension is the shear stress when sediment resuspension is imminent; as such it is a fundamental parameter for understanding and computing the flux of sediment resuspension.

The critical shear stress of sediment resuspension for different depths, sediment densities, and wave conditions in Lake Taihu was determined using a laboratory wave flume for 36 groups of experiments. The experiment was conducted for water depths of 30, 40, and 50 cm and sediment densities of 1.289, 1.290, and 1.338 g/cm^3 (mean of 1.300 g/cm^3 is consistent with surface sediment).

Sediment resuspension affected by waves was computed according to Liu (1966), who considers the thickness of the infiltration layer to be smaller than wavelength, and sediment resuspension to belong to the surface layer. The critical wave height and water depth for sediment resuspension were computed using the following equation:

$$H_0 = M\sqrt{\frac{Lsh(2kh)}{\pi g}\left[\frac{\rho_s - \rho}{\rho} gD + A_2\delta\left(\frac{D}{D_0}\right)^y \frac{\varepsilon_K}{D}\right]} \tag{3.47}$$

where H_0 was critical wave height (m); m was a parameter decided by the experiential model; L was wave length (m); k was wave number; h was water depth (m); g was gravity acceleration (m/s^2); ρ_s was sediment density (kg/m^3); ρ was water density (kg/m^3), D was particle size of sediment (μm); D_0, ε_k were parameters; y was an index parameter; for A_2, δ a value of 0.02 was adopted.

$$h_0 = \frac{L}{4\pi} arcsh\left\{\frac{\pi g H^2}{M^2 L\left[\frac{\rho_s - \rho}{\rho} gD + A_2\delta\left(\frac{D}{D_0}\right)^y \frac{\varepsilon_K}{D}\right]}\right\} \tag{3.48}$$

where h_0 was critical water depth, and the coefficient y was 0.5.

The maximal horizontal bottom wave orbital speed could be inferred from the sediment critical wave parameters and water depth:

$$U_{\max(-h)0} = M\left[\frac{\rho_s - \rho}{\rho} gD + A_2\delta\left(\frac{D}{D_0}\right)^y \frac{\varepsilon_K}{D}\right]^{\frac{1}{2}} \tag{3.49}$$

The M value above was related to particle shape coefficient, and is a function of sediment particle frontal resistance coefficient (φx) and circumferential motion coefficient (φz). It can generally be determined by the following model (Liu, 1966):

$$M = 0.12\left(\frac{L}{D}\right)^{1/3} \tag{3.50}$$

Then the critical wave height for sediment resuspension could be expressed as

$$H_0 = 0.12\left(\frac{L}{D}\right)^{1/3}\sqrt{\frac{Lsh2kh}{\pi g}\left[\frac{\rho_s - \rho}{\rho} gD + 0.02\left(\frac{D}{D_0}\right)^{1/2} \frac{2.56}{D}\right]} \tag{3.51}$$

The factors influencing the critical wave height for sediment resuspension with water depth h and particle size D included sediment concentration, wave height, wavelength, and wave period. Equation (3.49) shows that the critical wave orbital speed for sediment resuspension also changed with the change of wave factors once water depth and particle size were determined. The critical wave heights for sediment resuspension for different-sized sediment particles and at different depths are shown in Table 3.4. Calculated values agreed well with the measured values, suggesting that the equations could be used for determining critical conditions of sediment resuspension.

Table 3.4 Comparison of critical sediment resuspension wave height measured in field surveys and calculated by equation (3.51), for different sediment particle diameters, water depths, wave periods, and wavelengths

Sediment particle diameter (cm)	Water depth (cm)	Wave period (s)	Wave length (cm)	Critical wave height H_0 (cm)	
				Measured	Equation (3.51)
	30	1.83	295	7.2	6.8
0.006	40	1.83	333	7.9	8.5
	50	1.56	298	10.3	10.7
	30	1.56	245	6.7	6.6
0.010	40	1.56	275	8.5	8.5
	50	1.38	252	10.4	11.1
	30	1.26	189	8.7	7.0
0.018	40	1.39	237	9.7	9.0
	50	1.57	300	11.5	10.8
	30	1.33	202	7.6	7.2
0.025	40	1.22	198	11.8	9.9
	50	1.24	215	12.8	12.9
0.015	15	2.00	236	5.5	4.1
	36	1.00	143	10.5	10.1
0.020	120	4.00	1302	21.0	22.3
0.050	36	1.00	143	11.9	11.4
0.0017		1.20		9.3	9.79
	30	1.49		8.7	9.51
		1.68		8.5	9.50
		1.1		13.5	14.62
	40	1.50		12.0	12.32
	50	1.1		19	20.83
		1.24		15.6	17.07

Based on equations (3.51) and (3.53), the wave height and critical shear stress for bottom sediment resuspension for different water depths and wave periods in Lake Taihu were determined ($D_{50} = 0.017$ mm) (Table 3.5). Data for August–October 1992 and for August–November 1993 gave the mean wave height as 5–40 cm and mean wave period as 1.2–2.3 s (Qiao et al., 1996).

$$h_o = \frac{L}{4\pi}\operatorname{arcsh}\left\{\frac{\pi g H^2}{0.12\left(\frac{L}{D}\right)^{1/3} L\left[\frac{\rho_s-\rho}{\rho}gD+0.02\left(\frac{D}{D_0}\right)^{1/2}\frac{2.56}{D}\right]}\right\} \tag{3.52}$$

$$U_{\max(-h)_o} = 0.12\left(\frac{L}{D}\right)^{1/3}\left[\frac{\rho_s-\rho}{\rho}gD+0.02\left(\frac{D}{D_0}\right)^{1/2}\frac{2.56}{D}\right] \tag{3.53}$$

In Lake Taihu, for sediment with median particle size of 0.017 mm and density of 1.3 g/cm^3, the mean critical shear stress was about 0.037 N/m^2 (this value was calculated from the last column in Table 3.5). Another study of a large shallow lake, Okeechobee Lake, in Florida (USA), showed that for sediment of similar size and

Table 3.5 Critical conditions for sediment resuspension in Lake Taihu ($D_{50} = 0.017$ mm)

Water depth (m)	Wave period (s)	Critical wave height (cm)	Critical shear stress (N/m^2)
0.5	1.2	17.95	0.036
0.5	2.4	14.67	0.036
1.0	2.4	21.02	0.026
1.5	2.4	30.48	0.036
2.0	2.4	42.56	0.032
2.5	2.4	58.91	0.040
3.0	2.4	81.80	0.053

density, the critical shear stress was $0.032\,N/m^2$ (Mehta, 1991), which is close to the value we obtained for Lake Taihu.

The critical shear stress of $0.037\,N/m^2$ for Lake Taihu corresponded to a wind speed of 4 m/s, according to data for July 1992 (Fig. 3.14). At this time, the shear stress at the water–sediment interface was mainly caused by waves. When wave disturbance exceeded the critical condition for sediment resuspension, a large amount of sediment would be suspended, as was observed in Meiliang Bay in February 1998 (Fig. 3.15).

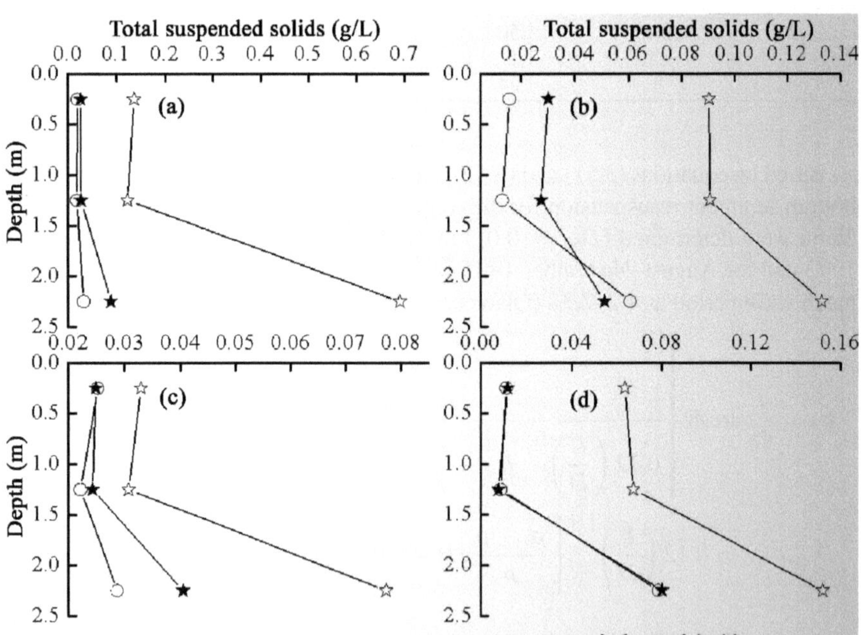

Fig. 3.15 Relationship between wind speed, depth, and suspended solid concentration at four sites in Meiliang Bay February–May 1998 (**a**, site 2; **b**, site 4; **c**, site 6; **d**, site 7) (Qin et al., 2000). See separate comments

In Lake Taihu, sediment resuspension is mainly caused by waves, and Glenn & Grant (1987) estimated that for a shallow lake 70% of driven strength comes from waves. Waves and lake currents can be considered together in an equation for computing lake bottom stress (Sheng & Lick, 1979). This bottom stress can cause sediment resuspension if it is larger than the sediment resuspension critical shear stress, which is mainly affected by particle size and density of the sediment. In Lake Taihu, there was no significant variation in particle size of the sediment with depth or region; critical shear stress is mainly determined by sediment density.

In the eutrophic regions of Lake Taihu, there was a 2- to 3-cm-thick boundary layer at the sediment–water interface. Composed of 95% water and 5% organic particulate matter or fine sand, the layer formed a sediment series of increasing density together with the underlying sediment. Thus, the critical stress for sediment resuspension increases with depth because the critical stress is mainly decided by sediment density, resulting in a decrease of erosion rate with increasing sediment depth. This result has been also shown by another study (Amos et al., 1992).

References

Amos, C. L., G. R. Daborn, H. A. Christian, A. Atkinson & A. Robertson, 1992. In situ erosion measurements on fine-grained sediments from the Bay of Fundy. Marine Geology 108: 175–196.

Blom, G., E. H. S. Van Duin, R. H. Aalderink, L. Lijklema & C. Toet, 1992. Modelling sediment transport in shallow lakes: interactions between sediment transport and sediment composition. Hydrobiologia 235: 153–166.

Drago, M., L. Iovenitti, 2000. σ-Coordinates hydrodynamic numerical model for coastal and ocean three-dimensional circulation. Ocean Engineering 27: 1065–1085.

Durance, J. A. & D. G. Hughes, 1983. North Sea dynamics. Berlin: Springer-Verlag.

Glenn, S. M. & W. D. Grant, 1987. A suspended sediment stratification correction for combined wave and current flows. Journal of Geophysical Research 92: 8244–8264.

Hawley, N., 2000. Sediment resuspension near the Keweenaw Peninsula, Lake Superior during the fall and winter 1990–1991. Journal of Great Lakes Research 26: 495–505.

Hu, W. P., P. M. Pu & B. Q. Qin, 1998a. A three-dimensional numerical simulation on the dynamics in Taihu Lake, China (I): the water level and the current during the 9711 typhoon process. Journal of Lake Sciences 10: 17–25 (In Chinese with English abstract).

Hu, W. P., P. M. Pu & B. Q. Qin, 1998b. A three-dimensional numerical simulation on the dynamics in Taihu Lake, China (II): the typical wind-driven current and its divergence. Journal of Lake Sciences 10: 26–34 (In Chinese with English abstract).

Hutter, K. 1983. Hydrodynamics of Lakes. International Center for Mechanical Sciences, CISM Courses Lectures No. 286. New York: Springer.

Jiang, W., T. Pohlmann, J. Sündermann & S. Z. Feng, 2000. A modeling study of SPM transport in the Bohai Sea. Journal of Marine Systems 24: 175–200.

Krone, R. B., 1962. Flume studies of the transport of sediment in estuarial shoaling. Final Report. Hydraulic Engineering Laboratory and Sanitary Engineering Research Laboratory, University of California, Berkeley, 120.

Liang, R. J. & J. H. Zhong, 1994. A three-dimensional numerical simulation of wind-driven water current in Taihu Lake. Journal of Lake Sciences 6: 289–297 (In Chinese with English abstract).

Lick, W., 1994. The flocculation, deposition, and resuspension of fine-grained sediments. In: DePinto, J. V., W. Lick & J. F. Paul (eds), Transport and transformation of contaminants near the sediment–water interface. London: Lewis Publishers, 35–57.

Chapter 4
Sediment–Water Exchange and Its Significance

4.1 Sedimentation Rate and Distribution of Sediment

Guangwei Zhu

Throughout the history of lakes, inorganic and organic matter, nutrients, and pollutants from within the lake, from rivers, and from the air in the watershed deposit new layers of sediments through physical, chemical, and biological processes such as flocculation and deposition. At the same time, sediments release nutrients to the water; this is the internal source of nutrients in lakes. The distribution and properties of sediments are closely related to the distribution, growth, and population sizes of hydrophytes and benthos.

Therefore, study of the distribution and properties of lake sediments helps us to understand the hydrological characteristics, ecosystem, and pollution status of lakes and their catchments. Such studies play an important role in quantifying the development of lake ecosystems, through understanding lake sediments, the laws governing exchange of materials across the sediment–water interface, and the effects of environmental factors, such as hydrodynamic processes, on this exchange of materials.

The role of sediments in oceans, lakes, and rivers as reservoirs of nutrients and pollutants is one of the most important topics of aquatic environmental studies. However, most studies have been focused on seasonal variation of the conversion between "sources" and "sinks" of sediments, the effects of sediments on adsorption and deposition of pollutants, and measurement of fluxes of nutrients and pollutants across the sediment–water interface using geochemical balance techniques. Few studies have considered the effects of hydrodynamic processes on the exchange of materials across the sediment–water interface and on the development of aquatic environments. In large and shallow Lake Taihu, hydrodynamic processes could strongly influence the release and/or adsorption of nutrients to sediments. Sediment resuspension driven by wind waves and lake currents has important effects on the

Zhu, G.W.
State Key Laboratory of Lake Science and Environment, Nanjing Institute of Geography and Limnology, Chinese Academy of Sciences, 73 East Beijing Road, Nanjing 210008, P. R. China
e-mail: gwzhu@niglas.ac.cn

Qin, B. Q. (ed.), *Lake Taihu, China*,

exchange of materials across the sediment–water interface, on nutrient cycles, and on algae blooming in the lake. In recent years, the exchange of materials across the sediment–water interface in Lake Taihu has been studied from a hydrodynamic perspective because of the effects of hydrodynamic processes on both deterioration and restoration of lake ecosystems. Some important achievements have been made.

Lake Taihu has numerous bays and can be divided into many ecosystem types. Areas differ from each other in deposition conditions for suspended particulates, environmental conditions, and exchange of materials across the sediment–water interface. Therefore, the distribution, sedimentation rates, and deposition situation of sediments in different areas of the lake must be understood to study the exchange of materials across the sediment–water interface in the whole lake.

Since the 1970s, Nanjing's Institute of Geography and Limnology (Chinese Academy of Sciences) has made numerous investigations on the bottom sediment properties of Lake Taihu and has surveyed the sedimentation history of the lake using deep-boring, shallow-boring, and sub-bottom profilers. Previous investigations showed that sediments could be divided into two categories: (1) primeval sediments, which could be further divided into two types [(1a) loess-like silt clay, which is wind-deposited in the middle and lower reaches of the Yangtze River, transported by rivers, and redeposited in the Lake Taihu basin, and is common in the bottoms of Lake Taihu, the Yellow Sea, and the East China Sea; (1b) silt-like deposits in rivers and lakes, distributed along river courses or in shallow depressions on loess-deposition lands]; and (2) modern sediment, mainly including silt and clay, that is distributed in most of the lake areas, participates in material exchange across the sediment–water interface and other activities, and has an important effect on eutrophication in the lake (Sun & Wu, 1987). In recent years, the studies on sediments in Lake Taihu have focused on sedimentation rate and the distribution of sediment in relation to managing the lake environment.

4.1.1 Sedimentation Rate

Since the 1980s, the deposition rates of surface sediment in different areas of Lake Taihu have been measured many times using dating methods such as ^{210}Pb, ^{137}Cs, and ^{14}C dating (Sun & Wu, 1987; Chang & Liu, 1996; Qu et al., 1997; Xue et al., 1998). There have been great differences in deposition rates of sediment in different areas, especially in sediments formed in the past 100 years, because Taihu is a shallow lake with a large area, numerous bays, and complex ecosystem types. It must be pointed out that parts of the lake bottom might be erosion areas, while other parts may be depositing areas because of the shallowness and spatial heterogeneity of the lake.

4.1.1.1 Sedimentation Rate in the Past 15,000 Years

Qu et al. (1997) used ^{14}C to date sediment layers formed in the past 15,000 years using scintillation at the WT1 core in the south area of Mashan Hill in western Lake Taihu. The relationship between date and depth could be obtained from the four ^{14}C dates, specifically, $5,936 \pm 44$ years B.P. (50–60 cm) (B.P., before present), $7,899 \pm$

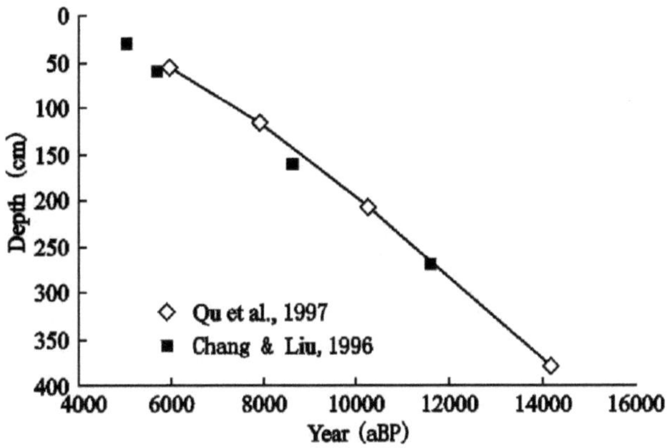

Fig. 4.1 Dates of sediment in a core from western Lake Taihu by the ^{14}C method

110 years B.P. (108–124 cm), 10, 299± 123 years B.P. (202–212 cm), and 14, 188± 865 years B.P. (374–384 cm) (Fig. 4.1). "Soft sediment" (which is easy to penetrate with a bamboo pole) at a depth of 50 cm was deposited 5,000 years ago, whereas at 100 cm it was deposited about 7,000 years ago, and at 2 m in western Lake Taihu, it was deposited about 10,000 years ago; this means that there was lowland with water-logged depressions 14,000 years B.P. (Xue et al., 1998). Chang & Liu (1996) showed that the age of surface sediments in Eastern Taihu Bay was 15,885 ± 170 years B.P. Other dating work showed that the history of loess sediments might be up to 11,240 years B.P. This sediment was not rock, but Quaternary loess, formed 14,000 years ago.

4.1.1.2 Sedimentation Rate in the Past 100 Years

The sedimentation rate in the past 100 years has been obtained using ^{210}Pb and ^{137}Cs dating methods at more than ten sampling locations (Table 4.1).

Sedimentation rates measured in 2002 were higher than in most previous years (Table 4.1) because sampling locations in 2002 were closer to the lakeshore than in previous years and were affected by human activities. For example, the sampling location in Wuli Bay was less than 200 m from the lakeshore. In contrast, the locations in northeastern Meiliang Bay, at Jiapu in southwestern Lake Taihu (with a large amount of sand input), and in Xukon Bay between Dongshan and Xishan Island, were 2.5 km, 3 km, and 4 km from the lakeshore, respectively. The data from the four locations in 2002 were all higher than those from the same locations in 1982, showing a recent trend towards increasing sedimentation in Lake Taihu. Moreover, it was estimated that the amount of sedimentation was about 2.35×10^6 t based on the sedimentation rates collected before 1990 (Sun & Huang, 1993). This value was much higher than that measured in 1954.

Table 4.1 Sedimentation rates of modern sediments in Lake (L.) Taihu

Locations	Year	Date range	Mean sedimentation rate (mm/yr)	Mean sedimentation fluxes (g/(cm^2 · yr))
Southeastern L. Taihu	1982	1927–1982	1.83	0.13
Northwestern L. Taihu	1982	1901–1982	1.11	0.08
Northeastern L. Taihu	1986	1920–1986	2.99	0.23
Dapu River mouth	1986	1885–1986	1.66	0.14
Meiliang Bay	1988	1931–1988	1.80	0.13
Gonghu Bay	1989	—	0.76	0.06
Northwestern L. Taihu	1992	—	0.60	0.05
Southern Meiliang Bay	2001	—	1.90	0.16
Wuli Bay	2002	—	3.1	/
Northeastern Meiliang Bay	2002	1922–2000	3.6	0.25
Jiapu, southwestern L. Taihu	2002	1911–2000	2.8	0.40
Xukou Bay	2002	1887–2000	2.5	0.34

Altogether, sedimentation rates ranged from 0.6 to 3.6 mm/yr, with a mean value of 2.1 mm/yr (Table 4.2), within the range for other large lakes in China and elsewhere (Table 4.2).

Previous dating data showed that silt deposition was extremely unstable. Sediments were disturbed violently, leading to coarsened sediments, decreased clay content, and decreased element content. For example, the quantity of sediments on the bottom of Gonghu Bay was recorded daily using sediment traps (Table 4.3). The maximum sedimentation during the survey was 9.23 kg/(m^2 · yr) on September 1, with another exceptionally high rate of 8.17 kg/(m^2·yr) recorded on September 29. The annual sedimentation rate at the same location was 0.76 mm/yr, with a mean value of 0.6 kg/yr or 0.0016 kg/(m^2·d). This value was ten times that obtained from sediment traps, which means that sedimentation rates can be extremely variable in shallow lakes, depending on weather conditions; bottom particles can be repeatedly disturbed, resuspended, transported, and redeposited in different places during strong wind-wave events.

There is a significant increasing trend of sedimentation in recent years in Eastern Taihu Bay. The sediment was from river and marsh deposits before 6,000 years

Table 4.2 Sedimentation rates in large lakes in the world (Sun & Huang, 1993)

Name of the lake	Depth (m)	Sedimentation rates (mm/yr)
Lake Taihu, China	2.6	0.6–3.6
Lake Poyang, China	16	2.0–2.5
Lake Dongting, China	5	25–35
Lake Chaohu, China	5	2.4
Lake Biwa, Japan	104	1.3–1.6
Lake Michigan, U.S.A.	400	1.8
Lake Constantine, Germany	300	1.2–1.4
Lake Tanganyika, Africa	1000	1.0–1.4
Lake Van, Turkey	450	0.4–0.9

Table 4.3 Daily sedimentation rates from traps in Gonghu Bay in September 1989

Date	Weight of sediments (g)	Mean deposition rate (kg/(m^2 · yr))	Date	Weight of sediments (g)	Mean sedimentation rate (kg/(m^2 · yr))
Sept. 1	9.34843	9.23	Sept. 16	0.29778	0.3
Sept. 2	5.09176	5.09	Sept. 17	0.35435	0.36
Sept. 3	0.65154	0.65	Sept. 18	0.65292	0.65
Sept. 4	0.14414	0.14	Sept. 19	0.30980	0.31
Sept. 5	0.08597	0.09	Sept. 20	0.17408	0.17
Sept. 6	0.13475	0.13	Sept. 21	0.09509	0.10
Sept. 7	0.09204	0.29	Sept. 22	0.11430	0.11
Sept. 8	0.08305	0.08	Sept. 23	0.02320	0.02
Sept. 9	0.08758	0.09	Sept. 24	0.13640	0.14
Sept. 10	0.09077	0.09	Sept. 25	0.03588	0.04
Sept. 11	0.09960	0.10	Sept. 26	0.01509	0.01
Sept. 12	0.06373	0.06	Sept. 27	0.00809	0.01
Sept. 13	0.06967	0.07	Sept. 28	0.02646	0.03
Sept. 14	0.09035	0.09	Sept. 29	8.16604	8.17
Sept. 15	0.09726	0.10	Sept. 30	0.86514	0.87

B.P. (Chang & Liu, 1996). Before the 1980s, the mean sedimentation rate in southeastern Taihu Bay was 0.11 g/(cm^2·yr), which is slightly lower than that in the northeastern bay, 0.299 g/(cm^2·a) (Sun & Huang, 1993) (Table 4.4).

In 1992, the mean sedimentation rate ranged from 0.17 to 0.45 cm/yr (Table 4.5). The annual mean quantity of sediment deposited on the bottom of Eastern Taihu Bay since 1920 showed that the sedimentation rate was increasing in Eastern Taihu Bay (Table 4.6).

Table 4.4 The deposition fluxes in Eastern Taihu Bay measured in 1982 and 1986 (Sun & Huang, 1993)

Location	Period	Sedimentation rates g/(cm^2 · a)	Mean sedimentation rates g/(cm^2 · a)
Southeastern Taihu Bay	1955–1982	0.145	0.11
	1945–1955	0.083	
	1977–1986	0.214	
Northeastern Taihu Bay	1970–1977	0.649	0.299
	1945–1970	0.140	
	1920–1945	0.192	

Table 4.5 Sedimentation rates measured at five sites in Eastern Taihu Bay in 1992

Locations	Entrance	Dongjiaozui	Mouth of Taipu River	Western shore	Eastern shore
Mean sedimentation rate (cm/yr)	0.17	0.23	0.10	0.19	0.16

Table 4.6 Mean annual sedimentation in Eastern Taihu Bay since 1920

Periods	1920–1940	1940–1970	1970–1991
Sedimentation rates ($t/(km^2 \cdot a)$)	1065.36	1357.62	7549.80
Annual sedimentation quality (10^4 t)	13.85	17.65	20.15

Moreover, according to data for 1999 obtained from the Taihu Basin Administration, Ministry of Water Resources, P.R. China, the average thickness of siltation in Eastern Taihu Bay was greater than 1 m, of which 98% was silt. From 1991 to 1998, the depth of new silt was about 0.2–0.3 m and the mean sedimentation rate was 0.25–0.38 cm/yr in the area from Dongjiaozui (the mouth of Eastern Taihu Bay) to the mouth of the Taihu River. The great amount of silt on the bottom of Eastern Taihu Bay showed that there was a trend towards bogginess, which requires attention for environmental protection and preservation of water resources.

4.1.2 Soft Sediment Distribution

With deterioration of the water environment and frequent algal blooms occurring, treatment of eutrophication has been undertaken gradually in recent years. Numerous investigations on sediment have been made because the distribution and physicochemical properties of sediments are closely related to the internal sources of nutrients. The fundamental purposes of such investigations are to understand the depth and properties of sediment and its nutrient content, to determine the potential release of nutrients from internal sources, and to discuss the environmental effects of ecological dredging and its practicability. The Nanjing Institute of Geography and Limnology, Chinese Academy of Sciences, has investigated the distribution of sediment at 108 sites in Lake Taihu.

Based on these data, the distribution of "soft" sediment, specifically, the sediment that was easy to penetrate by a hard bamboo pole, was mapped out for the whole lake (Fig. 4.2) (Zhu et al., 2006). A hard bottom was hypothesized as a sediment thicknesses of less than 0.1 m depth.

The investigations showed that the area where soft sediment thickness exceeded 0.1 m was 1,692 km^2, occupying 72.4% of the total area of the lake. The general distribution of sediment shows that there is more sediment in the western lake than in Eastern Taihu Bay, and more sediment is present in the littoral regions than in the centre (Fig. 4.2). The mean thickness of sediment in regions with more than 0.1 m sediment was 0.87 m in the area of 1,692 km^2; the mean depth of sediment was 0.65 m in the whole lake area (2,338.1 km^2). Moreover, the greater sediment thickness in the western and northern parts of the lake was significantly different than in the east, with a maximum sediment thickness of more than 5 m in the deepest region.

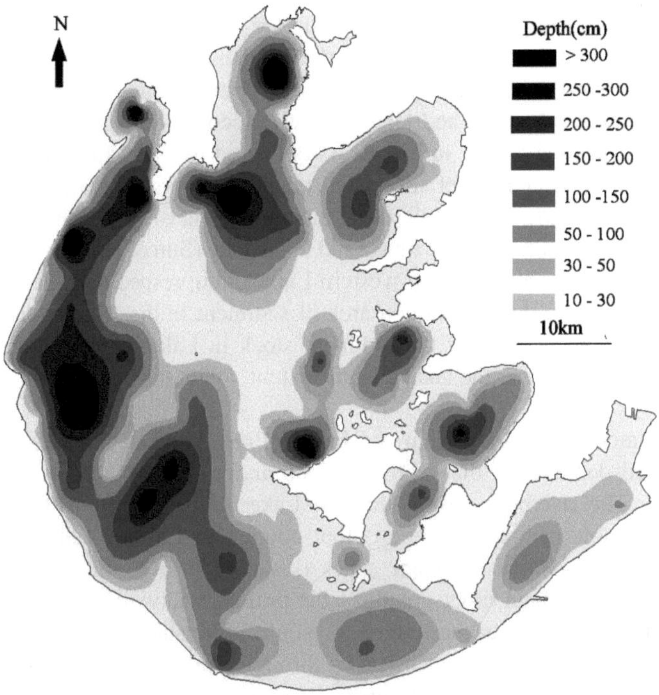

Fig. 4.2 Distribution of soft sediment in Lake Taihu (Zhu et al., 2006)

In the sediment investigations, sediment thickness was measured using a hard bamboo pole penetrated by hand into the sediment by a strong worker. Thus, sediment thickness in this investigation refers to the "soft" sediment layer, which includes the soft soil formed in the Quaternary period (Quaternary silt clay, under the modern sediments of Lake Taihu). Our results showed that the mean thickness of sediment was 0.87 m in the regions covered with sediment, and the sediments were formed 6,500 years ago, based on data obtained using ^{14}C dating.

Although Taihu is large and shallow and its sediment is frequently disturbed by wind waves, generally only the surface 30 cm of sediment has strong effects on the quality of the overlying water. Nevertheless, the thicker sediments must be considered if dredging is carried out, because such sediment would then be exposed to the sediment–water interface after the surface mud (generally 30–50 cm deep) was removed. Moreover, the resuspension potential, the nutrient content and chemical form, and the hardness of the deeper sediment must be considered if the surface sediment were removed, because these factors would have potential effects on water quality improvement and restoration of macrophytes. Therefore, understanding the distribution of sediment has important environmental significance in terms of treatment of lake environments.

The factors affecting sediment distribution are complex. The principal determinants are the activity of rivers entering the lake (most rivers entering Taihu are located in the west), the effects of the ancient river channel, and hydrodynamic factors, such as wind waves and lake currents. According to the investigation on changes in suspended matter in 139 rivers delivered to Lake Taihu during 1986–1990, the maximum content of suspended matter in rivers was 645 mg/L and the minimum was 13 mg/L. Of the 139 rivers, the drainage networks of the Tiaoxi River and the Yili River contributed the most sand to the lake (Sun & Huang, 1993). These rivers are located in southwest and western Lake Taihu, respectively, locations that roughly overlap with the distribution area of sediment in the western lake (this is the biggest region in terms of the sediment stock in Lake Taihu). Thus, particulates carried by rivers into the lake, to some extent, affect the distribution of surface sediments.

The sediment core study (Sun & Huang, 1993) showed a series of ancient riverbeds and depressions that were silted up and covered by the loess layer on the bottom. The riverbeds passed through Lake Taihu, extending from west to east, and are now merged with the output courses in Eastern Taihu. The distribution of the three main sediment zones is shown in Fig. 4.2. The first zone, from Dapukou, the main input mouth of the drainage networks of the Yili River, extends to the northeast, reaches the region south of Mashan Hill, and then divides into two branches; the first branch passes through western Meiliang Bay, extends to the north, and goes into the Liangxi River in the northern lake; the second branch extends to the east, goes into Gonghu Bay, and connects to the Wangyu River). The second zone extends from Zhushan Bay to the east, passes through Mashan reclaimed land, goes into Meiliang Bay, joins with the first zone, and then goes into the Liangxi River. The third zone is formed by extensions of the Yili and Tiaoxi Rivers (which join in the east of Daleishan Island), extends to the east along the north edge of Xishan Island, goes into Xukou Bay, and then joins with the old Xujiang River. The sediment region that joins Xiaomeikou north of Xishan, Ducun, and the paleochannel of the Wusongjiang River overlaps extensively with the sediment zone in the north of Xishan Island in Fig. 4.2. It also is possible that the merged sediment region was waterlogged lowland in ancient times.

The centre of Lake Taihu is almost all covered by hard earth, which may be the result of erosion by lake currents. Lake Taihu is significantly affected by the southeast monsoon in summer. Being a large shallow lake with a long wind fetch, it is characterized by strong wind waves and lake currents. According to long-term studies conducted by the Nanjing Institute of Geography and Limnology, Chinese Academy of Sciences, current velocity in the centre of Taihu increased with increased wind velocity (Ma & Cai, 2000). It is difficult for suspended matter to deposit and accumulate in the centre of the lake, and hard bottom areas are created there (Sun & Huang, 1993). Although many rivers of different sizes enter the lake, no deltas are formed at the mouths of these rivers. This observation implies that the hydrodynamic effects of lake currents are strong, because alluvial deposits were washed away and transported by the currents.

4.2 Physicochemical Properties of Sediments

Guangwei Zhu

4.2.1 Particle Size Distribution of Surface Sediments

The particle size distribution of lake sediment may reflect inputs into the basins, the lake ecosystems, and the sediment resuspension potential. Since the 1980s, studies have been undertaken on particle size distribution of the sediments, using techniques such as surface sampling, shallow boring, and deep boring.

The results showed that most surface sediment was silt clay, 60–80% of which was silt sand (0.01–0.1 mm diameter) and 20–40% of which was clay (<0.01 mm diameter). The size composition of sediment is shown in Table 4.7.

In June 2001, investigations on particle size distribution of sediments in some major regions were conducted by the Nanjing Institute of Geography and Limnology. Sediment cores were sampled at four sites: site 1, Dapukou (31°18′25″ E, 119°56′37″ N); site 2, Xiaomeikou (31°05′05″ E, 120°06′04″ N) in the southwestern lake; site 3, Eastern Taihu Bay (31°01′18″ E, 120°27′14″ N); and site 4, central Meiliang Bay

Table 4.7 Particle size composition of sediment in Lake Taihu measured in 1980 (Sun & Huang, 1993)

Regions	Properties	Particle size distribution (% per region)				
		Fine sand (> 1 mm)	Coarse silt (1–0.1 mm)	Fine silt (0.1–0.01 mm)	Clay (< 0.01 mm)	Median diameter (mm)
Northern Meiliang Bay	Silt clay	4.57	4.35	48.33	42.78	0.012
Southern Meiliang Bay	Silt clay	—	6.59	56.54	36.87	0.012
Northwestern L. Taihu	Silt clay	—	11.59	54.35	34.8	0.015
Centre of L. Taihu	Silt clay	6.44	4.93	46.19	32.06	0.016
Southern L. Taihu	Silt clay	8.04	8.31	41.59	32.0	0.165
Western L. Taihu	Silt sand	5.88	5.94	72.65	15.59	0.031
Southwestern L. Taihu	Silt sand	7.31	10.69	63.72	16.28	0.023
Eastern Taihu Bay	Silt sand	10.14	48.55	31.53	9.78	0.027

Zhu, G.W.
State Key Laboratory of Lake Science and Environment, Nanjing Institute of Geography and Limnology, Chinese Academy of Sciences, 73 East Beijing Road, Nanjing 210008, P. R. China
e-mail: gwzhu@niglas.ac.cn

Table 4.8 Particle size composition of sediments in Lake Taihu measured in 2001

Location	Properties	Particle size distribution (% per site)				Median diameter (mm)
		Fine sand (> 1 mm)	Coarse silt (1–0.1 mm)	Fine silt (0.1–0.01 mm)	Clay (< 0.01 mm)	
Dapukou	Silt clay	—	0.88	70.95	28.17	0.015
Xiaomeikou	Silt clay	—	0.45	62.21	37.35	0.013
Eastern Taihu Bay	Silt clay	—	0.85	64.36	34.79	0.015
Meiliang Bay	Silt clay	—	1.06	55.17	43.77	0.011

(31°29′00″ E, 120°10′03″ N). The sediment cores were stratified 2 cm by 2 cm, and then the size composition of sediments was measured using a particle size analyzer.

The surface sediments (10 cm thick), composed of silt clay, were mainly made up of silt sand and clay, which together constituted more than 98% of the total (Table 4.8).

There was no significant difference in the median size of sediment particles among the four sites in Lake Taihu in 2001, similar to a result obtained in the early 1980s (Sun & Huang, 1993). The surface sediment in Xiaomeikou was much sandier than that in Dapukou in the 1980s investigation; however, the 2001 investigation did not find a significant difference between the two sites, which may be because the sampling sites were nearer to the centre of the lake and were slightly affected by the Tiaoxi River in 2001. The median size of sediment particles was 0.015 mm in the upper 10 cm of sediments in Eastern Taihu Bay (site 3), mainly because rivers with high sand contents entered the eastern bay through the southern lakeshore, which shortened the water residence time significantly. The sand content of the sediments of Eastern Taihu Bay was significantly greater than in northern Lake Taihu. Thus, the minimum median particle size was found in Meiliang Bay, with the longest water residence time there.

The particle size distribution curves for sediments at the four sites were similar (Fig. 4.3). Sediments in Meiliang Bay were fine. Sediments in Dapukou and Eastern Taihu Bay were slightly smaller, whereas sediment in Xiaomeikou was intermediate size.

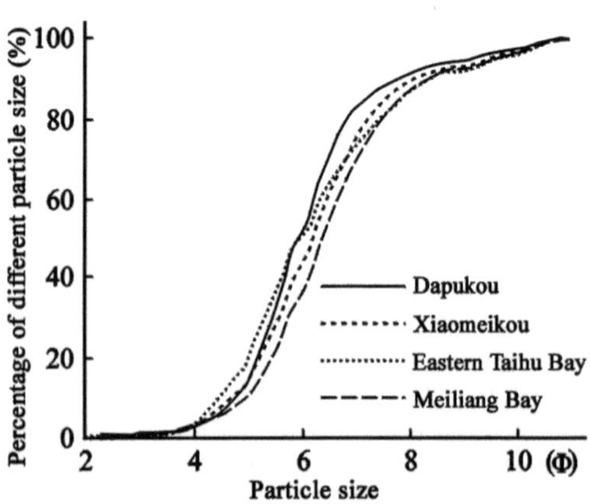

Fig. 4.3 Particle size distribution of surface 10 cm sediments in Lake Taihu. $\Phi = -\log_2 D$; D is grain size (mm). The larger the value of Φ, the smaller the particle size

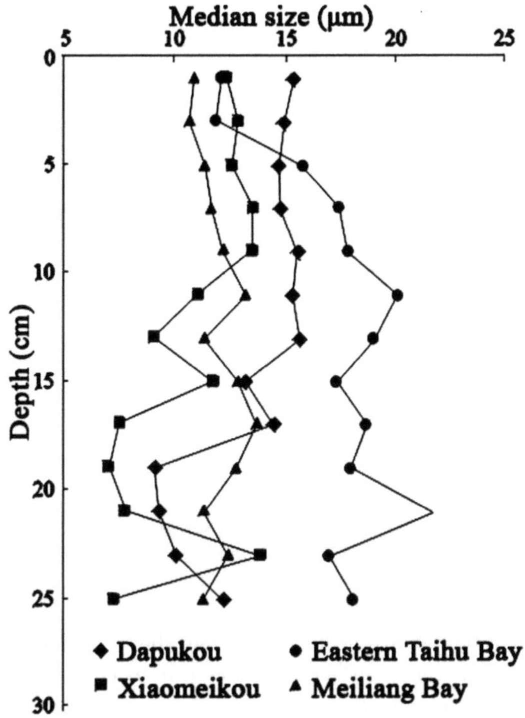

Fig. 4.4 Vertical profiles of median sediment size

The lake regions were significantly different in terms of the vertical distribution of sediment size (Fig. 4.4). In Eastern Taihu Bay, the median size of sediments below the surface 4 cm layer at all four sites reached the maxima at relatively shallow depths, which shows great changes have occurred in the lake sediments recently. The vertical change in median sediment size was slight in Meiliang Bay, because there has been no significant change in hydrological characteristics at this location in the past 50 years. The median size of the surface sediment was larger than that of the underlayer in Dapukou and Xiaomeikou, which might be related to increasing human activities and the increasing intensity of soil erosion in the catchments.

The size curve of sediments in the surface 2 cm layer was compared with that in the 18–20 cm layer (Fig. 4.5). In Dapukou and Xiaomeikou, sediments in the surface 2 cm were significantly larger than those in the 18–20 cm layer; in Eastern Taihu Bay, the reverse was true. In Meiliang Bay, the size curve of the surface 2 cm layer was not significantly different from that of the deeper layer.

4.2.2 Water Content and Porosity of Sediments

Water content and porosity of sediments are important parameters that reflect sediment resuspension potential. The higher water content, the smaller porosity, and the more easily sediment is resuspended by wind waves. During April and May 2002,

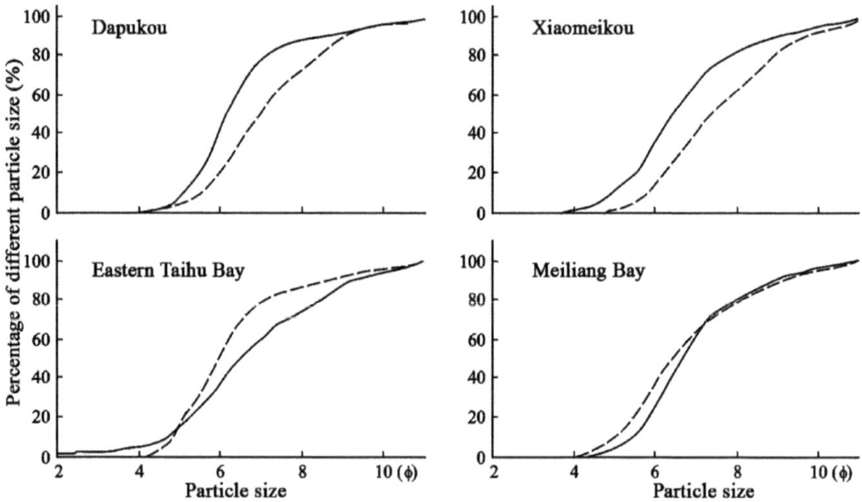

Fig. 4.5 Particle size distribution of surface 2 cm (*solid lines*) and deeper 18–20 cm (*dotted lines*) sediments. $\Phi = -\log 2D$, D is grain size (mm)

investigations were made at more than 100 sampling sites in the main regions of Lake Taihu to further understand the water content and porosity of the sediments. Soft sediment was collected from nearly 50 sites (Fig. 4.6). Hard bottom was found in central Lake Taihu, a section of the littoral belt, and southeastern Meiliang Bay.

Vertical profiles of water content and porosity of sediments in the seven lake regions are shown in Figs 4.7 and 4.8 (in the latter, porosity is the ratio of wet

Fig. 4.6 Sampling sites for water content of sediments in Lake Taihu. *1*, Northern Lake Taihu; *2*, Western Lake Taihu; *3*, Southwestern Lake Taihu; *4*, lake centre; *5*, Xukou Bay; *6*, Southern Lake Taihu; *7*, Eastern Taihu Bay

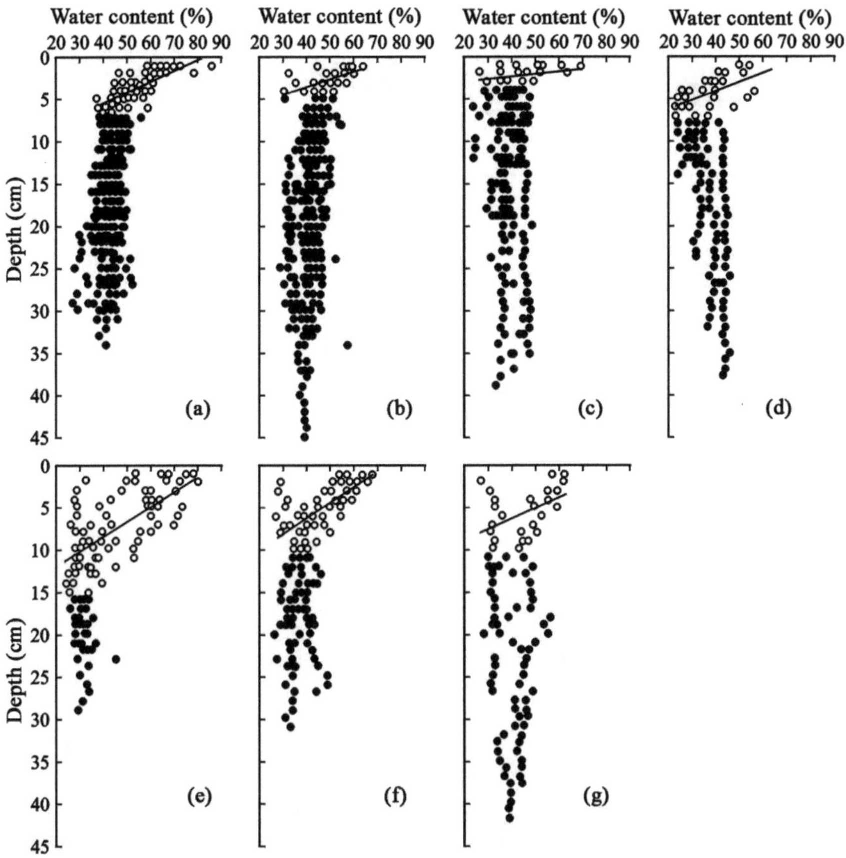

Fig. 4.7 Vertical profiles of water content in sediment of seven regions of Lake Taihu. (**a**) Northern Lake Taihu; (**b**) Western Lake Taihu; (**c**) Southwestern Lake Taihu; (**d**) Southern Lake Taihu; (**e**) Eastern Taihu Bay; (**f**) Xukou Bay; (**g**) Lake centre

sediment mass to its volume, in g/cm^3). Water content decreased significantly with increasing depth from sediment surface to bottom; water content decreased from more than 60% to less than 50%. Porosity increased from about 1.2 to about 1.6 g/cm^3. Although a transition in water content with depth was seen in all seven regions, the depth at which it occurred differed among sites, ranging from 4 cm to 15 cm (Fig. 4.7).

The greatest change in water content and porosity with sediment depth was in Eastern Taihu Bay, where the transition zone was the deepest (at 12 cm for water content, and 13 cm for porosity). This condition was related to both the abundance of organic matter and high water content in the sediments.

In general, the water content of the sediment determined its porosity, which is an important parameter reflecting compaction and resuspension potential of sediments. Not only water content, but also porosity, has decisive effects on the critical shearing stress for sediment resuspension.

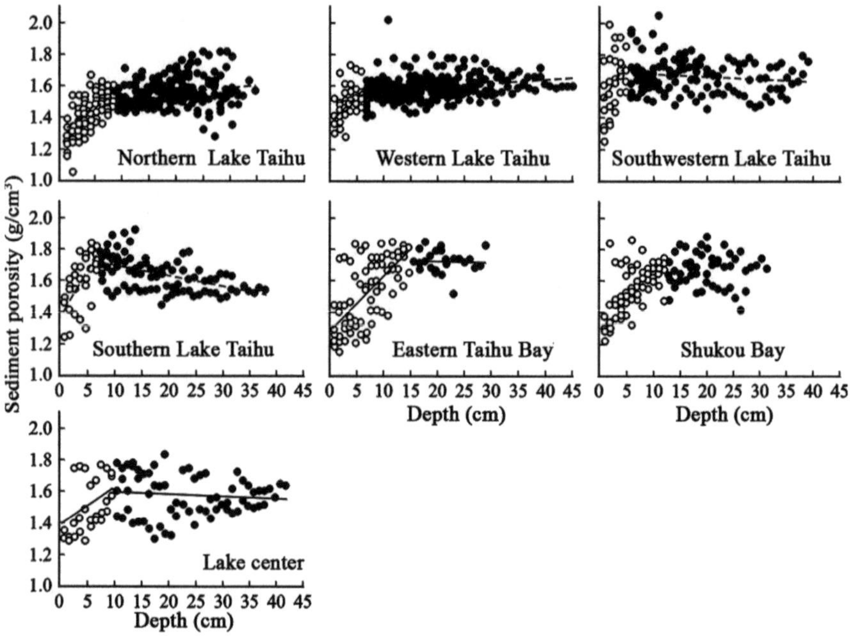

Fig. 4.8 Depth profiles of sediment porosity of seven regions of Lake Taihu

Therefore, identification of the transition depth of sediment water content is important in understanding the exchange between sediments and the overlying water and the nutrient cycling in shallow lakes. The transition depth of the water content of surface sediments (about 4–15 cm) should be given more attention in studies on the release of chemicals from the sediments, because the sediment above the transition depth may be the most active layer contributing to nutrient and pollutant exchange between sediments and the overlying water. Moreover, the different water content in sediments could affect microorganism communities in surface sediments and the conservation of living things such as plankton (for example, *Microcystis*), and then proceed to affect the biogeochemical cycles of nutrients.

4.2.3 The pH and Eh of Sediments

The pH and *Eh* (oxidation-reduction potential) of sediments have important effects on the formation and decomposition of minerals, and on the release, transportation, adsorption, and accumulation of elements. Moreover, the pH and *Eh* of sediment are also related to the growth and species distribution of aquatic vascular plants and benthic animals.

Analyses of sediment cores from eight sites with relatively deep mud showed that pH varied from 6.70 to 8.35 (Fan et al., 1998). In general, pH of the sediments decreased with increasing depth of sediment.

Pore water is defined as the water stored in the spaces between sediment particulates in contact with each other at the bottom of a lake. Pore water is affected by factors such as the ion exchange between solid- and liquid-phase sediments, the dissolution of solid-phase components in sediments, the dispersion of solid-phase components into liquid because of chemical decomposition, and the separation of some minerals or salts that do not readily dissolve in the pore water. Thus, the chemical properties of pore water are different from those of the overlying water and sediments.

The pH and *Eh* values of the surface overlying water and pore water sampled in Meiliang Bay and Eastern Taihu Bay using Peeper pore water samplers in July, September, and November 2002 are shown in Figs. 4.9 and 4.10.

The pH of the pore water varied with the seasons. The pH values of the lower water layer in pore water were less than or similar to those at the sediment–water interface and increased with increasing depth during July to November, with a trend of increasing rate. The acidity of sediments was reduced gradually, and the alkalinity increased with time; the transition from H_2CO_3 to CO_3^{2-} in the pore water carbonate system was controlled by the dissolution and removal of $CaCO_3$ and $MgCO_3$ from the sediments. The spatial and temporal trends of pore water pH were similar in Meiliang Bay, a typical algae-dominated lake, and in Eastern Taihu Bay, a typical submerged macrophyte-dominated lake. Comparison of the pH of the overlying water showed that the pH increased gradually within the 10 cm of overlying water in Meiliang Bay, yet there was almost no change in Eastern Taihu Bay. However, pH decreased when the season changed from summer to winter, in both Meiliang Bay and Eastern Taihu Bay, which was related to the abrupt decrease in production of

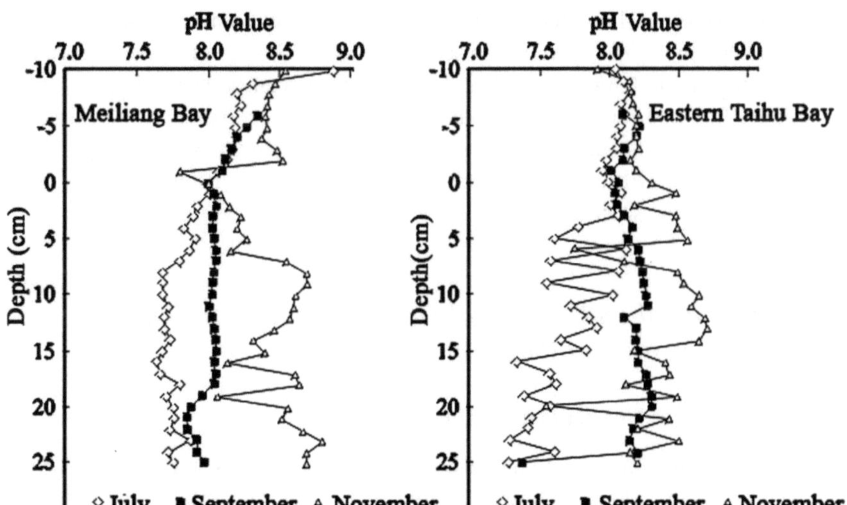

Fig. 4.9 pH of overlying water and pore water in Meiliang Bay and Eastern Taihu Bay in 2002. (Negative depth means overlying water; positive depth means pore water in sediment core) (Zhang Lu, unpublished data)

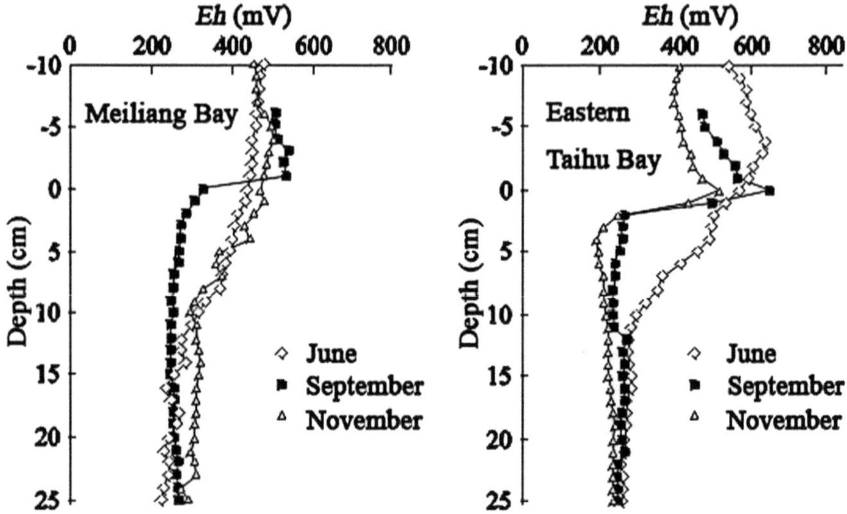

Fig. 4.10 *Eh* of overlying water and pore water in Meiliang Bay and Eastern Taihu Bay in 2002 (negative depth means overlying water, positive depth means pore water in sediment core) (Zhang Lu, unpublished data)

aquatic plants and algae in the overlying water and to the decrease in photosynthetic capacity.

The *Eh* values of overlying water and pore water decreased sharply in the interface between sediments and water, reaching a steady state (about 200 mV) at a depth of 3–10 cm (see Fig. 4.10). That result was significantly different from the change in the vertical distribution of *Eh* values in sediment, which showed that no lowest redox value occurred at the 10 cm depth. The differences between *Eh* values in pore water in Meiliang Bay and Eastern Taihu Bay showed that the redox states in sediments and pore water of Lake Taihu might be controlled by different systems (Fig. 4.10). The most important difference between the two bays is the difference of organic matter content in the sediment.

4.2.4 The Spatial and Temporal Distribution of Nutrients in Sediments and Pore Water

4.2.4.1 Contents of Nutrients in Sediments

Numerous investigations on the nutrient contents of sediments in Lake Taihu have been made since the 1960s, with several large-scale studies undertaken in the 1990s. An overview of the results is given in Table 4.9.

During the five decades since 1960, there have been sustained increases in the mean content of total nitrogen (TN) and total phosphorus (TP) in the sediment (Table 4.10). For example, TN in the surface sediment increased by 44.6% during

Table 4.9 The nutrient contents of sediment in Lake Taihu from 1960 to 1999 (%) (Fan et al., 2000a)

Year	Organic matter		Total nitrogen		Total phosphorus	
	Range	Mean value	Range	Mean value	Range	Mean value
1960	0.54–6.23	0.68	—	0.067	—	0.044
1980	0.24–2.78	1.04	0.022–0.147	0.065	0.037–0.067	0.052
1990–1991	0.57–15.10	1.90	0.049–0.558	0.080	0.040–0.107	0.056
1995–1996	0.31–9.04	1.70	0.022–0.45	0.094	0.039–0.237	0.058
1997–1999	0.31–15.73	1.83	0.022–0.618	0.092	0.028–0.280	0.060

Table 4.10 Phosphorus content of sediment samples from Meiliang Bay and Wuli Bay, showing difference in content of seven forms of phosphorus (μg/g) (Zhang et al., 2001)

Sample	Ex-P	Al-P	Fe-P	Ca-P	Oc-Fe-P	Oc-Al-P	Or-P
Meiliang Bay (I)	1.91	0.65	204.79	180.13	2,263.46	2.11	2,589.9
Wuli Bay (II)	2.63	0.64	300.74	181.58	2,042.44	5.98	1,652.1

Ex-P, exchangeable phosphorus; Al-P, aluminum-binding phosphorus; Fe-P, iron-binding phosphorus; Ca-P, calcium-binding phosphorus; Oc-Fe-P, occluded Fe-binding phosphorus; Oc-Al-P, occluded aluminum-binding phosphorus; Or-P, organic phosphorus.

the 15 years between 1980 and 1995–1996, and TP increased by 31.8% during the 35 years between 1960 and 1995. The results indicate that serious man-made pollution has affected the sediments in local areas of Lake Taihu.

In January 2003, a core of sediment was sampled from central Meiliang Bay, Lake Taihu. After stratifying the cores (each layer was 2 cm thick), TN and TP contents in each layer were analyzed. The results are shown in Fig. 4.11 (Zhu et al., 2006). The contents of nutrients in the top 5 cm layer was significantly higher than in the underlayers.

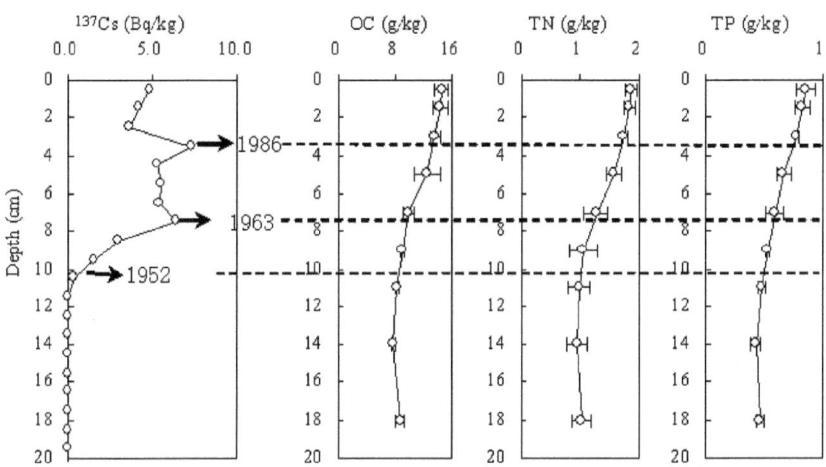

Fig. 4.11 Vertical profiles of organic carbon (OC), total nitrogen (TN), and total phosphorus (TP) in central Meiliang Bay, Lake Taihu

4.2.4.2 Vertical Distribution of Nutrient Content in Pore Water

In 2002, the P content in pore water in northern Meiliang Bay, Dapukou, southeast central lake, and Eastern Taihu Bay was investigated again (Fig. 4.12). In Meiling Bay, the P content in pore water in the surface 12 cm layer was significantly higher than that in the other three areas (Fig. 4.12). In Dapukou, the P content in pore water in the surface 10 cm also was rather high and then decreased between 10 and 18 cm. In the southeast central lake, the P content in pore water in the surface 12 cm layer was less than those in deeper layers. In Eastern Taihu Bay, the P content in sediment pore water was low.

These results for 2002 wcrc similar to those obtained from the analyses of the distribution of N and P in pore water in some lake areas in 1998 (Fan et al., 2000b). This phenomenon might be related to the depth of P exchange between pore water and overlying water. In large shallow lakes, resuspension often happens because of frequent disturbance by wind waves, and the depth of pore water exchange with the overlying water (or infiltration exchanging) is greater than that in deep lakes or in small lakes disturbed slightly by wind waves. However, for Lake Taihu, the specific depth of frequent exchange between overlying water and pore water is unknown. In the southwest central area, with the most violent disturbance driven by wind waves, there was no significant difference in the P content in pore water in the surface 12 cm layer (see Fig. 4.12), indicating that there might be active exchange between the surface 5–15 cm sediment layer and the overlying water.

4.2.5 Forms of Phosphorus in Sediments

In June 2001, 30-cm-deep sediment cores were taken in northern Meiliang Bay (A), Dapukou (B), southwest of central Lake Taihu (C), and Eastern Taihu Bay (D), and then air dried, ground, and passed through a 200-mesh sieve for analyses. The P forms were analyzed using the sequential extraction method to fractionate P in sediments (Ruttenberg, 1992), which was improved by Li et al. (1998a).

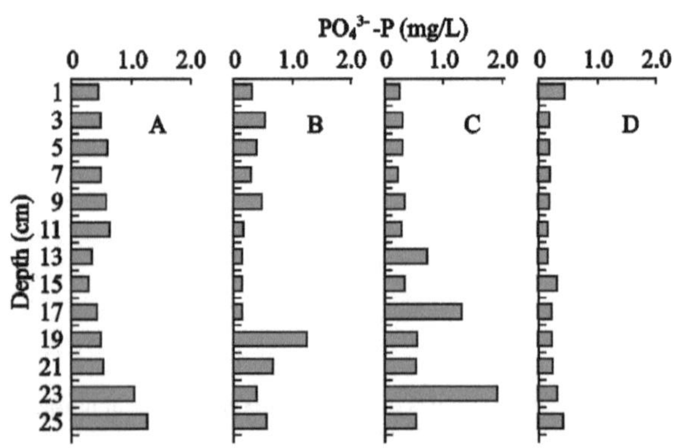

Fig. 4.12 Vertical profiles of PO_4^{3-}-P in sediment pore water of different lake regions (2002). (**A**) Northern Meiliang Bay; (**B**) Dapukou; (**C**) southeast of the central lake; (**D**) Eastern Taihu Bay

4.2.5.1 The Phosphorus Forms in Sediments

There were significant differences in the distribution of the seven P forms in the sediment cores from the four areas of Lake Taihu (Fig. 4.13). Concentrations of exchangeable P were commonly ranked as Eastern Taihu Bay <Meiliang Bay < Dapukou ≈ the southwestern central lake. The P content in sediments in Eastern Taihu Bay was significantly lower than those in the other three lake areas. At different sediment depths in Eastern Taihu Bay, there was little difference in the exchangeable P content, whereas in Meiliang Bay, the exchangeable P content in the upper sediment layers exceeded that in lower layers. These results contrasted with those in Dapukou and the southwest central lake, where the exchangeable P content decreased with depth within the surface 10 cm and then increased gradually at greater depths. The significant differences in P distribution characteristics in different lake areas were related to differences in biological conditions and hydrodynamics. In Eastern Taihu Bay, a grassy region of the lake with abundant plant roots, the macrophytes have a high demand for P in the sediments and water in the fast-growing season. Thus, much of the active P is adsorbed by the plants.

Previous studies in soil science, plant nutrition, and sediment geochemistry showed that both exchangeable P, and P fixed with Al and Fe in the sequential extraction method, were phosphorus with releasing potential, whereas other forms of P do not have such releasing potential. Based on the extraction (see Fig. 4.13), the content of P fixed with Al in sediments was low in different areas (all <1 mg/kg, and less than the content of exchangeable P). Therefore, the content of P fixed with Al in sediments could not contribute much to eutrophication. The distribution of P fixed with Fe was similar to that of exchangeable P; specifically, in Eastern Taihu Bay, the P content was significantly lower than that in the other three lake areas, and its vertical distribution was similar to that of exchangeable P. This result showed that aquatic plants could effectively inhibit the activity of P, and that it is unlikely that there would be a surge of P release from sediments in the grassy lake areas, even if there were large wind waves. The fact that P content occluded in sediments had irregular vertical distribution in the four sampling cores, combined with the fact that P content in the upper 10 cm layer exceeded that in deeper layers (Fig. 4.13), showed that occluded P was transformed into relatively stable calcium-P and organic-P during the early diagenesis of sediments.

The P in biogenic debris derives mainly from organism residues in the sediments. The P content in biogenic debris was similar to that of calcium-P, and the content in Eastern Taihu Bay was significantly higher than those in any other lake areas (see Fig. 4.13), which might be related to blooming macrophytes and the long-established pen culture of fish. In the other lake areas, the P contents in biogenic debris were not significantly different from each other and changed little vertically. Most calcium-P comes from calcium phosphate minerals such as hydroxyapatite and superphosphate, which are difficult to dissolve; these minerals, as the end products formed during the early diagenesis of sediments, are relatively stable. The extraction results showed that calcium-P content increased with increasing sediment depth in all lake areas except Meiliang Bay. In B, C, and D sampling cores (Fig. 4.13), the

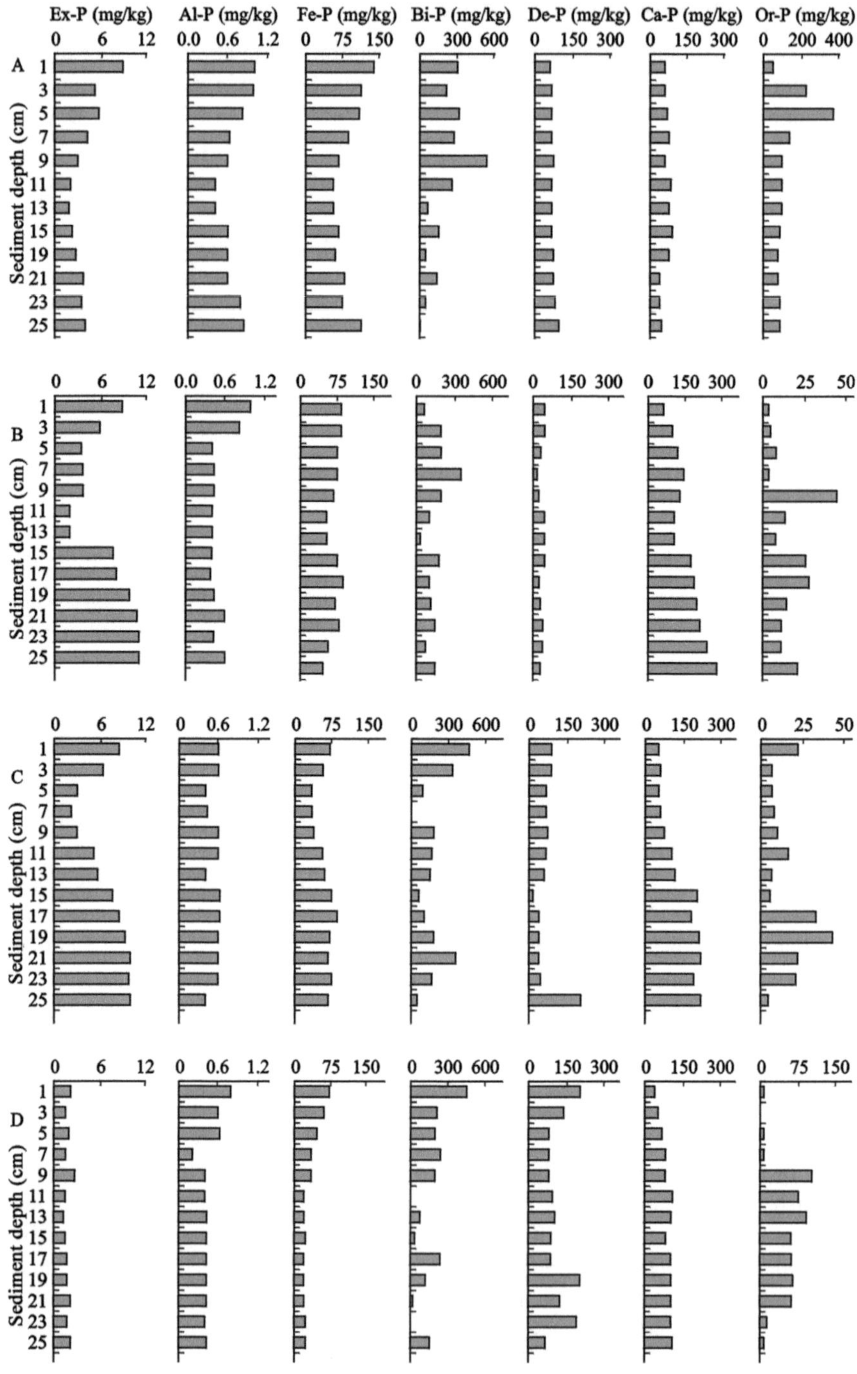

Fig. 4.13 Vertical profiles of seven P forms in the sediment of Lake Taihu. (**A**) Northern Meiliang Bay; (**B**) Dapukou; (**C**) southwest of central Lake Taihu; (**D**) Eastern Taihu Bay

correlation coefficients between the contents of calcium-P and the depth of sediments were 0.927, 0.929, and 0.865, respectively ($P < 0.001$). In the core of north Meiliang Bay, there was no correlation between calcium-P content and sediment depth, but these values were significantly positively correlated within the upper 16 cm layer ($R = 0.800$, $P = 0.017$, $n = 8$). The lack of correlation below 16 cm might be attributed to the significant differences in characteristics of the water environment in Meiliang Bay recently compared to the past. The results showed that calcium-P is the main P form in Lake Taihu.

The vertical distribution of organic P was irregular in three of the four lake areas (Fig. 4.13). The exception was in northern Meiliang Bay (core A in Fig. 4.13), where organic P content initially increased with depth (within the 5–6 cm layer, it reached 371.7 mg/kg, about 40% of the total), and then decreased at 8 cm, and remained rather stable below this depth.

In general the content of organic P in sediments was low in Dapukou, the southwest central lake, and in Eastern Taihu Bay. Especially, the organic-P content in the surface 8 cm was low, and peaked below 8 cm depth. In Eastern Taihu Bay, the low organic-P content in the surface layer contrasted markedly with the high organic matter content in the surface layer.

4.2.5.2 The Factors Controlling P Activity in Sediments

The main sedimentation mechanisms of phosphorus are adsorption with $CaCO_3$ or adsorption on ferric oxide colloids (Hartley et al., 1997; Hongve, 1997). The P generally passes though the process of release-deposition many times before being permanently deposited on the lake bottom (Hupfer et al., 1995).

Numerous factors affect P release in sediments, including biological (such as bacterial activity, the mineralization of organic matter, biological disturbances), chemical (such as *Eh*, pH, the percent of P fixed with Fe, nitrate content), and physical factors (such as resuspension of sediments) (Boers et al., 1998). The factors controlling P activity in sediments become more complex in shallow lakes than in deep lakes because of frequent wind-wave disturbance, numerous human activities, and complicated aquatic ecological factors.

Typically, P in sediment pore water passes into the overlying water through molecular diffusion and resuspension of surface sediments and is used by living organisms. Of the different P forms, P as PO_4^{3-}-P was the most active form in sediments. Exchangeable P primarily is deposited in the surface layer through physical adsorption, and is easily dissolved into water when P concentrations in the water change, maintaining a concentration balance between adsorption and dissolution with a relatively high activity. The P that is fixed with Al and Fe primarily is physically and chemically adsorbed on the surface of ferric oxide (or aluminum oxide) colloids; its amount can vary greatly because ferric oxide is easily affected by the redox potential, which is easily disturbed by the environment, and is the major source of potentially active phosphorus.

The foregoing analyses of the four sample cores (see Fig. 4.13) showed that the relationships between P concentrations in pore water, exchangeable P, P fixed with

Fe, other forms of P, and physical and chemical indexes were complex. In Meiliang Bay, P in sediment pore water only was positively correlated with P in biogenic debris and was negatively correlated with calcium-P. In Dapukou, P in sediment pore water was only positively correlated with loss on ignition (LOI) of the sediments. In the southwest of the central lake, P in sediment pore water was positively correlated with P fixed with Fe. In Eastern Taihu Bay, there were no significant correlations between P in sediment pore water and any of the other parameters investigated. These results agree with those obtained in 1998 (Fan et al., 2000b). In particular, redox potential, which can significantly affect P release in deep lakes, was not related to P in sediment pore water in the four sample cores in shallow Lake Taihu. Thus, redox potential had no decisive effects on the P distribution in sediment pore water in Lake Taihu.

In Meiliang Bay, the concentrations of exchangeable P in sediment cores were all significantly correlated with P fixed with Fe, P fixed with Al, LOI, and water content, showing that P fixed with Fe, P fixed with Al, and P adsorbed by organic colloids were the major active forms of P. However, in Dapukou, the concentrations of exchangeable P in sediment cores were not significantly correlated with P fixed with Fe or Al, which might be attributed to the polluted water from the Yili River. In the southwest of the central lake, exchangeable P in sediment cores was significantly correlated with P fixed with Fe and LOI. In Eastern Taihu Bay, concentrations of exchangeable P were not significantly correlated with any indexes listed, which also showed the individuality of this grassy bay relative to other kinds of bays.

The correlations showed that both biological and hydrodynamic factors play key roles in P geochemistry in Lake Taihu, which is a typical large shallow lake.

4.3 Experimental Studies on Nutrient Exchange at the Sediment–Water Interface

Lu Zhang and Wenchao Li

4.3.1 Phosphorus Release and Sorption Under Dynamic Disturbance Conditions

In shallow lakes, sediment–water interfaces are not stable. The shear forces caused by wind and wave disturbance directly influence the sediment–water interface, resulting in resuspension of the sediment. Disturbance of the water also significantly influences nutrient release from the sediment. Static release simulations (as described above in Part 1 of this section) can only reflect the biological and chemical changes in the release, as shown by the phosphorus release simulation experiments of Wu et al. (1998) and Yin et al. (1994) on the sediments of West Lake and Wuli Bay, Lake Taihu. To develop more realistic simulations, Zhang et al. (2001) have conducted preliminary studies on nutrient release in dynamic disturbance simulations.

4.3.1.1 Phosphorus Release

Zhang et al. (2001) selected two sampling sites: (I), at Meiliang Bay (31°28′31″ N; 120°11′39″ E), and (II), at eastern Wuli Bay (31°30′55″ N; 120°15′04″ E).

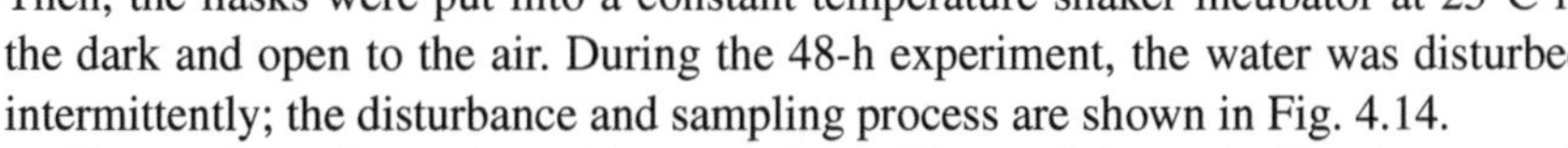

About 1 kg sediment was taken from the surface, and 2 L overlying water was collected and filtered through a Whatman GF/C membrane; 5 g fresh sediment was put into a 250-mL conical flask, to which 100 mL filtered overlying water was added. Then, the flasks were put into a constant temperature shaker incubator at 25°C in the dark and open to the air. During the 48-h experiment, the water was disturbed intermittently; the disturbance and sampling process are shown in Fig. 4.14.

The results indicate that under natural conditions of dynamic disturbance, sediments have both sorption and desorption effects on phosphate, and these two actions happen simultaneously (Fig. 4.14). Whenever the release of phosphate under dynamic disturbance conditions exceeds sorption by the sediments, the apparent release of the phosphate by the sediments is actually the difference between release and sorption.

Zhang, L.
State Key Laboratory of Lake Science and Environment, Nanjing Institute of Geography and Limnology, Chinese Academy of Sciences, 73 East Beijing Road, Nanjing 210008, P. R. China
e-mail: luzhang@niglas.ac.cn

For the surface sediments in central Meiliang Bay and Wuli Bay with the 60 rpm disturbance, the nutrient release was not obvious; instead, some sorption was observed. During the static period, between 12 h and 24 h, the release curve was almost level. When the disturbance strength was increased to 140 rpm after 24 h, there was marked release from sediment of both sites, with maximum release concentration reaching three times that seen at 60 rpm.

Although we cannot define the exact relationship between the simulated disturbance under experimental conditions and the actual disturbance caused by wind, waves, and biological events under natural conditions, this does not hamper our research on the behavior of phosphorus in the sediment under disturbance conditions. The resuspension release curves of the surface sediments in the two samples begin to drop after reaching peak values, which is similar to the situation in Lake Ge (Fan, 1995). The reason for this result may be that phosphorus release from resuspended sediments enters into an "exhausted" state after a certain degree of release to the water under dynamic disturbance conditions, when a dynamic balance of phosphorus release and sorption is reached.

The time taken to reach maximum release (T_{max}) was different for sample I (Meilieng Bay) and sample II (east Wuli Bay). Thus, T_{max} (I) > T_{max} (II), which may relate to the forms of phosphorus in the sediment (see Table 4.10). Exchangeable phosphorus can enter the overlying water more easily than organic phosphorus, so it has less influence on the time for release balance than organic phosphorus. Aluminum-binding phosphorus and iron-binding phosphorus can transform into dissolved phosphorus through chemical hydrolysis, with high reaction speeds and low content in sediment samples, so these are not the main factors affecting the phosphate release balance. The organic form of phosphorus can be transformed biologically, but this is a slow process. However, organic phosphorus is significant in this experiment and may be one of the main factors affecting the phosphate balance.

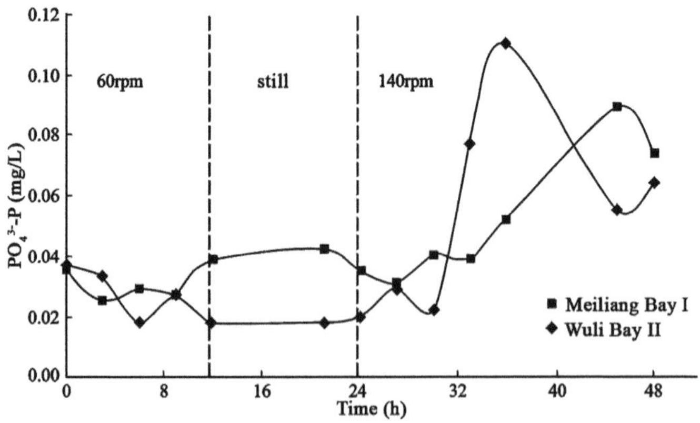

Fig. 4.14 PO_4^{3-}-P release from Meiliang Bay (■) and Wuli Bay (♦) sediments in simulated disturbance conditions (Zhang et al., 2001)

Phosphorus in other forms, such as occluded phosphate, is unlikely to enter the overlying water through hydrolysis under disturbance conditions. Because organic phosphorus content in Meiliang Bay sediment was higher than in Wuli Bay sediment, the time taken to reach release balance in the Meiliang Bay sample (I) was longer than that in the Wuli Bay sample (II), which was reflected in the lag in reaching maximum release.

4.3.1.2 Phosphorus Sorption Experiment

Zhang et al. (2001) conducted phosphorus sorption experiments on sediments under dynamic disturbance conditions, using the above-mentioned samples. Because the monthly mean phosphate concentration of Lake Taihu is 0.001–0.146 mg/L over a year [1], and the maximum was 0.29 mg/L (in Meiliang Bay), a value of 0.2 mg/L was adopted as the mean peak value for phosphate concentration. To simulate this in the experiments, lake water was mixed with a solution of 50 mg/L sodium dihydrogen phosphate ($NaH_2PO_42 \cdot H_2O$) to get a solution of 0.2 mg P/L as the overlying water in the incubation system. First, 5 g wet sediment was put into 250-mL conical flasks, then 100 mL filtered overlying water was added, and the preparation was incubated at 25°C, open, in the dark, at 100 rpm. The concentration of PO_4^{3-}-P was measured ten times during the 24-h incubation.

The sorption experiment under dynamic disturbance conditions indicated that the phosphate concentration drops quickly within the first 30 min of the experiment (Fig. 4.15). The concentration in the Meiliang Bay sample decreased from 0.215 to 0.013 mg/L (94%), and the concentration in the Wuli Bay sample decreased from 0.229 to 0.050 mg/L (78%). Thereafter, the decrease in phosphate concentration slowed down, and approached a stable concentration after 4 h, which indicates the sampling sites at Meiliang Bay and Wuli Bay have strong sorption activities under simulated disturbance conditions. The balanced concentrations during the 24-h incubation in systems with sediment from both the sampling sites approached 0.020 mg/L under the same disturbance conditions.

The sediments demonstrate marked adsorption under simulated disturbance conditions. The main reason for this may be the destruction and recreation of biological and chemical balances at the sediment–water surface, indicating significant changes to the adsorption behavior of the phosphorus in the sediment when disturbed. In large shallow lakes such as Lake Taihu, disturbance to the sediment–water interface is fairly frequent. In such circumstances, the release, sorption, and fate of sediment transformation is much more complex than in deep lakes that are less vulnerable to disturbance.

---

[1] *Data source: Taihu Laboratory for Lake Ecosystem Research, Chinese Academy of Sciences, 1998, Annual Report.*

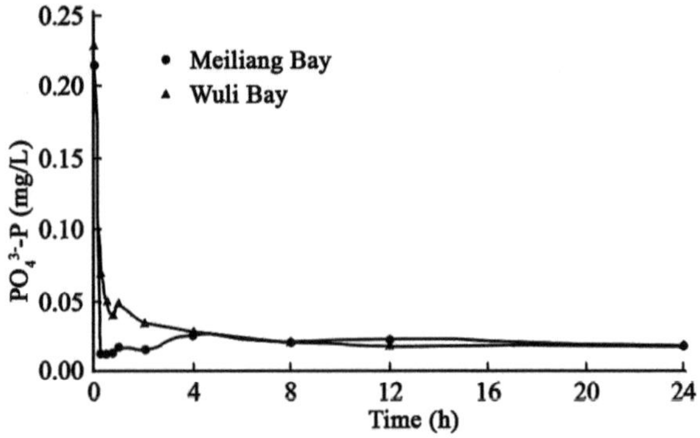

Fig. 4.15 Adsorption of P by sediments from Meiliang Bay and Wuli Bay under simulated conditions of dynamic disturbance (Zhang et al., 2001)

4.3.1.3 Phosphorus Adsorption Capacity of Sediment

The discharge of sewage may cause sudden increases in phosphate concentration in some areas. Capacity measurement experiments enable us to better understand how the sediments can reduce the peak phosphate concentrations. Experiments on the phosphorus adsorption capacity (PAC) of sediment were conducted on the aforementioned sediment samples (Zhang et al., 2001). Four phosphorus concentration gradients in the overlying water were chosen: 0.1, 0.2, 0.5, and 1.0 mg/L. The samples were placed in a constant temperature (25°C) shaker incubator at 100 rpm, to shake the sediment–water mixture to obtain the adsorption balance. The samples then were incubated for 24 h. The phosphorus adsorption capacity of resuspended sediments was calculated through comparison of the concentrations of phosphate before and after the incubation.

During the PAC experiments, there was no significant difference of phosphate concentration in the overlying water with sediment from Meiliang Bay and from Wuli Bay, which proves the sorption capability of the two samples were similar (Fig. 4.16). The phosphorus adsorption capacity (PAC, mg/g dry sediment) of surface sediment under simulated disturbance conditions can be obtained by the following formula:

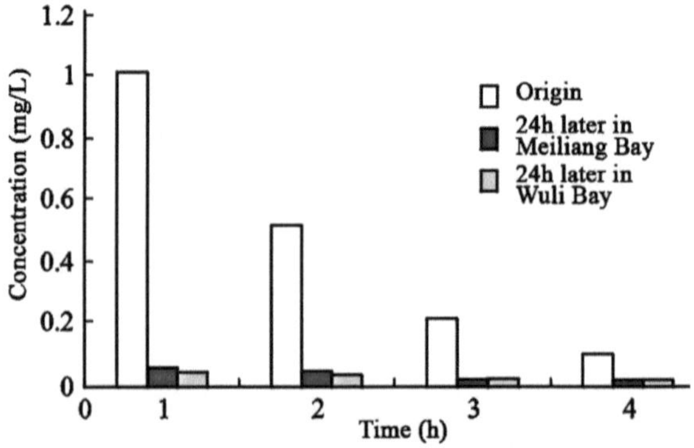

Fig. 4.16 Phosphate concentration in phosphorus adsorption capacity (PAC) experiments with sediments from Meiliang Bay and Wuli Bay (Zhang et al., 2001)

$$PAC = \frac{(C_0 - C_t) \times V}{\mathrm{W} \times (1 - \mu)}$$

In the formula, C_0 refers to the initial concentration of phosphate of the overlying water sample (mg/L), C_t refers to the concentration of phosphate after 24 h disturbance (mg/L), V refers to volume of water samples (L), W refers to wet weight of the sediments (g), and μ refers to the water content of the sediments (%).

The results of the PAC experiment are shown in Table 4.11.

Research on nutrient release and adsorption capacity of sediments in Lake Taihu indicates that the effects on phosphate release from surface sediments are significant. Under conditions of minor disturbance, release of nutrients from sediment is difficult to observe. However, when the disturbance strength is increased, marked release of nutrients from sediment is evident. The maximum release concentrations at 140 rpm are 0.110 and 0.089 mg/L for sediments from Wuli Bay and Meiliang Bay, respectively. These values are almost three times higher than those at 60 rpm conditions. The results indicate disturbances in natural conditions, such as winds, waves, and biological processes, have a significant effect on release of phosphate.

The effect of disturbance on phosphate adsorption by sediment also is marked. Based on these results, release and adsorption of phosphorus by sediments under dynamic disturbance conditions exist simultaneously. Adsorption of phosphorus by sediment (or suspended sediment) may be higher than release of phosphorus from sediment when the disturbance of the water–sediments system is strong enough. Which process dominates depends on the joint effects of factors such as phosphate concentration and disturbance intensity. Release will be the main activity if the disturbance is strong and the concentration of phosphate in the overlying water is low enough. However, sorption will be the main activity if the disturbance is strong and external phosphate concentration is higher than the balance concentration. The surface sediment acts as a natural buffer system for phosphate; whenever the phosphate concentration in the water is low, the sediment will release phosphorus and the release strength will increase with disturbance strength; in contrast, if the phosphate concentration in the water is high, the sediment will act as an adsorbent. The release process of phosphorus is comparatively slow, so the peak release value in summer usually lags behind wind activity, but the sorption process is more rapid.

Table 4.11 Initial and final phosphorus concentration in overlying water and the calculated phosphorus adsorption capacity (PAC) of sediments (Zhang et al., 2001)

	Meiliang Bay sediment				Wuli Bay sediment			
C_0 (mg/L)	1.01	0.514	0.213	0.104	1.01	0.514	0.213	0.104
C_t (mg/L)	0.048	0.046	0.021	0.017	0.041	0.032	0.019	0.016
PAC (10^{-3} mg/g)	42.2 ±0.4	20.9 ±0.3	8.4 ±0.2	3.8±0.1	50.0±0.5	25.3 ±0.3	10.3 ±0.2	4.7 ±0.1

4.3.2 Phosphorus Adsorption Saturation of Sediment from Eastern Taihu Bay

In the lake ecosystem, the sediment has a great capacity for phosphorus storage, yet this capacity is finite. With eutrophication of lakes, the amounts of phosphorus in the sediment approach saturation. The buffering capacity for phosphorus of the ecosystem decreases with increasing phosphorus pollution, as evidenced by a steep rise in phosphorus content and algae blooms. The relationship between phosphorus adsorption saturation and eutrophication, as well as the buffering capacity of sediment to external phosphorus pollution, can provide important data for controlling eutrophication.

Li et al. (1998b) have conducted research on the phosphorus adsorption saturation (PAS) of the sediments of Eastern Taihu Bay. They collected sediment samples at ten sites in Eastern Taihu Bay (Fig. 4.17). Sites 9 and 10 were located in the Eastern Taihu Bay section of the Taihu Laboratory for Lake Ecosystem Research, Chinese Ecosystem Research Network. Site 9 was located in an enclosure for the pen-culture fishery, and site 10 was located in a separate fishery pond near the lake.

Surface sediment was collected with a core sampler, with an inner diameter of 90 mm and a length of 1,000 mm. The surface 100 mm layer of sediment was separated for the experiment. Subsequently, 60 L clean lake water was filtered with a plankton net (no. 25) into a 90-L white plastic barrel, placed in a dark room (a total of ten barrels were used for the experiment). Approximately 2 kg sediment was collected from the surface layers (100 mm) from the above-mentioned 10 sites, then washed and filtered with a 100 − μm nylon sieve, separated into the ten barrels, and then more filtered water was added to bring each barrel to 65 L. The barrels were kept in the dark to minimize algal growth and their use of phosphate. Aeration was held at levels that maintained oxidation of the system, promoted aerobic degradation of organic matter, and facilitated phosphorus release and the phosphorus balance in the system. The sediments were disturbed each day by strong aeration to thoroughly suspend them. Water was regularly sampled and centrifuged, and then the phosphate content was

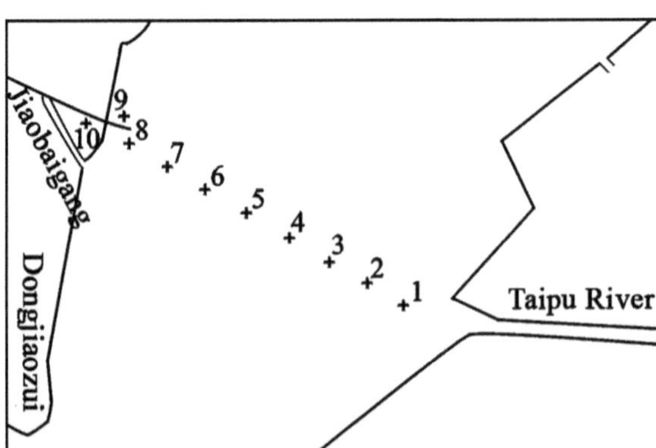

Fig. 4.17 Locations of sampling sites for phosphorus adsorption saturation experiment

determined. The ten experimental systems of water and sediments were incubated at 25°–30°C in darkness, with continuous aeration, and intermittently oxygenated to create disturbance.

The experiment ran from June 20 to September 3, 1997. After 20 days, the systems were stable (Fig. 4.18). On July 13, 6.5 mL phosphorus standard solution (buffer made with KH_2PO_4 and Na_2HPO_4, pH = 7.0; P content, 1 mg/mL) was added to each of the barrels. Additional phosphorus standard solution was added to each barrel on July 13 (20 mL), July 25 (6.5 mL), and July 30 (20 mL). After adding phosphorus, the balance of the system was determined. At the end of the experiment, the sediment was air dried and weighed, and the total phosphorus content and Olsen-P content were determined.

In the first 5 days, the phosphate concentration rose rapidly from the initial 7 μg/L to 18–43 μg/L, and then continually decreased to 1–23 μg/L at day 20, a typical sediment P release curve. Phosphorus release from the sediment occurs mainly because the original biological and chemical balances are disturbed, particularly the microbial system at the sediment–water interface, which breaks the phosphorus cycle. Thereafter, a new balance is created because oxidation of the sediment, increase in adsorption of phosphorus, and phosphorus release into water and readsorption by sediment result in decreased phosphorus concentrations in the water. On day 20, the addition of 6.5 mg phosphorus caused slight fluctuations of the phosphate concentrations. On day 23, the addition of 20 mg phosphorus caused

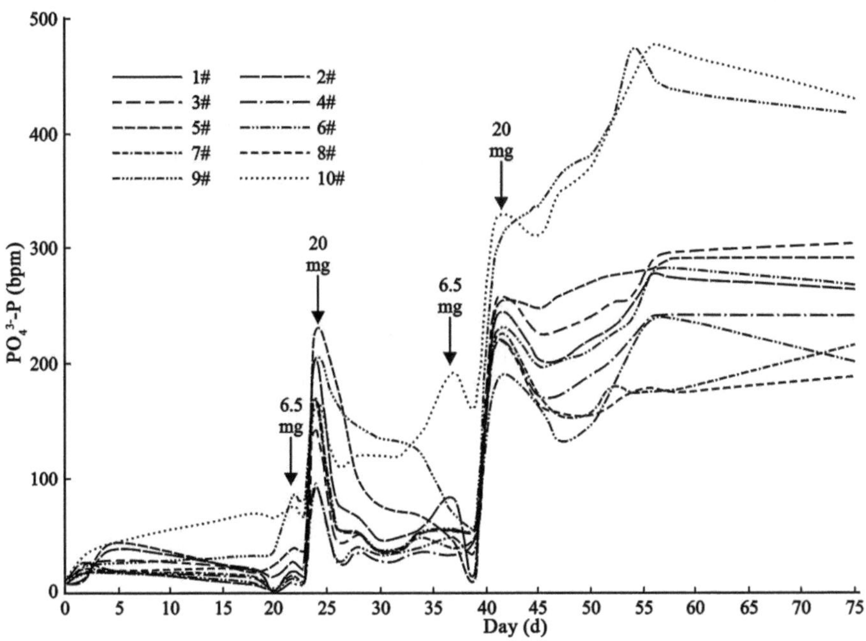

Fig. 4.18 Temporal changes in water phosphate concentration in ten systems in the phosphorus addition experiment (Li et al., 1998b)

rapid increases in phosphate concentrations to 92–225 μg/L. Thereafter, the concentrations dropped quickly to 30–60 μg/L. On day 40, the addition of 20 mg phosphorus again resulted in a rapid rise in phosphate concentration, this time reaching a new maximum of 184–253 μg/L. The concentrations did not return to their previous levels, but reached new levels of 200–300 μg/L 35 days after the second addition of 20 mg phosphorus.

Although the reactions of sediment samples 9 (from the fish pen enclosure) and 10 (from the fishery pond) were somewhat similar to those of other samples, they differed in their reaction to additions of external phosphorus. At day 20, the PO_4^{3-}-P concentrations in the water had reached 35 and 60 μg/L for samples 9 and 10, respectively (see Fig. 4.18). The stable concentrations after phosphorus additions were markedly higher than those of the other systems, with final PO_4^{3-}-P concentrations of 417 and 431 μg/L for samples 9 and 10, respectively. This result indicates that the phosphorus saturation is greatly decreased by the addition of fish foods where these sediment samples were taken. The initial and final amounts of phosphorus in the sediment and phosphate adsorption saturation (PAS) for each of the ten experimental systems are given in Table 4.12.

In general, the content of exchangeable phosphorus in the sediment of Lake Taihu is lower than 10 mg/kg (Zhang et al., 2001; Zhu et al., 2006). Table 4.12 shows that the addition of phosphorus was 53 mg. The dose was equal to 52–172 mg P per kilogram sediment, which is several times higher than the exchangeable phosphorus content of Lake Taihu sediment. With the large load of active phosphorus, the sediment still adsorbs more than half of the additional phosphorus, which indicates that its adsorption potential is fairly high. The main adsorption action can be attributed to oxidation of the sediment and to products such as active ferric oxide that adsorb the active phosphorus in the water. However, with such a large load of external phosphorus, although the sediment is thoroughly oxidized, it cannot completely overcome the effects of external phosphorus, and reaches maximal phosphorus adsorption capacity.

Table 4.12 Initial and final phosphorus contents and PAS in ten sediment samples (Li et al., 1998b)

Sediment sample	Dry weight (g)	Initial P content (mg)	Final P content (mg)	P addition (mg)	P adsorption (mg)	Adsorption (%)	PAS (mg/kg)
1	556	458	489	53	31	58	56
2	611	557	593	53	36	68	59
3	435	385	419	53	34	64	78
4	455	418	456	53	38	72	84
5	512	427	463	53	36	68	70
6	316	331	371	53	40	75	127
7	308	283	322	53	39	74	127
8	536	486	527	53	41	77	76
9	732	558	585	53	27	51	37
10	1021	1248	1274	53	26	49	25

Initial concentration of phosphate was 7 mg/m^3.

Sediments used in the experiment were collected from the lake bottom with the area of about 0.02 m^2. The addition of 53 mg phosphorus was equal to a phosphorus pollution load of 2.65 g/m^2, which was similar to the actual external phosphorus pollution load on Eastern Taihu Bay during 7 months. The experiment demonstrates that the PAS of sediment in Lake Taihu is finite, so additional methods of phosphorus export from the lake ecosystem are necessary. The phosphorus content in Eastern Taihu Bay is consistent with conditions of eutrophication. In addition, the phosphorus carried by the water from western Lake Taihu cannot be well controlled, so in addition to water exchange, proper biological export should be considered; also, phosphorus pollution in Dongshan Peninsula should be reduced to control and prevent further eutrophic deterioration. In addition, the research shows that fish pen-culture causes significant phosphorus pollution. The phosphorus content of the sediment and the PAS of the sediment in the fish pen-culture zone in the northwest of Eastern Taihu Bay are higher than those in other areas of the lake.

4.4 Conceptual Model of Internal Nutrient Loading

Boqiang Qin, Guangwei Zhu, and Lu Zhang

4.4.1 Effects of Wind-Wave Disturbance on Nutrient Concentrations in the Water

4.4.1.1 Field Investigations

Field investigations were conducted on five occasions in February–March and July 1998, June and September 2001, and May 2003.

The two investigations in 1998 were conducted inside Meiliang Bay (sampling site no. 1) and at the mouth of the bay (sampling site no. 4) (Fig. 4.19); both sites were in pelagic areas, but with a greater depth of soft sediment at site 4 than at site 1. These studies focused on the vertical profiles of nutrient and total suspended solid (SS) concentrations under different wind conditions. Wind speed, wind direction, water temperature, transparency, and lake current were measured in situ. Samples from different water depths were taken to analyze total organic matter (TOM) content, different N and P forms, and SS concentration. In February–March, three different wind intensities were observed: 2 m/s (February 26), 5 m/s (February 24), and 6.5 m/s (March 11), which represented light, moderate, and strong wind conditions, and water samples were collected from three depths: 0.25, 1.25, and 2.25 m (water depth was 2.50 m). On July 23, 25, and 30, wind speeds were all 2–4 m/s (light–moderate), and water samples were collected from five depths: 0.25, 0.63, 1.25, 1.88, and 2.25 m.

In June 2001, a total of four sediment cores, each 26 cm long, were collected from site 2 at central Meiliang Bay, site 5 at Dapukou in the northwest, site 6 in the southwest, and site 7 in the east part of the lake (see Fig. 4.19). Each core was sliced at 2-cm intervals, and measurements made of the oxidation-reduction potential (*Eh*), porosity, specific gravity, size, moisture content, organic matter content, total nitrogen (TN), and total phosphorus (TP). Pore water from each core was obtained by centrifugation, and interstitial TN and TP concentrations were measured by a spectrophotometer after alkaline potassium persulfate digestion.

From September 8 to 11, 2001, an observation platform was built at site 1 in Meiliang Bay, and a capacitive wave recorder and a SonTek Doppler current meter were deployed to obtain time-series measurements of waves and currents. In situ wind speed, water temperature, transparency, and water depth were also measured. Meanwhile, water samples from seven different depths (0, 1.0, 1.5, 2.0, 2.25, 2.50, and 2.7 m; bottom depth was 2.75 m), were taken three times a day (at 13:00, 17:00,

Qin, B.Q.
State Key Laboratory of Lake Science and Environment, Nanjing Institute of Geography and Limnology, Chinese Academy of Sciences, 73 East Beijing Road, Nanjing 210008, P. R. China
e-mail: qinbq@niglas.ac.cn

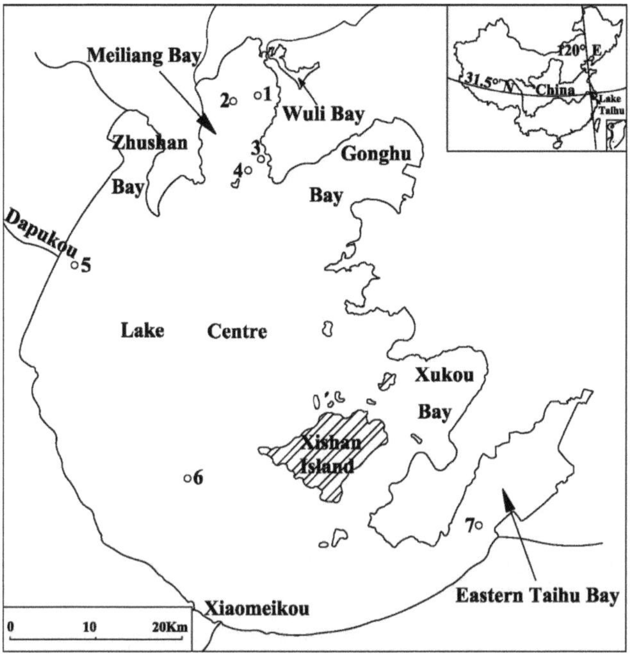

Fig. 4.19 Location of field sampling sites in Lake Taihu

and 21:00 on September 8 and at 9:00, 13:00, and 17:00 on September 9 to 11) to monitor changes in SS and nutrient concentrations.

In May 2003, samples were taken at site 3 in southern Meiliang Bay, approximately 100 m from the bank, where the sediment was about 20 cm thick. For 3 days until the afternoon of May 6, there had been light winds and rain. From the evening of May 6, the wind speed increased, and on May 7, speeds exceeded 8 m/s from 9:00 to 17:00 and 12 m/s at 13:00. The prevailing west wind and longer wind fetch created strong dynamic disturbance in the lake. On May 8, the wind speed decreased to 4–6 m/s and subsequently dropped below 3 m/s. On May 10, the east wind prevailed and the waves subsided in the observed area. Water samples were taken at 1300–1400 on May 7 and 10 from 0.2, 0.9, 1.4, and 1.9 m (bottom depth was 1.95 m). Measurements were made of SS, loss of ignition (LOI), TP, total dissolved phosphorus (DTP), and soluble reactive phosphorus (SRP).

4.4.1.2 Vertical Profiles of Nutrient Concentrations in the Sediment and Overlying Water

The vertical profiles of the mean nutrient concentrations in the sediment pore water and the water column in Meiliang Bay on September 8–11, 2001, are shown in Fig. 4.20. The data show that the total dissolved nitrogen (DTN) concentration was ten times higher in the sediment pore water than in the overlying water, and that

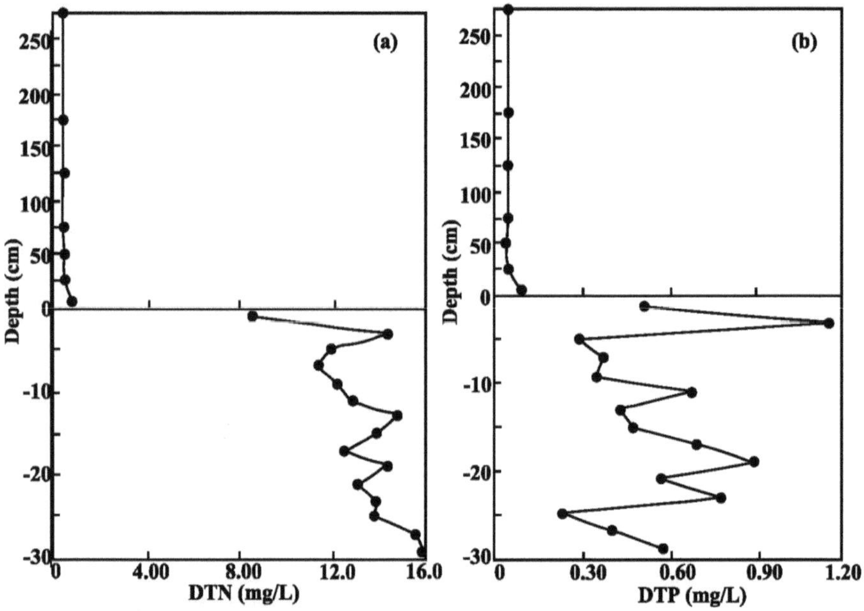

Fig. 4.20 Vertical concentration profile of (**a**) of total dissolved nitrogen (DTN) and (**b**) total dissolved phosphorus (DTP) in the water column and sediment pore water at site 2 in September 2001. Data are means for September 8 to 11 (4 days, 3 measurements/day)

the DTP concentration was seven to eight times higher; this concentration gradient caused the release of nutrients from the sediment pore water into the water column. In the sediments, concentrations of DTN varied and generally increased with depth, whereas concentrations of DTP varied without any apparent trend with depth.

The vertical profiles of six water quality parameters in the water column at sites 1 and 4 in Meiliang Bay on July 23, 25, and 30, 1998 are shown in Fig. 4.21. Total dissolved nitrogen (DTN) concentration in lake water increased with increasing water depth. SS concentration and its organic matter content (SS-OM) were typically higher near the lake bottom than near the surface of the water column. In contrast, TN, TP, and Chl a concentrations were higher in the upper layer than near the bottom of the water column. These vertical distributions showed that the dissolved nitrogen and phosphorus will be affected by nutrient release from the sediments, but that TN, TP, and Chl a will be influenced by the phytoplankton and suspended organic detritus. Thus, nutrient concentrations vary greatly by water depth in this shallow lake.

4.4.1.3 Relationship Between Nutrient Concentration in the Overlying Water and Wind Speed

The column mean values of eight water quality variables from seven depths in September 2001, at site 2, were used to determine correlations of the variables

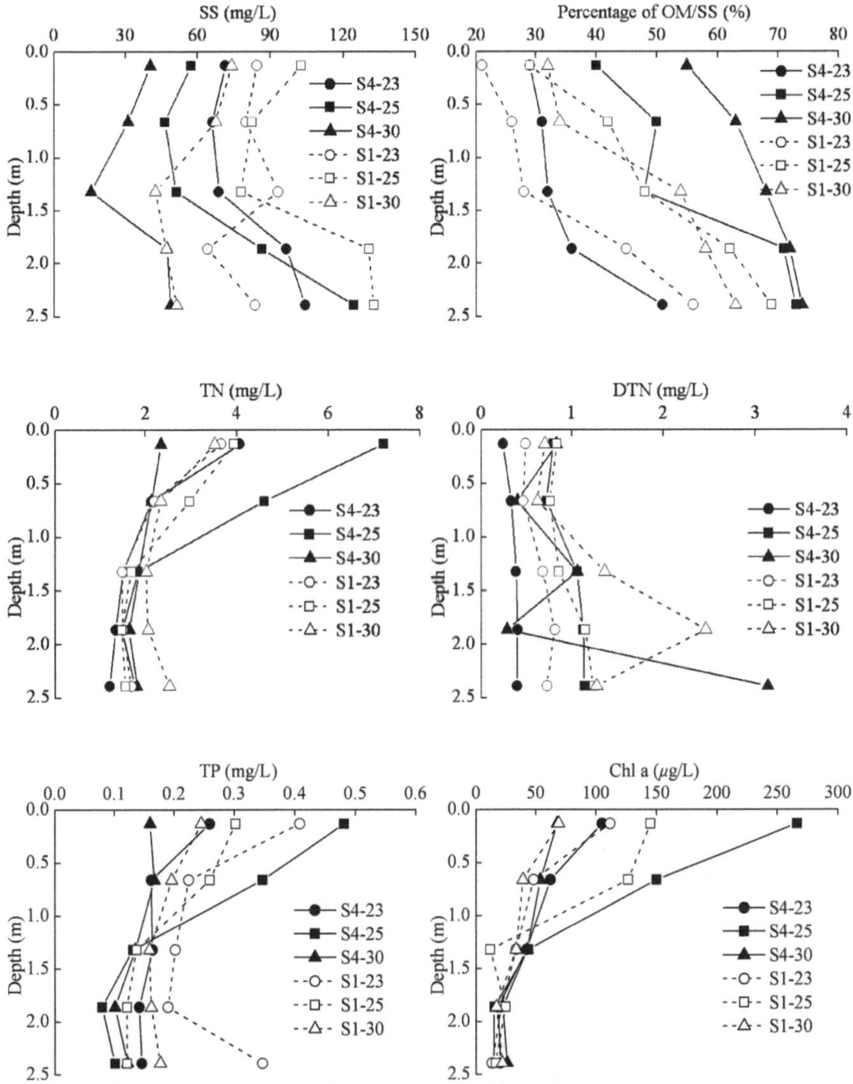

Fig. 4.21 Vertical profiles of suspended solids (SS) and organic matter content (SS-OM), total nitrogen (TN), total dissolved nitrogen (DTN), total phosphorus (TP), and chlorophyll *a* (Chl a) content in the water column at sites 1 and 4 in Meiliang Bay on July 23, 25, and 30, 1998

with wind speeds (Fig. 4.22). Data for the lowest depth of the water column was excluded from calculation of the mean concentration, since SS concentration was much higher near the sediment-water interface than that in upper layers, and was affected by sampling. The excluded data should not affect the qualitative analysis of the release of nutrient under dynamic disturbances.

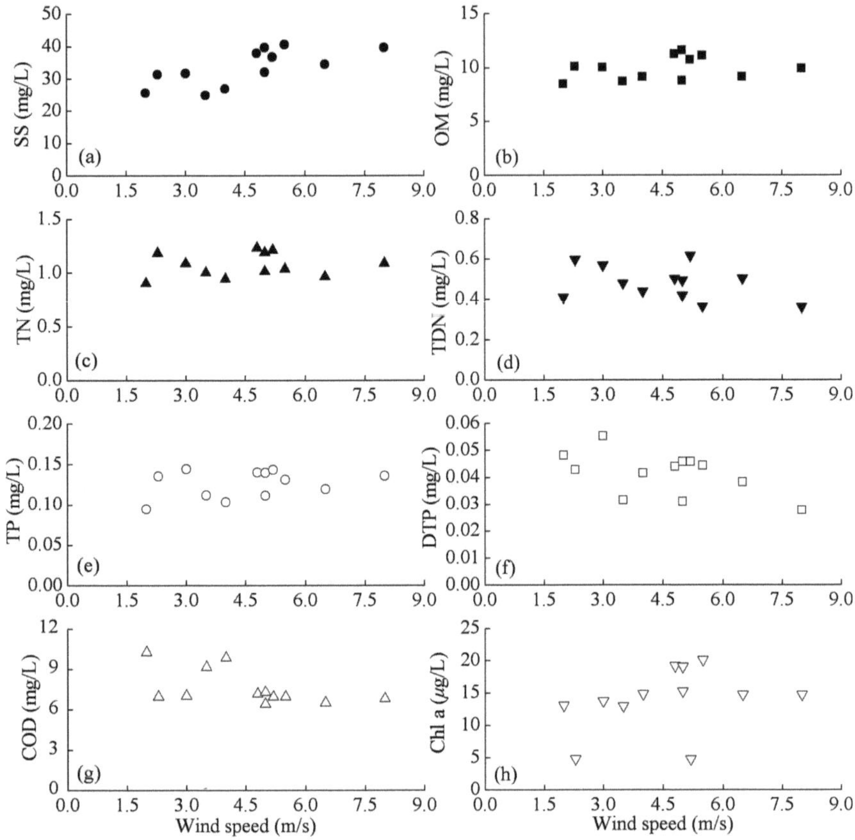

Fig. 4.22 Relationships between column mean values of SS (**a**), OM (**b**), TN (**c**), DTN (**d**), TP (**e**), DTP (**f**), chemical oxygen demand (COD) (**g**), and Chl a (**h**) with wind speed in Meiliang Bay (site 2) in September 2001

The concentrations of SS, TP, and Chl a increased with wind speed (Fig. 4.22a,e,h, respectively). In contrast, DTP and COD decreased with wind speed (Fig. 4.22f,g, respectively). The increased dynamic disturbances associated with increasing wind speed brought more SS and SS-OM into the water column by means of resuspension, and as a result, the TN and TP content of the water would also greatly increase. However, soluble nutrients such as DTN and DTP did not show a marked increase with increasing of wind speed, probably because of sorption of particles.

From May 6 to 11, 2003, at site 3 in Meiliang Bay, data were collected on concentrations of SS, LOI of organic matter (which denotes the organic matter content of the suspended solids in lake water), TP, and DTP under different wind conditions (Fig. 4.23). Concentrations of SS and TP were high at high wind speeds (7–8 m/s) and low at low wind speeds (<3 m/s), which was the direct result of increased suspension of sediments under strong wind stresses (Fig. 4.23). However, LOI of the suspended solids was low at high wind speed and high at low wind speed. No

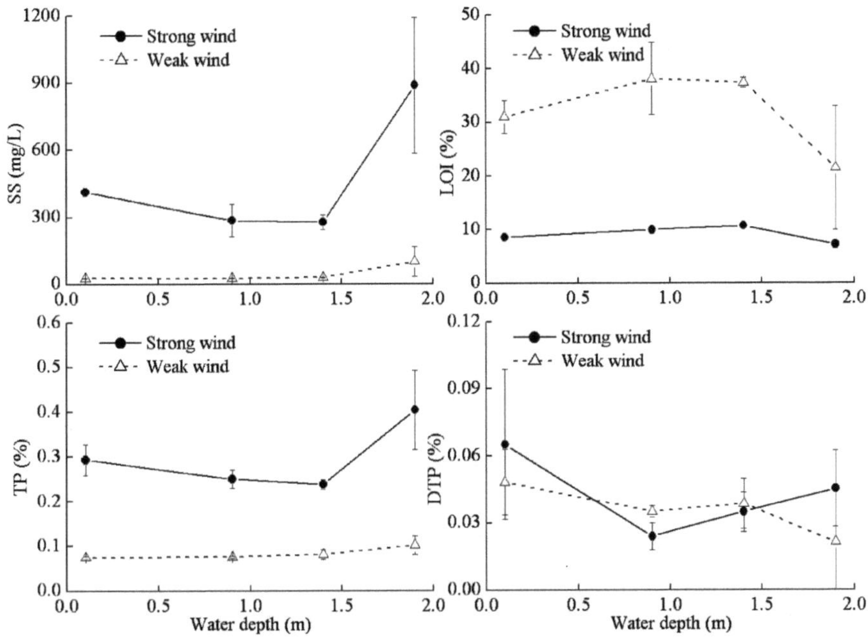

Fig. 4.23 Relationship between concentration of suspended solids (SS), loss on ignition of SS (LOI), total phosphorus (TP), and dissolved total phosphorus (DTP) in lake water with wind strength (strong wind >8 m/s; weak wind <3 m/s) and depth at site 3 in Meiliang Bay in May, 2003

marked changes in DTP contents in lake water were noted at different wind stresses, similar to the situations at another site, site 2, in Meiliang Bay in September 2001, as noted earlier.

4.4.2 Internal Loading of Nutrients

A variety of mechanisms can transport nutrients from the sediment to the overlying water; for example, diffusion, wind-induced water turbulence, bioturbation, release of sediment gases, and removal by algae and rooted aquatic plant harvest (Wetzel, 2001). In Lake Taihu, the first two mechanisms are the main types of nutrient release. Diffusion of nutrients from the sediment to the overlying water occurs in calm conditions, and is temperature dependent, with faster diffusion occurring at higher temperatures. In calm conditions, the sediment–water interface becomes more reducible than in strong mixing conditions. Nutrient release following sediment resuspension takes place during wind-induced water turbulence. This latter mechanism of nutrient release, which can involve significant amounts of nutrients, is more efficient than the former diffusive release. Under field conditions, these two mechanisms of nutrient release alternate.

Nutrient concentrations in pore water increase with sediment depth in shallow lakes (Qin et al., 2004). The constant weak oxidized conditions across the sediment–water interface favor decomposition of organic matter in the sediment. At the same time, dynamic disturbance erodes the surface sediment particles and releases dissolved nutrients to the water column. Some dissolved interstitial nutrients, especially SRP, may be adsorbed by ferric iron and other oxides. In calm conditions, the suspended sediment particles and the adsorbed nutrients may settle to the bottom of the lake and be subjected to another suspension during the next wind disturbance (Qin et al., 2004). During calm periods, nutrient release from the sediment is controlled by diffusion associated with nutrient gradients from the dissolved interstitial regions to the overlying water. During strong wind-wave activity, there is not only a great increase in TP and TN concentrations in the water column, but also a possible decrease of soluble nutrients in the water. Thus, in a shallow lake such as Lake Taihu, there is a totally mixed mode of nutrient release; this is distinct from the situation in a deep lake, where nutrient release is only from diffusion.

4.4.2.1 Nutrient Exchange Flux Across the Sediment–Water Interface Under Static Conditions

The daily exchange rates and annual released amounts of nutrients at three different temperatures in static conditions are given in Table 4.13 (NH_4^+-N) and Table 4.14 (PO_4^{3-}-P). Generally, in the lower water temperatures, some areas of the lake, such as southern Mashan, the west littoral area, and the southwest centre, serve as a "sink" rather than a "source" for nutrient exchange, meaning that relatively low temperatures reduce the production of ammonium in the sediments. In contrast, release of PO_4^{3-}-P did not increase with temperature. The annual amount of nutrient release in static conditions was considerable, equal to some 10,000 t NH_4^+-N and 900 t PO_4^{3-}-P, or 20–30% and 30–35% of the annual external load of nitrogen and phosphorus, respectively.

4.4.2.2 Estimates of the Total Nutrient Release Under Dynamic Conditions

Estimates for soluble nutrient release are complicated, because sediment suspension was usually accompanied by soluble nutrient release and adsorption, and the adsorption and suspension processes for SS in sediments are related to the oxygen and Fe concentrations in the water.

Here we describe an estimate for release of DTN and DTP from pore water, and of TN and TP from sediments under one distinct dynamic process. It was assumed that dissolved nutrients released from pore water were not adsorbed by suspended solids, and that there was no sedimentation. In the wind-wave disturbance process, the dissolved nutrients in sediment pore water were mixed in the overlying water column and were bioavailable without any inactivation reaction. Therefore, the internal loading of soluble nutrients comprises the total dissolved nutrients in pore water in all the eroded sediments. Similarly, the internal loading of TN and TP comprises the total nitrogen and phosphorus in all eroded sediments. Based on the

Table 4.13 Flux of $NH_4{}^+$-N at the sediment–water interface of Lake Taihu, at three different temperatures under static conditions (Fan et al., 2004)

District of the lake	Area of soft sediment (km^2)	Flux of $NH_4{}^+$-N exchange at different temperatures ($mg/(m^2 \cdot d)$)			Loading of ammonium (t/yr)
		5°C (75 d)	15°C (120 d)	25°C (170 d)	
Eastern Wuli Bay	3.4	78.8 ± 27.6	−113.9 ± 52.4	13.1 ± 7.5	−18.8 ± 10.0
Western Wuli Bay	1.1	33.7 ± 11.8	66.2 ± 30.5	187.0 ± 106.6	46.5 ± 24.9
Xiaowanli	19.2	23.1 ± 8.1	17.1 ± 7.9	40.7 ± 23.2	205.7 ± 105.6
Central Meiliang Bay	21.5	17.3 ± 6.1	131.7 ± 60.6	−11.2 ± 6.4	326.5 ± 142.6
Mashan Bay	21.2	94.9 ± 117.7	−44.8 ± 42.1	−77.2 ± 49.4	−241.3 ± 98.1
Gonghu Bay	74.8	13.2 ± 16.4	−92.0 ± 86.4	63.1 ± 40.4	51.6 ± 170.2
Zhushan Bay	24.9	−19.3 ± 23.9	−15.1 ± 14.2	56.9 ± 36.4	159.6 ± 66.9
West littoral area	247.6	−48.6 ± 17.0	−22.2 ± 10.2	62.4 ± 35.6	1, 067.0 ± 879.2
Southwest centrr	214.0	−27.1 ± 33.6	1.0 ± 1.0	39.6 ± 25.3	1, 030.8 ± 406.8
Southwest littoral	203.6	−3.9 ± 4.8	4.2 ± 4.0	45.6 ± 29.2	1, 621.7 ± 1, 033.2
Southern Mashan	101.8	−75.3 ± 93.4	79.3 ± 74.5	30.6 ± 19.6	923.3 ± 536.6
Central Lake Taihu	119.8	−13.6 ± 4.8	0.1 ± 0.1	20.7 ± 11.8	301.8 ± 198.7
Littoral of Xishan	162.2	−18.3 ± 22.7	55.2 ± 51.9	23.5 ± 15.0	1, 500.7 ± 1, 149.5
Xukou Bay	118.0	6.4 ± 7.9	−51.3 ± 48.2	19.7 ± 12.6	−274.4 ± 359.6
Eastern Taihu Bay	289.7	28.7 ± 10.0	64.5 ± 29.7	18.0 ± 10.3	3, 868.9 ± 1, 809.5
Mean value	–	−16.0 ± 17.6	12.6 ± 6.9	34.1 ± 20.8	—
Sum	1631.8	—	—	—	10, 569.6 ± 6, 991.4

Note: Minus means the contents of ammonium decrease when exchanged with sediments.

Table 4.14 Flux and loading of phosphate at the sediment–water interface of Lake Taihu at three different temperatures under static conditions (Fan et al., 2006)

District of the lake	Area of soft sediment (km^2)	Exchange flux of phosphate at different temperatures ($mg/(m^2 \cdot d)$)			Loading of phosphate (t/yr)
		5°C (75 d)	15°C (120 d)	25°C (170 d)	
Eastern Wuli Bay	3.4	2.95 ± 1.99	−0.40 ± 0.34	0.50 ± 0.50	0.9 ± 0.7
Western Wuli Bay	1.1	5.20 ± 3.51	3.13 ± 2.61	2.11 ± 2.10	1.2 ± 1.0
XiaoWanli	19.2	2.87 ± 1.94	−3.25 ± 2.72	−3.72 ± 3.70	−15.5 ± 15.6
Central Meiliang Bay	21.5	1.20 ± 0.81	−0.13 ± 0.11	−0.57 ± 0.57	−0.5 ± 1.0
Mashan Bay	21.2	2.57 ± 0.84	0.29 ± 0.10	−0.80 ± 0.28	1.9 ± 0.6
Gonghu Bay	74.8	−0.08 ± 0.03	4.02 ± 1.35	0.63 ± 0.22	43.6 ± 14.8
Zhushan Bay	24.9	−2.08 ± 0.68	2.42 ± 0.82	2.54 ± 0.88	14.1 ± 4.9
West littoral area	247.6	3.93 ± 2.65	1.60 ± 1.34	0.60 ± 0.60	145.9 ± 114.2
Southwest centre	214.0	1.33 ± 0.43	−2.83 ± 0.95	−0.66 ± 0.23	−75.2 ± 25.8
Southwest littoral	203.6	2.19 ± 0.71	5.67 ± 1.91	1.37 ± 0.47	219.4 ± 74.0
Southern Mashan	101.8	−3.88 ± 1.26	2.49 ± 0.84	0.29 ± 0.10	5.7 ± 2.3
Central Lake Taihu	119.8	1.44 ± 0.97	−1.76 ± 1.47	0.78 ± 0.78	3.7 ± 3.5
Littoral of Xishan	162.2	2.65 ± 0.87	3.13 ± 1.05	1.44 ± 0.50	133.1 ± 44.9
Xukou Bay	118.0	3.37 ± 1.1	0.22 ± 0.08	1.49 ± 0.52	63.0 ± 21.2
Eastern Taihu Bay	289.7	0.24 ± 0.16	3.59 ± 3.00	4.41 ± 4.39	358.1 ± 334.0
Mean value	—	1.52 ± 0.78	1.78 ± 0.96	1.32 ± 1.05	—
Sum	1631.8	—	—	—	899.4 ± 573.6

Note: Minus means the contents of ammonium decrease when exchanged with sediments.

wave flume experiments, the strong waves in Lake Taihu may erode the upper 10 cm of the sediment (Qin et al., 2004). Therefore, the hypothesis is that the upper 10 cm of sediment could be suspended in a strong enough hydrodynamic process.

Estimates for nutrient release from pore water were derived from calculations of sediment porosity, dissolved total nitrogen (TN), and dissolved total phosphorus (TP) concentrations in pore water (Table 4.15). The physicochemical characteristics (gross weight, organic matter content, and porosity) of four sediment cores taken in June 2001 revealed that the upper 5–10 cm of the sediment had been actively involved in exchange of sediment with the overlying water (Qin et al., 2004). As a result, estimates of nutrient release from sediments in one strong dynamic process were calculated assuming that suspension occurred in the upper 10 cm of the sediment. Spatial distribution of surface sediments was not uniform in Lake Taihu, and a detailed distribution of the depth of the loose sediment was presented by Luo et al. (2004).

Lake Taihu consists of three regions: Meiliang Bay, Eastern Taihu Bay, and the remainder of the lake (main Lake Taihu). The vertical distribution of the physicochemical characteristics of the sediments was calculated according to the four sediment cores. As a result, if the upper 10 cm of the sediments was taken in lake water and thoroughly mixed, the total release of nutrients from the sediment to lake water was calculated to be 23,059 kg DTN and 1,236 kg DTP in Meiliang Bay, 29,618 kg DTN and 1,514 kg DTP in Eastern Taihu Bay, and 477,668 kg DTN and 20,606 kg DTP in the main Lake Taihu (see Table 4.15); this was equivalent to an increase in TN of 0.12 mg/L and in TP of 0.005 mg/L in all of Lake Taihu when the upper 10 cm sediments were eroded into the overlying water column. Compared with the mean concentrations of TN of 2–4 mg/L and TP of 0.01–0.1 mg/L in the water column in Lake Taihu, strong sediment suspension (upper 10 cm sediments moved into the lake water) would result in an increase of DTN and DTP of 3–6% and 5–50%, respectively.

The actual release of nutrients from the sediments would be lower than these estimates, which are based on only physical factors, because the abundant aquatic plants in Eastern Taihu Bay will decrease the suspension and release of sediments. Moreover, adsorption and flocculation by particles may also decrease the concentration of dissolved nutrients, because the high SS concentration in the overlying water column has sufficient organic and inorganic material to adsorb the dissolved

Table 4.15 Estimates of internal loading of dissolved total nitrogen (DTN) and dissolved total phosphorus (DTP) from pore water of the upper 10 cm of sediments in three regions of Lake Taihu

	Meiliang Bay	Eastern Taihu Bay	Main Lake Taihu
Porosity (%)	45.3	45.2	40.2
DTN content (mg/L)	8.21	4.89	8.37
DTP content (mg/L)	0.44	0.25	0.36
Areas included (km^2)	62	134	1,431
N release from pore water (kg)	23,059	29,618	477,668
P release from pore water (kg)	1,236	1,514	20,606

nitrogen and phosphorus and resediment it to the lake bottom. Sometimes, some soluble nutrients, especially SRP, decreased in dynamic suspension conditions.

4.4.2.3 Estimate of Annual Nutrient Loading in This Large Shallow Lake

Estimates of nutrient release in the foregoing two release scenarios provide an opportunity to calculate the annual nutrient release. Observations of wind speed made by the Taihu Laboratory for Lake Ecosystem Research showed that Lake Taihu is located in the monsoon climate zone with prevailing northwest winds in winter and southeast winds in summer. In addition, observations of wave height versus wind speeds show that the strong wind could cause significant wave heights frequently. When wind speed is less than or equal to 2 m/s, significant wave height is just 3.5 cm and conditions are defined as "calm" because there is only rippling on the lake (Luo & Qin, 2003). When wind speed is $>2\,m/s$ and $<6\,m/s$, the significant wave height is 15 cm (1/10 of the time wave height is 60–70 cm), which is defined as "gentle" because this wind force aerates the overlying water well but does not cause significant sediment resuspension. When the wind speed is $>6\,m/s$, defined as "gusty," there is significant sediment resuspension. In the 2001 wind speed study, 45 days were "calm" (12% of the year), 298 days were "gentle" (82%), and 22 days were "gusty" (6%). Because instantaneous gusts happen frequently in some areas of Lake Taihu, the actual frequency of "gusty" is probably underestimated.

By use of the percentage of "calm" weather (12% of a year) and the internal loading obtained in the laboratory calculated in Table 4.15, the total releases of nitrogen (NH_4^+-N) and phosphate (PO_4^{3-}-P) are 1,272 t and 108 t in a year, respectively. For the "gentle" weather, no laboratory experiment was conducted to determine the release rate, so we used results from flume experiments (Luo et al., 2006; Sun et al., 2006). In "gentle" wind conditions, the sediment will not be markedly resuspended, but the water will be well aerated, and the shear stress between the water and sediment is about $0.019\,N/m^2$. In "gusty" conditions, the sediment will be eroded and resuspended significantly, with a shear stress of about $0.217\,N/m^2$. Correspondingly, during "gentle" conditions, the release rates of TN and TP are $1.92 \times 10^{-3}\,mg/(m^2 \cdot s)$ and $5.69 \times 10^{-4}\,mg/(m^2 \cdot s)$, respectively, whereas during gusty conditions, the release rate of TN and TP are $1.16 \times 10^{-2}\,mg/(m^2 \cdot s)$ and $2.14 \times 10^{-3}\,mg/(m^2 \cdot s)$ (Luo et al., 2006). Therefore, under "gentle" wind conditions, the daily internal loading of TN and TP should be 186 t and 55 t, respectively (Qin et al., 2006). Accordingly, with 298 days of "gentle" wind conditions in a year, the annual contributions from "gentle" wind conditions to TN and TP should be 55,430 t and 6,390 t, respectively (Qin et al., 2006). Calculated in the same way, the annual contributions of TN and TP from "gusty" wind conditions (22 days) would be 81,000 t and 21,000 t, respectively (Qin et al., 2006).

Investigations of external loading in Lake Taihu have given values ranging from 39,000 to 41,000 t for nitrogen and from 2,900 to 3,800 t for phosphorus (Sun & Huang, 1993). The amounts of nutrients released from the sediment are two to six times those from external loading, and the internal release during turbulence is much higher than during "calm" conditions, during which diffusion provides the main release of nutrients. For TN and TP release following sediment resuspension, even one "gust" on the lake causes sediment resuspension, resulting in massive nutrient release. Therefore, wind-wave hydrodynamic disturbance must play a key role on the internal release of nutrients in large and shallow lakes such as Lake Taihu.

Only a small percentage of the released nutrients will contribute to eutrophication, because most nutrient fractions are not easily utilized by the phytoplankton. Moreover, under dynamic conditions, nutrient release occurs near the sediment–water interface, and most of the particulate nutrients will resediment quickly. Additionally, as a result of flocculation and adsorption by suspended particles and ferric iron, a considerable proportion of the dissolved nutrients will precipitate again and be buried in the sediments. Thus, most of the nutrients released from the sediment to the overlying water do not participate in biogeochemical cycling in the water ecosystem during strong wind events. In spite of that, such significant internal nutrient sources as 137,702 t TN and 27,498 t TP should not be neglected in the view of eutrophication control in Lake Taihu

4.4.3 Conceptual Model of Internal Loading of Nutrients in Large Shallow Lakes Such as Lake Taihu

The nutrient (TN and TP) concentrations in the overlying water will greatly increase under wind-stressed disturbances. Laboratory experiments have shown that winds cause suspension and diffusion-induced nutrient increases in the overlying water that may be dozens of times greater than those occurring under calm conditions, when diffusion alone operates (Reddy et al., 1996). In Lake Arresø, Denmark, with a surface area of 41 km^2 and a mean depth of 2.9 m, wind-induced resuspension could increase nutrient concentrations in the water column by up to 20–30 times those in "calm" condition simulated in the laboratory (Søndergaard et al., 1992). Statistics indicate that SS and TP concentrations are strongly correlated with wind speed (Kristensen et al., 1992; Søndergaard et al., 1992). Similar phenomena also were noted during laboratory simulations using samples from Lakes Taihu, Gehu, and Xuanwu (Wang et al., 1994; Fan 1995; Zhang et al., 2001). Investigation in Meiliang Bay in Lake Taihu during 1998 revealed that SS would double in winds of 6 m/s over those in "calm" conditions, and would increase with depth.

The upper 5–10 cm of the sediments of Lake Taihu was not only actively involved in nutrient exchange across the sediment–water interface but also provided nutrient transport into the overlying water under dynamic disturbance and resuspension. Below a sediment depth of 5–10 cm, nutrients could be carried into the upper water through gradient-induced diffusion (Qin et al., 2004). A similar finding has been reported for Lake Apopka, a large shallow lake in Florida, United States (Reddy et al., 1996).

Nutrient exchange across the sediment–water interface is not only controlled by dynamic disturbances, but also by the oxidation and reduction potential (*Eh*); this is especially true for soluble nutrients such as SRP (Mortimer, 1971; Moore & Reddy, 1994). Oxidizing conditions favor formation of Fe- or Mn oxides in surface sediments, which can enhance the adsorption ability of soluble nutrients and inhibit nutrient release to the overlying water. Reducing conditions enhance release of dissolved nutrients to the overlying water (Gächter et al., 1988; Hohener & Gächter, 1994). For this reason, variations in dissolved oxygen at the sediment–water interface in a given site are good indicators of rates of nutrient release.

Wind-stressed sediment resuspension is common in shallow lakes, such as Lake Taihu, leading to an unstable sediment–water interface. In contrast, deep lakes have a stable sediment–water interface that is much less affected by wind stress. The constant dynamic disturbance in shallow Lake Taihu frequently transports additional oxygen to the sediment–water interface. Field observations revealed small spatial variations in dissolved oxygen (DO) concentrations in Lake Taihu, ranging from 9.0 to 9.5 mg/L (Sun & Huang, 1993; Yu, 1994). As a result, the *Eh* potentials across the sediment–water interface differed slightly, from -100 to -200 mV, indicating a weak reducing environment (Qin et al., 2004). The static release from sediments in shallow lakes is likely lower than that in deep lakes, which typically possess a more reduced sediment–water interface.

In shallow lakes such as Lake Taihu, accurate estimates for internal nutrient loading, especially for SRP, require further understanding of the behavior of metallic elements such as iron and manganese in water and in sediments. Dynamic disturbances could release nutrients from the sediments through suspension, leading to increased concentrations of SS, TN, and TP in the overlying water. At the same time, the release of soluble nutrients, especially SRP, will decrease because of reoxygenation in the water. The Fe content of sediments could have a great impact on the internal nutrient release (Jensen et al., 1992). Investigations on sediments from more than 100 lakes in Denmark found that total Fe and TP in sediments were positively correlated, whereas the TP in the overlying water was negatively correlated with Fe:P in the sediments (the higher the Fe:P ratio in the sediments, the lower the TP concentration in the overlying water) (Jensen et al., 1992). A Fe:P ratio of 15:1 in sediments could be a criterion for defining marked P release from the sediments. In Lake Taihu, the Fe:P ratio was between 20 and 40 (Zhu, 2003), indicating that under intensive dynamic disturbance there is a trend towards higher TP content with lower SRP content in the lake water. In such a case, most nutrients released following resuspension would not affect the ecosystem and eutrophication. However, in Lake Taihu, wind gusts and resuspension are so frequent, and the calculated annual nutrient release is so high, that the cumulative effects of nutrient release exert profound influences on the lake ecosystem, and these effects should be addressed in subsequent studies.

References

Boers, P. C. M., W. Van Raaphorst & T. D. Van der Molen, 1998. Phosphorus retention in sediments. Water Science and Technology 37: 31–39.

Chang, W. Y. B. & J. L. Liu, 1996. The origin and evolution of Taihu Lake: a 11 000 year journal. Acta Palaeontologica Sinica 35: 129–135.

Fan, C. X., 1995. Physicochemical characteristics of sediments in Gehu Lake and simulation of its phosphorus release. Journal of Lake Sciences 7: 341–350 (In Chinese with English abstract).

Fan, C. X., J. Ji & G. R. Sui, 1998. Characteristics of the sludge storage in Taihu Lake and the special distributions of its main physical and chemical properties. In: Cai, Q. M. (ed.), Study on the Environment and Ecosystem of Lake Taihu (I). Beijing: China Meteorology Press, 55–62 (In Chinese with English abstract).

Fan, C., Y. Liu & H. Chen, 2000a. Approach on estimating storage sludge in Lake Taihu and its distributing characteristics. Shanghai Environmental Science 19: 72–75 (In Chinese with English abstract).

Fan, C. X., L. Y. Yang & L. Zhang, 2000b. The vertical distributions of nitrogen and phosphorus in the sediment and interstitial water in Taihu Lake and their interrelations. Journal of Lake Sciences 12: 359–366 (In Chinese with English abstract).

Fan, C., L. Zhang, B. Qin, W. Hu, G. Gao & J. Wang, 2004. Migration mechanism of biogenic elements and their quantification on the sediment-water interface of Lake Taihu: I. Spatial variation of the ammonium release rates and its source and sink fluxes. Journal of Lake Sciences 16: 10–20 (In Chinese with English abstract).

Fan, C. X., L. Zhang, X. M. Bao, B. S. You, J. C. Zhong, J. J. Wang & S. M. Ding, 2006. Migration mechanism of biogenic elements and their quantification on the sediment-water interface of Lake Taihu: II. Chemical thermodynamic mechanism of phosphorus release and its source-sink transition. Journal of Lake Sciences 18: 207–217 (In Chinese with English abstract).

Gächter, R., R. S. Meyer & R. Mares, 1988. Contribution of bacteria to release and fixation of phosphorus in lake sediments. Limnology and Oceanography 33: 1542–1558.

Hartley, A. M., W. A. House, M. E. Callow & B. S. C. Leadbeater, 1997. Coprecipitation of phosphate with calcite in the presence of photosynthesizing green algae. Water Research 31: 2261–2268.

Hohener, P. & R. Gächter, 1994. Nitrogen cycling across the sediment-water interface in an eutrophic, artificially oxygenated lake. Aquatic Sciences 56: 115–131.

Hongve, D., 1997. Cycling of iron, manganese and phosphate in a meromictic lake. Limnology and Oceanography 42: 635–647.

Hupfer, M., R. Gächter & R. Giovanoli, 1995. Transformation of phosphorus species in settling seston and during early sediment diagenesis. Aquatic Sciences 57: 305–324.

Jensen, H. S., P. Kristensen, E. Jeppesen & A. Skytthe, 1992. Iron:phosphorus ratio in surface sediment as an indicator of phosphate release from aerobic sediments in shallow lakes. Hydrobiologia 235/236: 731–743.

Kristensen, P. M., M. Søndergaard & E. Jeppesen, 1992. Resuspension in a shallow lake. Hydrobiologia 228: 101–109.

Li, Y., D. N. Wu & Y. X. Xue, 1998a. A development sequential extraction method for different forms of phosphorus in the sediments and its environmental geochemical significance. Marine Environmental Science 17: 15–20 (In Chinese with English abstract).

Li, W., K. Chen, Q. Wu & G. Wang, 1998b. A preliminary study on phosphorus saturation of the top sediment in east Taihu Lake. Journal of Lake Sciences 10: 49–54 (In Chinese with English abstract).

Luo, L. C. & B. Q. Qin, 2003. Comparison between wave effects and current effects on sediment resuspension in Lake Taihu. Hydrology 23: 1–4 (In Chinese with English abstract).

Luo, L., C. Qin & G. Zhu, 2004. Sediment distribution pattern mapped from the combination of objective analysis and geostatistics in the large shallow Taihu Lake, China. Journal of Environmental Sciences 16: 908–911.

Luo, L. C., B. Q. Qin, G. W. Zhu, X. J. Sun, D. L. Hong, Y. J. Gao & R. Xie, 2006. Nutrient fluxes induced by disturbance in Meiliang Bay of Lake Taihu. Science in China Series D Earth Sciences 49(S1): 186–192.

Ma, S. W. & Q. M. Cai, 2000. Study on upwinding finite element numerical model for wind-driven current in shallow lakes. Advance in Water Science 11: 70–75 (In Chinese with English abstract).

Moore, J. P. A. & K. R. Reddy, 1994. Role of *Eh* and pH on phosphorus geochemistry in sediments of Lake Okeechobee, Florida. Journal of Environmental Quality 23: 955–964.

Mortimer, C. H., 1971. Chemical exchanges between sediments and water in the Great Lakes: speculations on probable regulatory mechanisms. Limnology and Oceanography 16: 387–404.

Qin, B. Q., W. P. Hu, G. Gao, L. C. Luo & J. S. Zhang, 2004. Dynamics of sediment resuspension and the conceptual schema of nutrient release in the large shallow Lake Taihu, China. Chinese Science Bulletin 49: 54–64.

Qin, B. Q., G. W. Zhu, L. Zhang, L. C. Luo, G. Gao & B. H. Gu, 2006. Estimation of internal nutrient release in large shallow Lake Taihu, China. Science in China Series D Earth Sciences 49(S1): 38–50.

Qu, W. C., B. Xue, Y. H. Wu, S. M. Wang, R. J. Wu, P. Z. Zhang & J. F. Chen, 1997. Record of paleoenvironmental evolution of Taihu Lake in the past 14000 years. Journal of Geomechanics 3: 53–61 (In Chinese with English abstract).

Reddy, K. R., M. M. Fisher & D. Ivanoff, 1996. Resuspension and diffusive flux of nitrogen and phosphorus in a hypereutrophic lake. Journal of Environmental Quality 25: 363–371.

Ruttenberg, K. C., 1992. Development of a sequential extraction method for different forms of phosphorus in marine sediments. Limnology and Oceanography 37: 1460–1482.

Søndergaard, M., P. Kristensen & E. Jeppesen, 1992. Phosphorus release from resuspended sediment in the shallow and wind-exposed Lake Arreso, Denmark. Hydrobiologia 228: 91–99.

Sun, S. C. & Y. F. Wu, 1987. Formation and evolution of the Taihu Lake and modern sedimentation. Science in China Series B Chemistry 12: 1329–1339 (In Chinese).

Sun, S. C. & Y. P. Huang, 1993. Lake Taihu. Beijing: Ocean Press 4–130 (In Chinese with English abstract).

Sun, X. J., G. W. Zhu, L. C. Luo & B. Q. Qin, 2006. Experimental study on phosphorus release from sediments of shallow lake in wave flume. Science in China Series D Earth Sciences 49(S1): 92–101.

Wang, T. J., R. Su, X. C. Jin, Z. L. Xia, Y. Lin & S. Q. Fang, 1994. The effects to water quality of phosphorus loading and its release in the sediments of urban eutrophic lakes. Environmental Science Research 7: 12–19 (In Chinese with English abstract).

Wetzel, R. G., 2001. Limnology: lake and river ecosystems, 3rd edition. San Diego: Academic Press.

Wu, G. F., X. C. Wu, & C. T. Jin, 1998. Preliminary studies on release of phosphorus from the sediment of West Lake, Hangzhou. China Environmental Science 18: 107–110 (In Chinese with English abstract).

Xue, B., W. C. Qu, Y. H. Wu, S. M. Wang & R. J. Wu, 1998. Sedimentological record of paleoenvironment of Taihu Lake in late-glacial to Holocene. Journal of Lake Sciences 10: 30–36 (In Chinese with English abstract).

Yin, D. Q., Q. R. Qin & H. Yan, 1994. Effects of environmental factors on release of phosphorus from sediments in Wuli Lake. Journal of Lake Sciences 6: 240–244 (In Chinese with English abstract).

Yu, Y. S., 1994. Sediments and their relationship to eutrophication in Lake Taihu, China. Memoirs of Nanjing Institute of Geography and Limnology (Nanjing Institute of Geography and Limnology, ed.). Beijing: Science Press 9: 48–61 (In Chinese with English abstract).

Zhang, L., C. X. Fan, B. Q. Qin & L. Y. Yang, 2001. Phosphorus release and absorption of surficial sediments in Taihu Lake under simulative disturbing conditions. Journal of Lake Sciences 13: 35–42 (In Chinese with English abstract).

Zhu, G. W., 2003. Hydrodynamics and the internal loading of phosphorus in shallow lakes. Postdoctoral report, Nanjing Institute of Geography and Limnology, Chinese Academy of Sciences 65–67 (In Chinese with English abstract).

Zhu, G. W., B. Q. Qin & L. Zhang, 2006. Phosphorus forms and bioavailability of lake sediments in the middle and lower reaches of Yangtze River. Science in China: Series D Earth Sciences 49(S1): 28–37.

Chapter 5
Physiological and Ecological Characteristics of Blue-Green Algae in Lake Taihu

5.1 Effect of Alkaline Phosphatase on Algal Blooms

Guang Gao

In aquatic ecosystems, 80–90% of organic matter is dissolved. Of this, only a small fraction, 10–20%, can penetrate the cell membrane andf be used directly by heterotrophic microbes. This part of the organic matter, which is highly variable, does not entirely fulfill a microbe's requirements for metabolism and growth. The other 80–90% of dissolved organic matter is composed of polymers that need to be hydrolyzed before being of any use to microbes. Therefore, the depolymerization and hydrolyzation of dissolved organic matter is regarded as an important potential tool for control of microbes because it limits their nutrient intake.

5.1.1 Alkaline Phosphatase in Lake Water

Since the effects of microbial enzymes in aquatic ecosystems were intensively studied by Overbeck in the early 1960s, alkaline phosphatase in particular has been the focus of many studies because it is related to the phosphorus cycle (Reichardt et al., 1967; Overbeck, 1991). Initially, it was reported that, at low dissolved reactive phosphorus concentration, alkaline phosphatase activity would rapidly increase; therefore, it was regarded as an index of phosphorus inhibition of phytoplankton (Kuenzler & Perras, 1965; Healey, 1978). Subsequently, it was found that alkaline phosphatase in water does not entirely reflect the phosphorus inhibition of phytoplankton. Other ecological factors, such as temperature, growth rate, N:P ratio, and heavy metals, all could influence alkaline phosphatase activity of phytoplankton. Furthermore, not all alkaline phosphatase in the aquatic environment is produced by phytoplankton (Healey, 1973; Cembella et al., 1984). Therefore, it is difficult to

Gao, G.
State Key Laboratory of Lake Science and Environment, Nanjing Institute of Geography and Limnology, Chinese Academy of Sciences, 73 East Beijing Road, Nanjing 210008, P. R. China
e-mail: guanggao@niglas.ac.cn

Qin, B. Q. (ed.), *Lake Taihu, China*,

evaluate phytoplankton phosphorus status by using alkaline phosphatase activity as the only indicator (Rhee, 1973; Wynne & Rhee, 1986).

5.1.1.1 Origin of Alkaline Phosphatase in Lakes

Dissolved extracellular enzymes in aquatic ecosystems are mainly released from microbes, microbial cell destruction or autolysis, zooplankton digestion of algal cells, and protozoans preying on bacteria. As early as the 1960s, the existence of dissolved free enzymes in freshwater lakes had been recognized (Reichardt et al., 1967), and this was subsequently confirmed by many experiments (Chróst et al., 1989). Phosphate is one of the limiting nutrients and is extremely important for primary production in freshwater. Its regeneration is mainly related to phosphatase. Therefore, the alkaline phosphatase activity of algae has received considerable attention.

Generally speaking, the alkaline phosphatase of freshwater ecosystems comes mainly from bacteria, phytoplankton, and micro- and macrozooplankton. Studies on the sources of alkaline phosphatase normally use filters of different pore sizes to separate the seston was into different size-classes, that is, bacteria and algae (Jamet et al., 1997). When this method is used, the fraction less than 0.2 μm is usually defined as dissolved, the fraction between 0.2 and 3 μm as bacteria, and the fraction larger than 3 μm as algae and organic detritus. Because the size of organisms varies greatly within communities, this classification by particle size is only approximate and does not represent a classification in the strict systematic sense. Some experiments indicate that the alkaline phosphatase activity in water can vary with the type of bacterial and algal populations present (Rivkin & Swift, 1979; Chróst et al., 1986). Sometimes, the enzyme comes mainly from bacteria (Stewart & Wetzel, 1982), and at other times mainly from algae (Berman, 1970). In addition, alkaline phosphatase can come from zooplankton itself, and not from their food (Boavida & Heath, 1983).

5.1.1.2 Characteristics of Alkaline Phosphatase

Alkaline phosphatase is composed of two subunits, each with a molecular weight of about 160,000 kDa and containing one tightly and one less tightly bound zinc atom. The former is used to maintain the integrity of the enzyme structure, and the latter is used in hydrolysis of enzymes. In each subunit, magnesium ions combine in different locations with the zinc atoms. Via their structural characteristics, magnesium ions can stimulate activity of the enzyme and accelerate its hydrolysis of the monoester of phospholipids. In contrast, the chelators of divalent ions, such as ethylenediaminetetraacetic acid (EDTA), are common inhibitors of this enzyme (Chróst, 1991). At physiological pH, the two subunits of alkaline phosphatase negatively interact; therefore, alkaline phosphatase shows only "half-of-the-site" reactivity. However in alkaline pH, alkaline phosphatase activity is consistent with Michaelis–Menten dynamics.

Alkaline phosphatase usually functions at pH above 7, with the optimum pH of 9–10. As an extraenzyme, its synthesis is dependent on phosphorus status and the biological community of the aquatic ecosystem. When phosphate concentration is low and restricts the growth of plankton in the environment, the alkaline phosphatase of bacteria and phytoplankton is induced, and its activity increases (Chróst & Overbeck, 1987); in contrast, when phosphate concentration is higher, the activity of alkaline phosphatase is inhibited.

The Michaelis constant, K_m, of alkaline phosphatase in lake water is quite small, generally 10^{-6} to 10^{-4} mol/L. The value of K_m represents the ability of the enzyme to bind to the substrate: the smaller the value, the higher the binding ability. Phytoplankton can increase the enzyme reaction speed and increase the binding ability (i.e., reduce the Michaelis–Menten constant), thus increasing the activity of the enzyme (Zhou & Zhou, 1997).

5.1.1.3 Alkaline Phosphatase Synthesis and Phosphorus

Various environment factors in aquatic systems not only influence microbial growth and metabolism but also influence their enzyme synthesis. Enzyme synthesis is determined by gene expression, and different enzymes have different adjustment and control mechanisms. The synthesis of an inducible enzyme is controlled by concentration of its metabolic product; therefore, its synthesis is determined by the concentration of the final product in the environment. The control gene signals for enzyme synthesis to begin when its final product concentration falls below some threshold in the environment. Conversely, a restraint gene signals for enzyme synthesis to stop when the final product accumulates in the cell or environment (Chróst & Overbeck, 1987). As with other inducible enzymes, alkaline phosphatase synthesis speed and activity vary with nutrient status and other environmental factors in a lake and are inhibited by the amount of hydrolyzed final product of PO_4^{3-} accumulated in the cell or environment.

Numerous publications note that high alkaline phosphatase activity corresponds with low phosphorus concentration in the environment. Therefore, it was generally assumed that alkaline phosphatase was directly induced or activated by low phosphorus concentration in the environment (Berman, 1970; Petterson, 1980); however, some experiments indicate that this hypothesis may not be correct. When, for example in Lake Plußsee in Germany, the concentration of phosphorus in the water was relatively high, microbes could store phosphorus as polyphosphate, with this fraction of phosphorus in the microbes reaching 23–34% (mean, 30%) of total particulate phosphorus. When this fraction of phosphorus decreased to 15% (corresponding to $PO_4^{3-} < 1$ μg/L), the microbes began rapid synthesis of alkaline phosphatase, and there were rapid increases in the activity of alkaline phosphatase (Chróst, 1991).

The synthesis of alkaline phosphatase in microbes probably is not directly induced by low external phosphorus concentrations. When phosphorus concentration decreases in the environment, microbes first utilize the PO_4^{3-} and polyphosphate stored in their bodies, rather than beginning to synthesize alkaline phosphatase. Pelagic microbes begin to synthesize alkaline phosphatase, which has specialized active hydrolyzed phospholipids to compensate for limited phosphate, only after the phosphorus stored inside the cell and outside the cell is almost consumed. Thus, the mechanism for induction of alkaline phosphatase production is likely controlled by phosphorus storage in the microbial cell, which is determined by the phosphorus concentration of the surrounding environment (Chróst, 1991).

When the PO_4^{3-} concentration in microbes or their surrounding environment increases to a certain extent, alkaline phosphatase activity is inhibited. This type of PO_4^{3-} competitive inhibition is mainly evident as competition with monophospholipids for active sites of the enzyme, sequentially increasing the K_m value of the enzyme, which decreases binding of enzyme and substrate, and finally slowing the enzyme reaction. This kind of competitive inhibition is reversible, and could be reversed through increased reactive substrate concentration; therefore, the maximum reaction speed of the enzyme does not change (Chróst & Overbeck, 1987). PO_4^{3-} is a strong competitive inhibitor for alkaline phosphatase in algae (Chróst & Overbeck, 1987); however, it is only a weak inhibitor for alkaline phosphatase in bacteria (Chróst, 1986).

5.1.2 Alkaline Phosphatase and the Phosphorus Cycle

In theory, the nutrients should satisfy the growth demands of phytoplankton in eutrophic waters; however, in the season of algal growth, low orthophosphate concentrations often occur in mesotrophic and eutrophic lakes (Schindler, 1977). Monthly field observation data from May 1998 to May 1999 for Lake Taihu (Fig. 5.1) have shown that even in highly eutrophic areas such as Meiliang Bay, dissolved organic phosphorus (DOP) and particulate phosphorus (sestonic P) are still the main forms of phosphorus. The concentration of dissolved inorganic phosphorus (DIP) was low, at most times less than 0.020 mg/L (in 84.6% of all samples), sometimes even below the detection threshold, and this accounted for only about 8% of total phosphorus (TP) (in 77% of all samples) (Fig. 5.1).

The nitrogen (N) to phosphorus (P) ratios were high, with only 10% of the annual N:P ratios being less than 15, and the maximum N:P ratio was 104 in April 1999. The annual mean N:P ratios all were above 15, ranging from 19 to 63, with the maximum also in April 1999 (Fig. 5.2).

The temporal and spatial distribution patterns of phosphorus in the water during this time were almost the same as those from 1997 to 2001, indicating that even in eutrophicated water phosphorus is possibly still a factor inhibiting phytoplankton growth. Especially in summer when algal blooms occur and DIP is absent, different forms of phosphorus can transform to DIP, providing a continual source of DIP for algae and becoming the key factor controlling algal blooms.

Filters with different pore sizes are used to separate the alkaline phosphatase activity (APA) in water: total (TAPA, unfiltered), dissolved (DAPA, passing through a 0.2-μm filter), and bacterial (BAPA, 0.2–3.0 μm). The phytoplankton alkaline phosphatase activity (PAPA) = TAPA − BAPA − DAPA.

Experiments have shown that alkaline phosphatase activity varies significantly among the different size fractions. The proportion of DAPA of TAPA was the high-

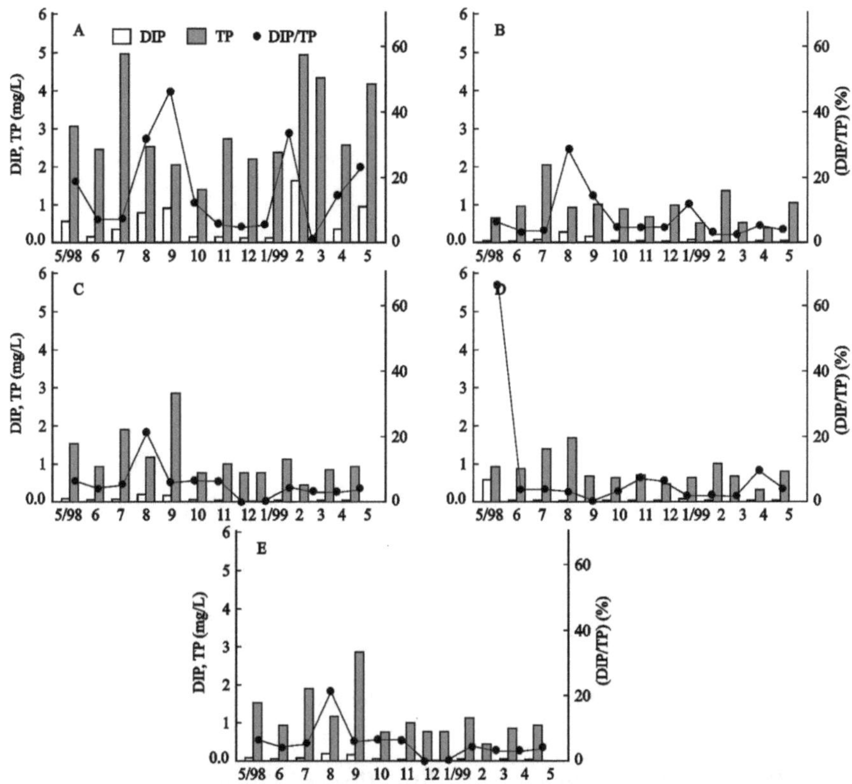

Fig. 5.1 Annual variation of DIP, TP concentrations, and DIP versus TP ratio in the water at five sites in Lake Taihu (May 1998–May 1999). *DIP*, dissolved inorganic phosphorus; *TP*, total phosphorus. (**A**) Site 0#, river mouth; (**B–D**) sites 2#, 3#, 5#, respectively, Meiliang Bay; (**E**) site 8#, open lake

est, ranging from 12% to 91% (mean, 57%); the proportion of PAPA ranged from 5% to 85% (mean, 39%); and the proportion of BAPA was only 1–12% (mean, 4%) (Fig. 5.3).

Although the proportions of DAPA, PAPA, and BAPA in TAPA varied markedly, there were no significant spatial differences between sampling sites, probably because Lake Taihu is a typical shallow lake, subject to strong wind disturbance, and thus the enzymes in the water would be fully mixed. By contrast, at each sampling site in Meiliang Bay, there were significant temporal differences in activity of DAPA, PAPA, and their proportions of TAPA [$P < 0.001$; analysis of variance (ANOVA)] and of BAPA ($P < 0.05$; ANOVA). All these parameters peaked in spring and summer (Fig. 5.3). In May, the PAPA could reach more than 80% of the total. By contrast, TAPA in the water peaked in April, but PAPA was only about 20% of TAPA.

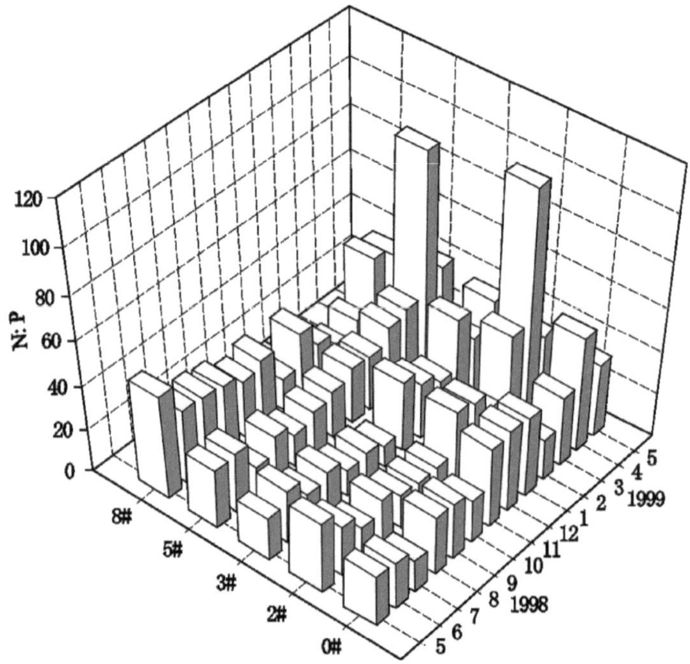

Fig. 5.2 Temporal and spatial variation in the N:P ratio for five sites in Lake Taihu

5.1.3 Relationship Between Alkaline Phosphatase and Algal Blooms

In oligotrophic and eutrophic waters, the external inorganic phosphorus concentration is usually low, especially during the summer algal bloom period, which often inhibits algal growth. At that time, the phytoplankton alkaline phosphatase could be induced, and organic phosphorus hydrolyzed and released as inorganic phosphorus for algal growth (Reichardt, 1971). In many lakes, dissolved organic phosphorus (DOP) is more than one order of magnitude higher than dissolved reactive phosphorus (DRP). Approximately 60% of the DOP could be used for algal growth after its hydrolysis by alkaline phosphatase (Hantke et al., 1996). The mechanism by which algae use organic phosphorus compounds may be the key process in algal competition.

5.1.3.1 Phosphorus Concentration Threshold in Lake Taihu

In general, when DRP concentration in the water is lower than 0.2 mg/L, alkaline phosphatase is induced in microbes (Wynne et al., 1991). In Lake Taihu, when PO_4^{3-} in water was below 0.01 mg/L, TAPA significantly increased ($P < 0.001$) (Fig. 5.4). When PO_4^{3-} was below 0.02 mg/L, PAPA significantly increased ($P < 0.05$), but when PO_4^{3-} was above 0.20 mg/L, PAPA significantly decreased

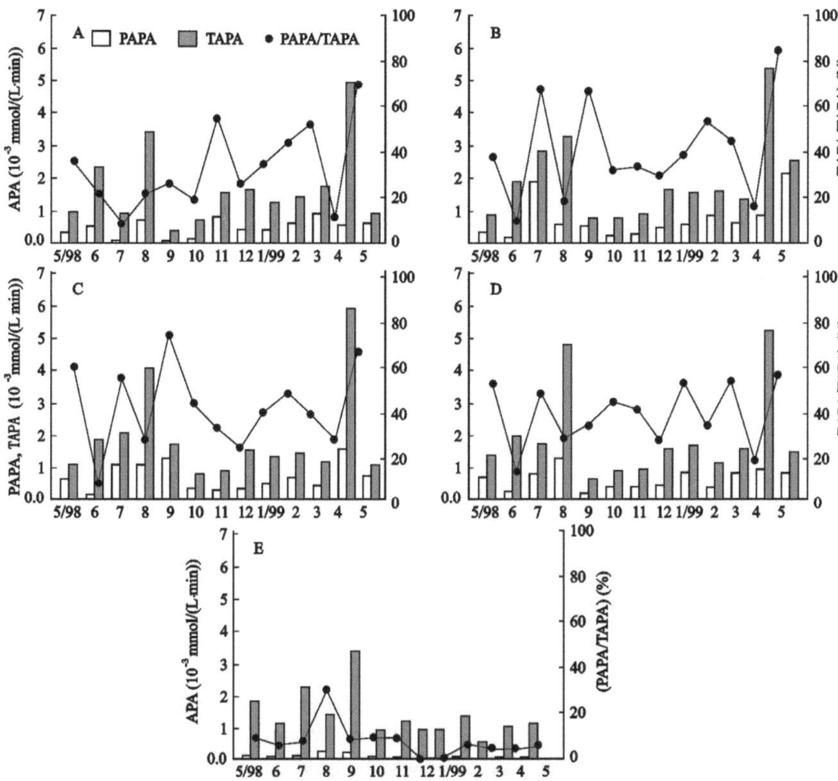

Fig. 5.3 Temporal changes of the different size fractions of alkaline phosphatase activity. *PAPA*, phytoplankton alkaline phosphatase activity; *TAPA*, total alkaline phosphatase activity. *Right axes* show the proportion of PAPA versus TAPA. (**A–E**) are same as in Fig. 5.1

($P < 0.001$). This finding indicates that, in Lake Taihu, a PO_4^{3-} concentration of 0.02 mg/L may be the threshold concentration for inducing alkaline phosphatase activity in phytoplankton (Gao et al., 2000); this is a much lower threshold than generally reported.

Experiments show that TAPA is negatively correlated with PO_4^{3-}. When PO_4^{3-} concentration was between 0 and 0.010 mg/L, TAPA was not significant different; however, when PO_4^{3-} exceeded 0.010 mg/L, TAPA was significantly reduced ($P < 0.001$). When PO_4^{3-} concentration in the water is relatively lower, alkaline phosphatase activity is increased, which would accelerate organic phosphorus decomposition and release, and the PO_4^{3-} concentration in the water would increase. With increasing PO_4^{3-} concentration, alkaline phosphatase activity would be inhibited and gradually decrease, which would reduce organic phosphorus decomposition and release. Utilization of PO_4^{3-} by bacteria and algae leads to a decrease of PO_4^{3-} concentration, after which alkaline phosphatase activity would finally gradually increase again (Fig. 5.5). During the experiment, this kind of cycle was most clearly demonstrated at low PO_4^{3-} concentration conditions (Gao et al., 2000).

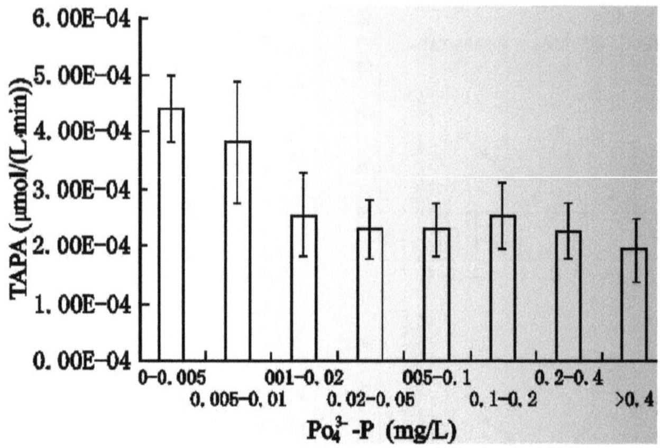

Fig. 5.4 TAPA variation at different $PO_4{}^{3-}$-P concentrations. Columns show total alkaline phosphatase activity with standard error bars

5.1.3.2 Relationship Between Phosphorus Cycling and Algal Blooms

As an essential element for algal growth, phosphorus influences almost all metabolic processes, especially photosynthesis. The optimal phosphorus concentration for algal growth in culture is much higher than in natural conditions. For *Microcystis aeruginosa*, which is predominant in Lake Taihu, the minimum phosphorus concentration for growth is about 0.5 $\mu g/m^3$ cell volume (Wetzel, 2001).

In lake water, the concentration of DIP, which could be used directly by algae, is usually lower than that, and 90% of TP is organic phosphorus. Organic phosphorus or polyinorganic phosphorus could be used as alternative phosphorus sources by algae. For this type of utilization, the algae either accelerate the synthesis of alkaline phosphatase or remove its inhibition (Berman, 1970; Jones, 1972).

This kind of organic phosphorus transformation to inorganic phosphorus is usually rapid, especially in summer, when there is high demand and low external load import, and recycling time can be as brief as 5–100 min (Wetzel, 2001).

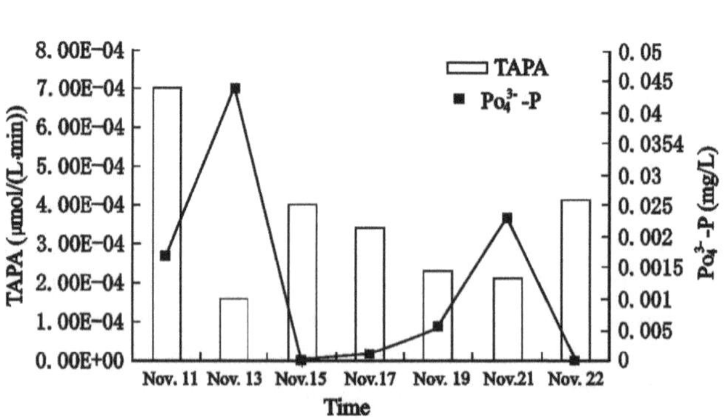

Fig. 5.5 Changes in TAPA relative to $PO_4{}^{3-}$-P concentration over time

During the period of summer algal blooms, the phosphate concentration in the water is far below the alkaline phosphatase induction threshold. The enzymes in the algae are induced, causing greatly increased alkaline phosphatase production and activity. As a consequence, other forms of phosphorus in the water continually decompose, and DIP is released, compensating for the PO_4^{3-} loss in the algal bloom. Phospholipid decomposition through the action of alkaline phosphatase is rapid, exceeding 0.05 mmol/h (Wetzel, 2001). This rapid decomposition of organic phosphorus leads to the release of phosphate, which supports algal growth. Especially when the external phosphorus load is reduced, this internal phosphate cycle becomes the key factor for the supply of phosphorus to algal blooms.

5.2 Relationship Between the Lake Taihu Microecosystem Structure and Algal Blooms

Guang Gao and Liuyan Yang

5.2.1 Microbial Food Webs, Organic Particles, and Algal Blooms

The food web, which is one of the fundamental and frequently used concepts in ecology, describes the complicated nutrient and trophic relationships among communities. The microbial food web model, put forward in the 1980s (Azam et al., 1983; Sherr & Sherr, 1988), considered nutrient and energy flow pathways in the microbial community and the trophic relationships among various microbes, for example, commensalism [phytoplankton produce dissolved organic matter (DOM) and bacteria consume DOM], competition (phytoplankton and bacteria compete for nutrients), and feeding (flagellates, ciliates, and microzooplankton provide nutrient and DOM feedback). In aquatic ecosystems, most of the dissolved organic carbon in phytoplankton metabolism can be used by larger consumers (such as zooplankton) via the microbial food web and retained in the pelagic zone for recycling. The microbial food web is an important concept and an area of focus in current freshwater ecology (Havens, 2001; Pomeroy, 2001).

5.2.1.1 Microbial Food Web in Lake Taihu

In the lake, the main components of the microbial food web are picoplankton (bacteria, algae, and microeukaryotes), heterotrophic or mixotrophic flagellates, and ciliates (Stockner & Porter, 1988). A simplified microbial food web model, nutrient relationships among its components, and release and use of organic carbon are shown in Fig. 5.6.

(1) Heterotrophic bacteria

The important role that heterotrophic bacteria play in the recycling of nutrients such as nitrogen and phosphorus is well recognized. The abundance, biomass, production, and respiration rates of pelagic bacteria in natural waters are highly variable, depending on primary production and the characteristics of the habitats.

In general, 15–25% of bacterial cell wet weight is organic carbon, 0.6–0.8% is phosphorus, and 2–4% is nitrogen. Pelagic bacteria in aquatic ecosystems are usually 0.2–1.0 μm in size, with a cell volume of 0.01–0.3 μm^3, although some cells exceed these dimensions (Porter et al., 1988). Cell volume of pelagic bacteria is markedly variable, depending on the species and the nutrient status. For example, the cell volume of small cocci is usually 0.06–0.08 μm^3 and that of medium bacilli

Gao, G.
State Key Laboratory of Lake Science and Environment, Nanjing Institute of Geography and Limnology, Chinese Academy of Sciences, 73 East Beijing Road, Nanjing 210008, P. R. China
e-mail: guanggao@niglas.ac.cn

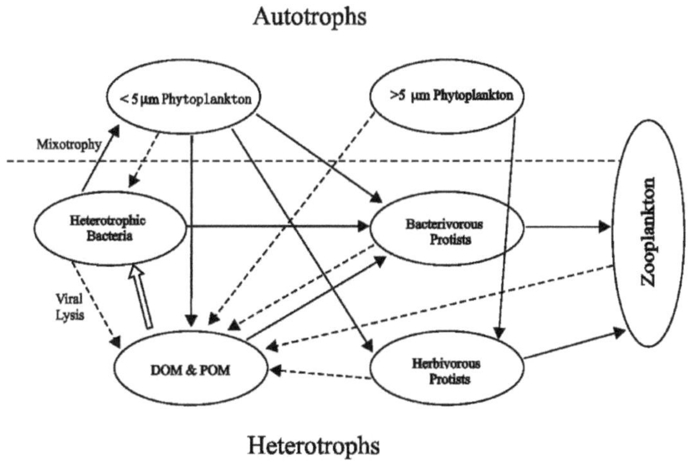

Fig. 5.6 The microbial food web model, nutrient relationships among components, and organic carbon release and use. *DOM*, dissolved organic matter; *POM*, particulate organic matter. (From Sherr & Sherr, 2000)

0.1–0.2 μm^3; however, a large bacillus could have a cell volume of 0.5–2.0 μm^3. In medium to highly productive waters, bacterial cell volume generally is in the range of 0.08–0.15 μm^3. In oligotrophic seawater, it is less than 0.1 μm^3, and in eutrophic waters, it sometimes exceeds 0.2–0.25 μm^3 (Sorokin, 1999).

As a large, shallow, eutrophic lake, Taihu exhibits great temporal and spatial differences in the distribution of heterotrophic bacteria biomass and production. This characterization has been investigated by using 4-6-diamindino-2-phenylindole (DAPI), direct counting through fluorescence microscopy (Porter & Feig, 1980), measuring bacterial volume in a LUCIA image analysis system, and applying the formula $CC = 218 \times V^{0.86}$ to convert bacteria cell volume to biomass (Posch & Arndt, 1996). The temporal and spatial distributions of pelagic bacteria abundance in different parts of Lake Taihu are shown in Fig. 5.7

There was a significant difference in bacterial abundance at three sites (one-way ANOVA, $P = 0.005$). At the river mouth, the annual mean value of bacteria abundance was 15.4 × 10^9 cell/L (range, 4.3–34.8 × 10^9 cell/L), which was 54.3% and 96.8% higher than that in Meiliang Bay (9.99 × 10^9 cell/L; range, 3.6–15.4 × 10^9 cell/L), and the open lake (7.83 × 10^9 cell/L; range, 2.7–16.5 × 10^9 cell/L), respectively.

The heterotrophic bacteria cell size also varied significantly in different parts of the lake (one-way ANOVA, $P < 0.0001$). Although the cell volume was generally small, in the river mouth, volume (mean, 0.080 μm^3; range, 0.004–3.596 μm^3) was about 70% larger than those in Meiliang Bay (mean, 0.047 μm^3; range, 0.003–2.964 μm^3) and in the open part of the lake (mean, 0.046 μm^3; range, 0.004–1.580 μm^3) (Fig. 5.8).

The carbon content of the bacterial cells also varied significantly among different parts of Lake Taihu (one-way ANOVA, $P < 0.0001$). In the river mouth, mean

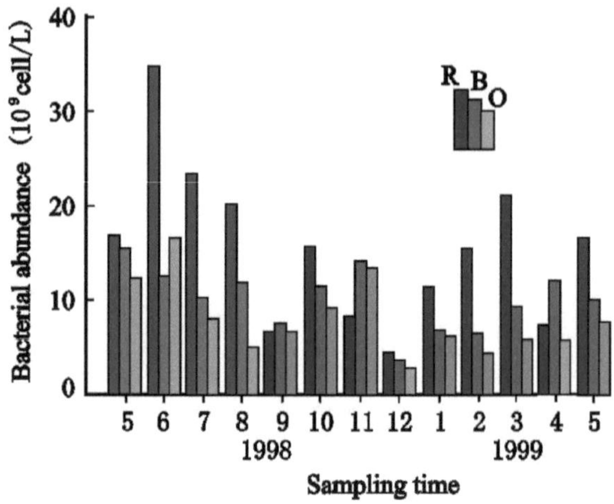

Fig. 5.7 The temporal and spatial distributions of bacterial abundance at three sites in Lake Taihu. *R*, *B*, *O*, river mouth, Meiliang Bay, and open lake, respectively

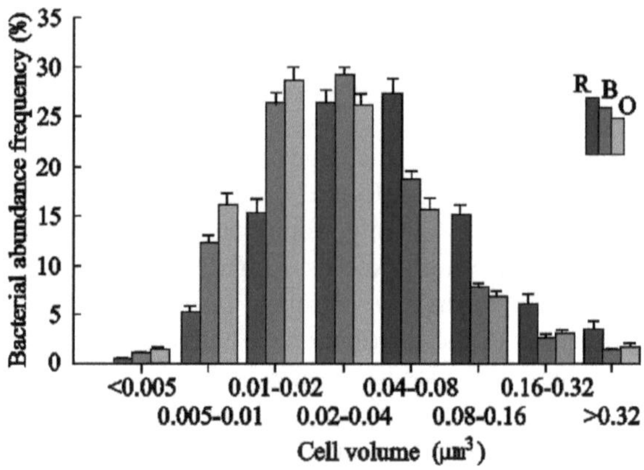

Fig. 5.8 The distribution of bacterial cell volume at three sites in Lake Taihu (mean ± SD). *R*, *B*, *O*, river mouth, Meiliang Bay, and open lake, respectively

carbon content was 23.4 ± 5.7 fg/cell (range, 17.6–40.0 fg/cell), which was about 60% higher than in Meiliang Bay (14.9 ± 2.1 fg/cell; range, 12.0–19.4 fg/cell) and in the open lake (14.5 ± 2.8 fg/cell; range, 10.3–19.6 fg/cell) (Fig. 5.9). Although there were some seasonal fluctuations, this variation was not statistically significant.

Bacterial biomass in different parts of Lake Taihu could be calculated from bacterial abundance and cell carbon content. Bacterial biomass was quite high, reaching 803.6 μg C/L at the river mouth (Fig. 5.10). Its distribution showed significant spatial differences (one-way ANOVA, $P < 0.0001$). In the river mouth, bacterial total biomass averaged 363.9 ± 224.5 μg C/L (mean ± SD), which was 1.5 and 2.5 times higher than in Meiliang Bay (148.1 ± 51.3 μg C/L) and in the open lake (113.1 ± 62.2 μg C/L), respectively.

The carbon cycle in the aquatic ecosystem is dominated by bacterial carbon production, which can contribute 20–40% of phytoplankton primary production in the water column of freshwater and ocean ecosystems (Ducklow et al., 1993). Through

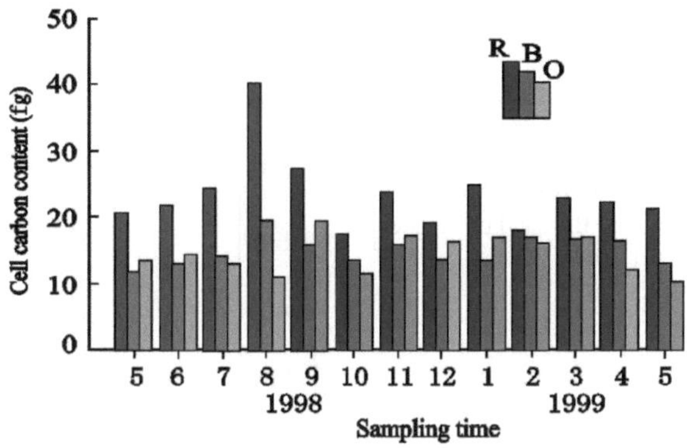

Fig. 5.9 The temporal and spatial distributions of bacterial carbon content in three sites in Lake Taihu. *R, B, O*, river mouth, Meiliang Bay, and open lake, respectively

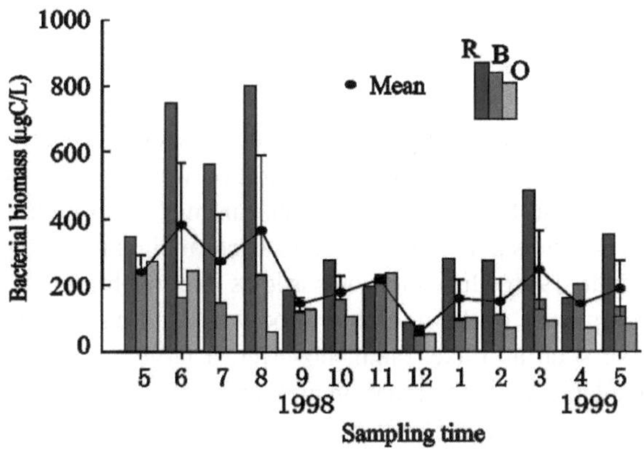

Fig. 5.10 The temporal and spatial distributions of bacterial biomass in three sites in Lake Taihu. *R, B, O*, river mouth, Meiliang Bay, and open lake, respectively

measurement of bacterial production, the bacterial biomass that could be transferred to a higher trophic level by predation can be determined. Bacterial production also is used to estimate the mean growth rate of the bacteria population and is an index for responses to external environmental fluctuations (Looij & Riemann, 1993).

In Lake Taihu, with abundant organic matter in water, heterotrophic bacteria usually have a high productivity, reaching 500–800 mg/($m^3 \cdot$ day) (wet weight), or about 100–200 mg C/($m^3 \cdot$ day) (Sorokin, 1999).

Bacterial production in different parts of the lake was measured by [^{3}H]thymidine and [^{14}C]leucine, as shown in Fig. 5.11.

Heterotrophic bacterial production in Lake Taihu is relatively high and shows significant temporal and spatial variation (paired t test, $P < 0.05$). In the river mouth, with a maximum of 1044.1 μg C/(L · h), the mean bacterial production was 292.5 ± 82.4 μg C/(L · h), which was 1.1 and 2.8 times higher than in Meiliang Bay [141.6 ± 40.1 μg C/(L · h)] and in the open lake [76.5 ± 11.3 μg C/(L · h)], respectively.

(2) Flagellates and ciliates

There are numerous pelagic flagellates and ciliates in Lake Taihu. These organisms feed on picoplankton and bacterioplankton, and also on microphytoplankton

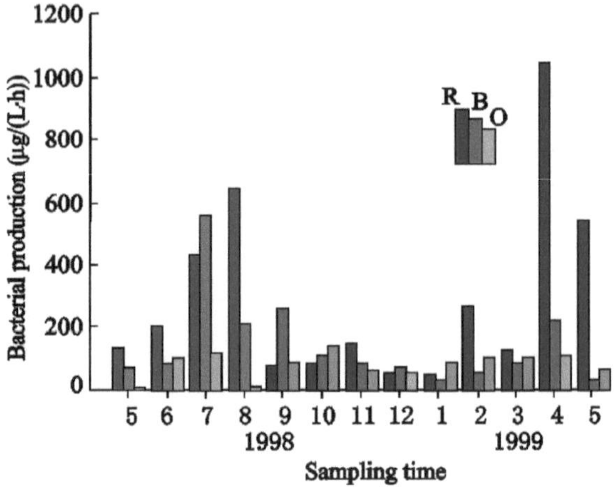

Fig. 5.11 Temporal and spatial variation in heterotrophic bacterial production in three different areas of Lake Taihu. *R*, *B*, *O*, river mouth, Meiliang Bay, and open lake, respectively

that may be even larger than the flagellate or ciliate cell. These organisms are important heterotrophs in the microbial food web (Sherr & Sherr, 1987).

Recent data show that ciliates can ingest sub-micrometer- and micrometer-level fine detritus particles, except for bacterioplankton and picophytoplankton (Posch & Arndt, 1996). Ingestion of particles over such a great range of sizes increases the complexity of the microbial food webs and particle decomposition. In a eutrophic lake such as Lake Taihu, however, a large amount of external organic matter is available, and the predation pressure in the microbial food web is mainly from zooplankton such as *Daphnia*. The nutrient recycling attributable to feeding by ciliates and flagellates was greatly reduced (Carney & Elser, 1990).

5.2.1.2 Relationship Between Algal Blooms and Organic Matter Metabolized in the Microbial Food Web

(1) Suspended particulate organic matter in the lake

In aquatic ecosystems, particulate organic matter (POM) usually appears as macroscopic aggregates. The macroscopic organic aggregations are detritus communities formed by algae and large amounts of bacteria and Protozoa. Such communities are important food sources for fish and zooplankton and provide unique chemical environments for microbes. Such aggregates have a relatively large volume and high sedimentation rate and form the main part of the downward flux of matter ("detritus energy pool"). At the same time, plenty of inorganic and organic nutrients also are included in these organic aggregates (Shanks & Trent, 1979; Herndl, 1992), which serve as carriers for POM vertical flux and can be used by microbes. Therefore, these macroscopic organic aggregates are extremely important in metabolic processes such as organic production, decomposition, and nutrient recycling.

(2) Formation and decomposition of suspended organic particles in the lake

Many types of dead organic matter in particulate or dissolved states form the detritus energy pool and can be produced externally or within the lake. The energy stored in this pool generally can be used directly by zooplankton and benthos or through the actions of microbes.

Organic aggregates in aquatic ecosystems are usually formed by a series of physical, chemical, and biological process. The complicated aggregation process is usually controlled by the following parameters: (1) concentration, density, size, and shape of particles in the water; (2) shear stress and sedimentation rate; and (3) particle agglutination after collision. Numerous experiments confirm that aquatic microbes play an important role in the process of organic aggregation (Biddanda, 1985; Alldredge et al., 1993).

Usually, part of the bacteria population (10–40%) exists in bacterial aggregations (5–20 μm). These aggregations are formed as a consequence of fast growth and do not need external particles as their core. Some bacterial plankton can also form aggregations around particles, such as dead phytoplankton cells and zooplankton faecal pellets. Some aggregations are covered by the mucus produced by bacteria and other plankton and form microscopic organic aggregates ($<$ 5–500 μm) or large-scale macroscopic organic aggregates ($>$ 500 μm). This kind of large-scale, congregated bacterial population is important to plankton population nutrient dynamics in aquatic ecosystems (Sorokin, 1999; Simon et al., 2002).

These detritus particles (organic aggregates) are all active centers of microbial activity in the water and on the sediment surface. In recently formed detritus particles, the microbial aggregates metabolize the organic matter in the particle; however, in the completely formed particles, microbial aggregates absorb DOM through the detritus-absorbing effect. Direct microscopic counts show that the microbes assembled in such organic detritus can account for more than 3% of total organic aggregation. Thus, total bacteria could reach 5–25 $\times$ 10^9 cells/g detritus (gross weight). Large quantities of microbes are present in organic detritus; for example, in 1 g (net weight) of organic detritus from decomposed diatoms, there can be 3 $\times$ 10^9 bacteria, 20 $\times$ 10^6 diatoms, 6 $\times$ 10^6 flagellates, 5 $\times$ 10^4 ciliates, and numerous Metazoa. Therefore, although the organic detritus is classified as "dead" organic matter, it is, in fact, an active part of the metabolism in aquatic systems (Sorokin, 1999).

(3) Rate of decomposition of organic matter in the lake

Organic matter decomposition is an important function in aquatic ecosystems. In general, there are three routes of organic matter degradation in lake water: ingestion, filter feeding, and absorption of DOM. Because bacterial enzymes can decompose dissolved organic macromolecules, and bacterial cells have an extremely high surface area to volume ratio, some bacteria can use organic matter concentrations as low as micromolar to nanomolar; therefore, bacterial activity strongly influences the concentration and composition of DOM.

The rate of organic decomposition in aquatic ecosystems is equal to the total respiration losses of all organisms. Because bacterial respiration is much higher than that of protozoa and zooplankton, heterotrophic bacteria are usually regarded as the main decomposers of organic matter (Sorokin, 1999).

The amount of material assimilated by bacteria can be estimated according to this formula (Sorokin, 1999):

$$A = P + Mb \tag{5.1}$$

where A is assimilated organic matter, P is bacterial production, and Mb is bacterial respiration rate.

Therefore, organic matter decomposed by bacteria:

$$(D) = Mb \tag{5.2}$$

$$P/A = K2 \tag{5.3}$$

K2 as the efficiency of bacterial assimilation of organic matter for its own growth.

For cultured and natural pelagic bacteria, K2 measured by isotopes has values of 0.4–0.5; however, the K2 value is apparently lower, about 0.25–0.35 (mean, 0.32), when there is external organic input.

The following formula can be derived from formulae (5.1) and (5.3) taken into (5.2), and by using 0.32 as the mean value for K2 in natural water:

$$D = Mb = P(1 - K2)/K2 = 0.68/0.32 \times P = 2.125 \times P \tag{5.4}$$

The rate of decomposition was calculated for three different parts of Lake Taihu (Table 5.1) by use of formula (5.4) and the bacterial production shown in Fig. 5.11. In the river mouth, which has much external organic matter input, the bacterial production and organic decomposition were much higher than other areas.

Table 5.1 Bacterial production and organic decomposition in three different parts of Lake Taihu

	Bacterial production (μg C/(L h))		Organic decomposition (μg C/(L h))	
	Mean ± SE	Range	Mean ± SE	Range
River mouth	292.5 ± 82.4	45.1–1044.1	621.6 ± 175.1	95.8–2218.7
Meiliang Bay	141.6 ± 40.1	26.2–555.9	300.9 ± 85.2	55.7–1181.3
Open lake	76.5 ± 11.3	1.1–138.7	162.6 ± 24.0	2.3–294.7

(4) Microbial food web and metabolism of Microcystis blooms

There is mucilaginous glue on the surface of *Microcystis* cells, and when algal blooms occur, large colonies (> 100 μm in diameter) commonly form. Attached to these colonies are usually numerous heterotrophic bacteria. These attached bacteria live in a microenvironment differing from that of the surrounding water (Worm & Søndergaard, 1998) because they are influenced by the members of the *Microcystis* colony and by the metabolism of other autotrophic and heterotrophic microbes.

In this microenvironment, the attached heterotrophic bacteria are closely connected to *Microcystis* metabolism. *Microcystis* provides an attachment place and organic carbon for the heterotrophic bacteria, which have extremely high metabolism. Through bacterial degradation, carbon and other nutrients are provided for the *Microcystis*. This metabolic association between these autotrophic and heterotrophic organisms could represent a shortcut of the nutrient cycle that not only increases the amounts of nutrients and energy that recycle but also increases the latent activity of whole microbes (Herndl, 1988).

In a pure *Microcystis* colony, the pH in the mucilage glue would increase as a consequence of photosynthesis (Richardson & Stolzenbach, 1995); however, in an algal bloom colony, the respiration of commensal heterotrophic bacteria could balance this chemical change, for example, O_2, CO_2, and pH produced by photosynthesis.

The vesicle of the *Microcystis* cell lets the colony float in water, preventing its sedimentation, and it also increases nutrient exchange between the colony and the surrounding water (Worm & Søndergaard, 1998).

The cells of the colony have one of three fates: (1) to disperse and produce a large amount of soluble organic matter and algal cells; (2) to congregate and form algal blooms with other microbes; or (3) to die and sink to the sediment. DOM is produced through these processes, which support bacterial growth (Becquevort et al., 1998). At present, data on the dynamics of attached and free-living bacteria indicate that phytoplankton, through the effects of microbes, form aggregations that provide highly dynamic and favorable microenvironments for bacteria. Availability of high-quality nutrients and easily decomposed matter stimulates the free-living bacteria and particle consumers (Herndl, 1988; Chróst, 1991).

5.2.2 Growth Characteristics of Bacteria Attached to Algae

Abundant bacteria are attached to decaying algal colonies; however, how they grow and how they affect degradation of algal blooms is still unknown (Becquevort et al., 1998), despite the extremely important ecological and biogeochemical roles such bacteria play. The close metabolic relationship between algal blooms and heterotrophic bacteria increases mineralization of organic matter in the water and transfers organic matter to a higher trophic level through the microbial food web, which allows more organic matter to be recycled and greatly reduces the sedimentation of organic matter.

During the growth of *Microcystis*, bacteria are often attached in its mucilage (Zhou et al., 1998) because *Microcystis* provides a suitable microenvironment for attached bacteria (Brunberg, 1999) and also provides organic matter for bacterial growth (Worm and Søndergaard, 1998). Meanwhile, the bacteria produce some nutrients and growth factors for *Microcystis*, but at the same time also compete for nutrients with *Microcystis* and can decompose it (Caiola, 1991; Zhao & Liu, 1996; Lian et al., 1999). The form of bacteria attached to *Microcystis* mucilage is similar to that of *Vibro* spp. associated with decomposed *Microcystis* (Caiola, 1991; Liu et al., 2000). During summer when *Microcystis* blooms, numerous cells die rapidly, and more bacteria than live cells surround the dead algal cells (Gu et al., 2000). It is assumed that the production of such large numbers of bacteria is related to death and decomposition of *Microcystis*. Furthermore, the adhesion of the attached bacteria to healthy blue-green algal cells is assumed to be benign, and possibly beneficial. The close relationship between blue-green algae and bacteria requires attention to the important effects of bacteria attached to blue-green algae when seeking to deal with eutrophication.

5.2.2.1 Isolation and Identification of Algae-Attached Bacteria in Lake Taihu

An important part of aquatic ecosystems are the abundant bacteria attached to the mucilage of algae. Four isolated strains of bacteria may have some relationship with growth of *Microcystis*. By comparisons of bacterial colony and individual shapes, staining characteristics, physiological and biochemical reactions, and molar ratio of cytosine and guanine in nuclcic acid, identifications were made according to Bergey's bacteria identification manual (*Bergey's Manual of Determinative Bacteriology*, 8th edition). Four strains have been recognized: three strains (D, X, Y2) are *Pseudomonas* sp. (Fig. 5.12), and another one (H) is a *Bacillus* sp. Reim et al. (1974) have reported that *Bacillus* sp. can lyse algae.

5.2.2.2 Bacillus sp.

To study the relationship between bacteria and growth of *Microcystis*, several variables such as temperature and concentrations of C, N, and P were selected as controlling factors. *Bacillus* sp. H, which needed growth factors, was studied, and a group of four factors, three levels, an orthogonal test (Table 5.2), and a single factor test were set up (Table 5.3). The nutritional relationship of the attached bacteria and their environment was analyzed.

Increasing temperature and concentrations of ammonia (N), organic carbon (C), and phosphate (P) are all favorable for growth of bacilli. Variance analysis showed that the influence of organic carbon was highly significant, and temperature was also significant; however, there were no significant effects of phosphate or ammonia. The

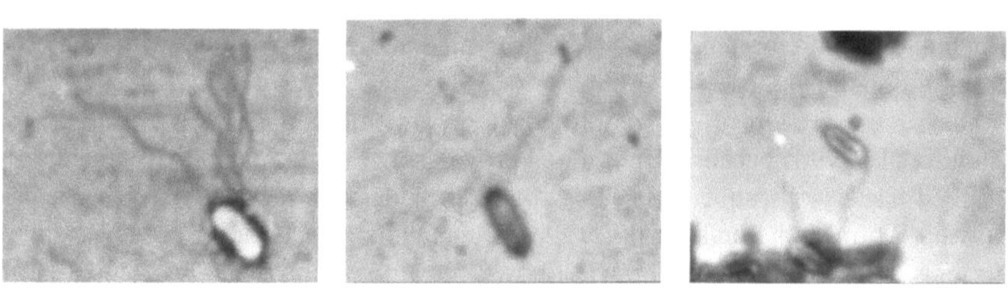

Fig. 5.12 The forms of three strains of *Pseudomonas* bacteria that attach to algae

Table 5.2 Factor levels in orthogonal test of attached bacteria growth and environmental factors

Factor level	Temperature (°C)	Ammonia (g/L)	Organic carbon (g/L)	Phosphate (g/L)
1	25	0.13	0.175	1.14
2	30	0.26	0.350	2.28
3	35	0.39	0.525	3.42

Table 5.3 Factor level in single-factor test of attached bacteria growth and environmental factors

Factor level	Temperature (°C)	Organic carbon (g/L)	Phosphate (g/L)	Nitrogen (g/L)	
				NH_4^+-N	Peptone-N
1	10	0.19	0.088	0.13	0.5
2	20	0.38	0.175	0.26	1.0
3	25	0.76	0.262	0.52	2.0
4	30	1.52	0.350	—	—
5	35	2.28	0.525	—	—

experimental environment most favorable for growth of *Bacillus* sp. H was temperature 25°–35 °C, glucose concentration 2.28 g/L, phosphorus 0.088 g/L, nitrogen (ammonia) 0.52 g/L, and peptone nitrogen 2.0 g/L.

Growth is shown by increasing optical density of the culture (a parameter measuring population density). Increased temperature reduced the delay and logarithmic phase at 25°–35 °C (Fig. 5.13), and increased organic carbon concentration prolonged the logarithmic growth phase (Fig. 5.14). Decreased phosphate concentration also reduced the delay phase and extended the logarithmic phase. However, phosphorus concentrations of 0.262–0.350 g/L supported stable growth of *Bacillus* sp. H (Fig. 5.15). Increased nitrogen concentration enhanced growth; at higher NH_4^+-N concentrations, the growth rate in the logarithmic phase was greater; at higher peptone-N content, the logarithmic phase was longer; and increasing the organic nitrogen concentration prolonged the logarithmic and delay phases (Fig. 5.16).

The growth of *Bacillus* sp. H is enhanced by an increase in either the ratio of C:N or that of N:P. Some publications show a ratio of N:P close to 7 in phytoplankton. Therefore, an N:P value higher or lower, meaning a lack of P or N, would limit algal growth. It is generally thought by limnologists that phosphorus is the main limiting factor for net organic carbon production in lakes (Qin & Zou, 1997). It is

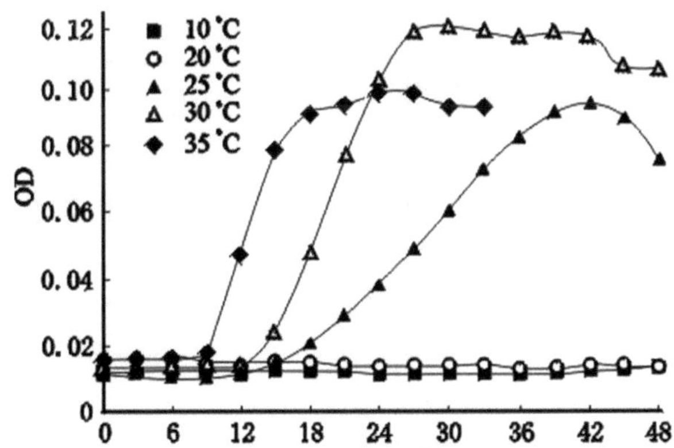

Fig. 5.13 Effect of temperature on growth of *Bacillus* sp. H. *OD*, optical density

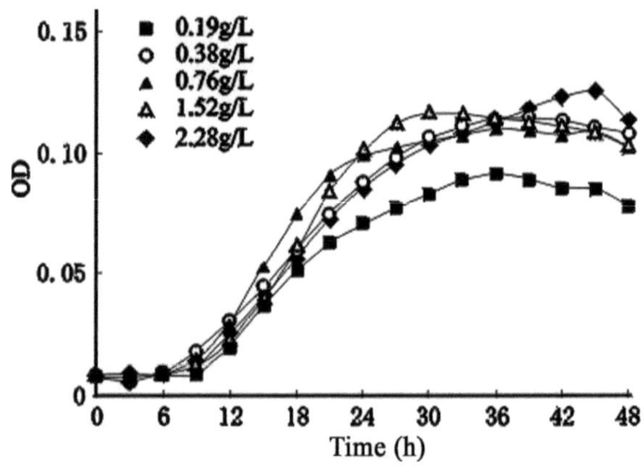

Fig. 5.14 Effect of organic carbon concentration on growth of *Bacillus* sp. H. *OD*, optical density

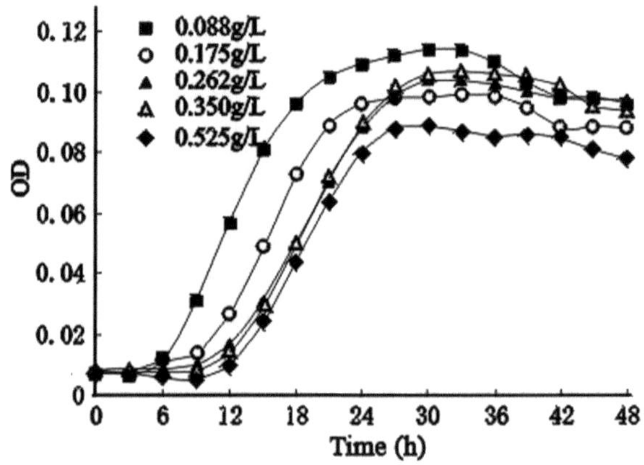

Fig. 5.15 Effect of phosphate concentration on growth of *Bacillus* sp. *OD*, optical density

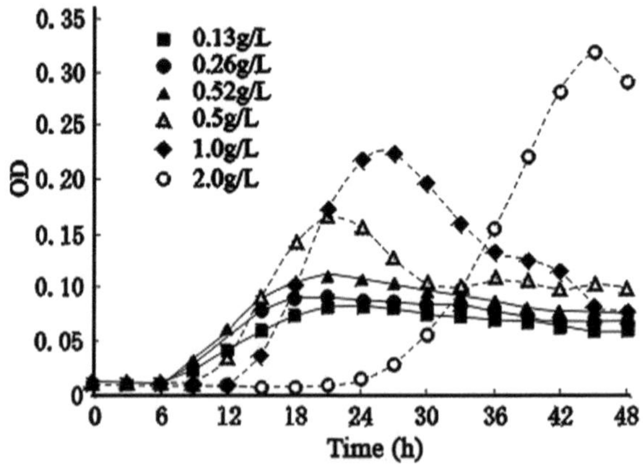

Fig. 5.16 Effect of nitrogen concentration and source on growth of *Bacillus* sp. H. *Solid lines*, ammonia; *dashed lines*, peptone-nitrogen; *OD*, optical density

assumed that when algae grow, phosphate concentration decreases, which can slow down algal growth. At that time, algae provide usable carbon and nitrogen for the attached bacteria.

5.2.2.3 Pseudomonas sp. X

Possible biological control of eutrophication requires sound understanding of phosphorus metabolism in lake ecosystems, especially that of algae and bacteria and the relationship between them. Phosphorus is the key element in eutrophication, and lowering the phosphorus content is an important step in reducing it. The influence of the attached bacteria on phosphorus release and transfer in *Microcystis* was studied using isotope tracer methods (Zhou et al., 1998; Gu et al., 2000; Liu et al., 2000). These authors also studied the influence of acetate on growth of the attached bacteria. The relationship between the *Microcystis* and its attached bacteria can be demonstrated by measuring the bacterial utilization of the *Microcystis* metabolic products.

Dakhama et al. (1993) reported that *Pseudomonas* sp. can dissolve algae; thus, study of the behavior of attached *Pseudomonas* sp. X can provide important information for control of algae.

(1) Absorption and utilization of different forms of phosphorus by Pseudomonas sp. X

The different forms and amounts of phosphorus, and the speed of their transformation, are important in determining the structure and function of aquatic communities. In the lake, the orthophosphate can be directly utilized by organisms, whereas other forms such as suspended and soluble organic phosphorus are potential sources for use by organisms (Wu et al., 1997). The role of different forms of phosphorus in the metabolism of Lake Taihu is shown in Fig. 5.17. Total phosphorus included different forms of phosphorus in the culture medium, which can be separated as soluble phosphorus and particulate phosphorus, or active phosphorus, acid water-soluble phosphorus, and organic phosphorus. Total organic phosphorus was total phosphorus minus the sum of total acid water-soluble phosphorus and active phosphorus. Particulate phosphorus was determined as total phosphorus minus total soluble phosphorus.

Most soluble phosphorus is absorbed and quickly used by bacteria (Fig. 5.18), but organic phosphorus was not taken up during the first 7 h in culture. If both inorganic and organic phosphorus are available, the former is used preferentially

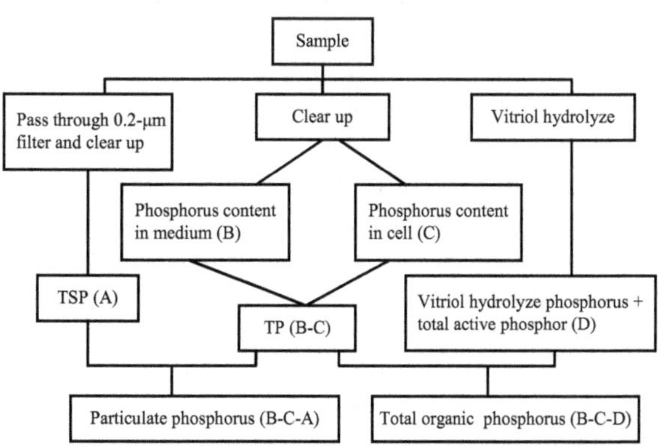

Fig. 5.17 The protocol of different forms of phosphorus analysis

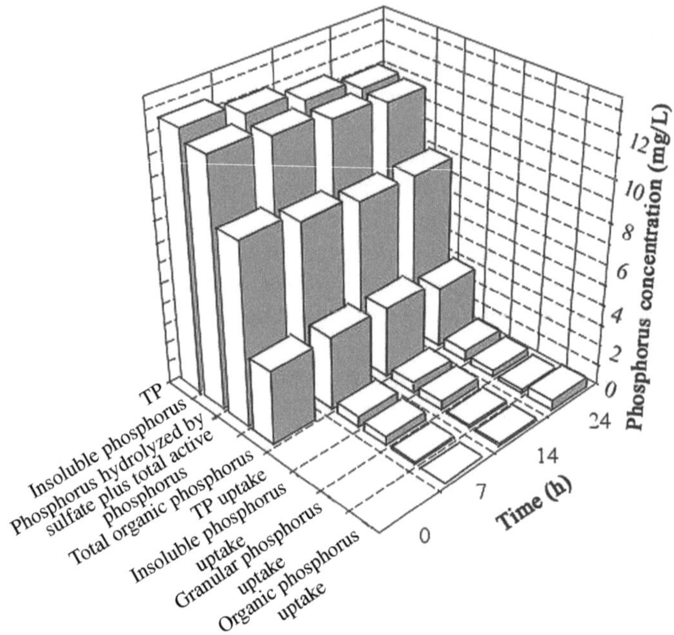

Fig. 5.18 Absorption and utilization of different forms of phosphorus by *Pseudomonas* sp. X in culture

by *Pseudomonas* sp. X. Particulate phosphorus was positively correlated with algal growth. It could be absorbed by algae directly and transformed to other forms that are easier to absorb (Wu et al., 1997). Particulate phosphorus could be used by *Pseudomonas* sp. X if calcium phosphate was added to the culture medium (Fig. 5.18).

(2) Influence of different forms and concentrations of phosphorus on growth of Pseudomonas sp. X.

Three different forms of phosphorus can be used by bacteria: inorganic phosphate, organic lecithoid, and lecithoid. Our results show that the concentration of inorganic phosphorus has little effect on growth of *Pseudomonas* sp. X (Fig. 5.19), and the bacteria can grow well in low concentrations of inorganic phosphorus. *Pseudomonas* sp. X can use adenosine triphosphate (ATP) as a source of organic phosphorus; however, at concentrations of ATP $= 1,000\ \mu g/L$ and $2,500\ \mu g/L$, growth was inhibited (Fig. 5.20). This inhibitory effect of ATP on the bacteria decreased with the decrease of the energy source in the culture medium (Fig. 5.21).

(3) Influence of the dissociation of attached Pseudomonas sp. X on phosphorus release in Microcystis

Dissociation of attached *Pseudomonas* sp. X can impact total phosphorus release from *Microcystis* and affect its growth (Fig. 5.22), and the effects vary according to the different growth phases of the bacteria-bearing *Microcystis*. In the lag and logarithmic phases, this bacteria restricts phosphorus release of *Microcystis*, which

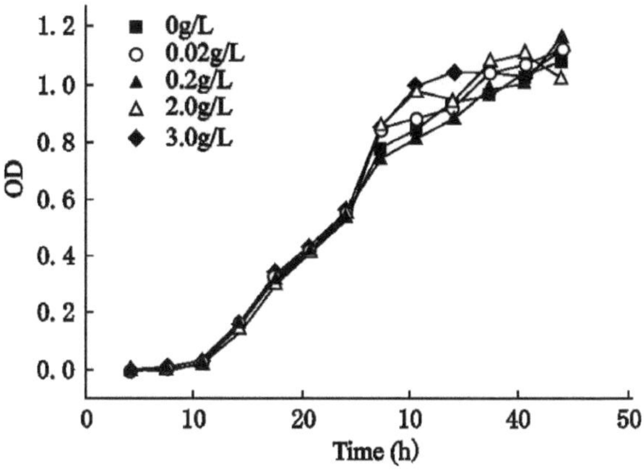

Fig. 5.19 Growth curve for *Pseudomonas* sp. X with inorganic phosphorus as the phosphorus source ($K_2HPO_4 : KH_2PO_4 = 1:1$). *OD*, optical density

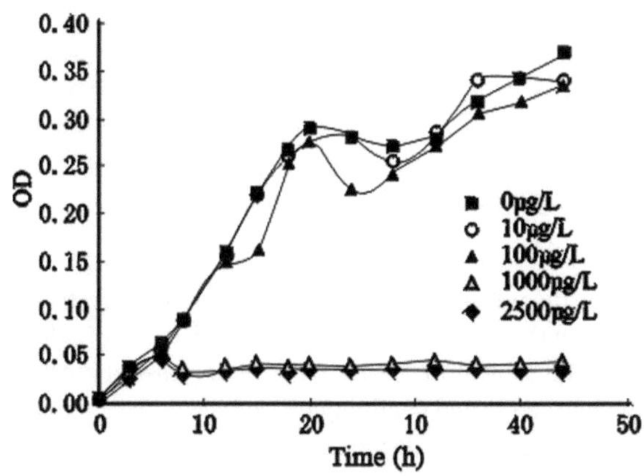

Fig. 5.20 Growth curve for *Pseudomonas* sp. X with ATP as the organic phosphorus source. *OD*, optical density

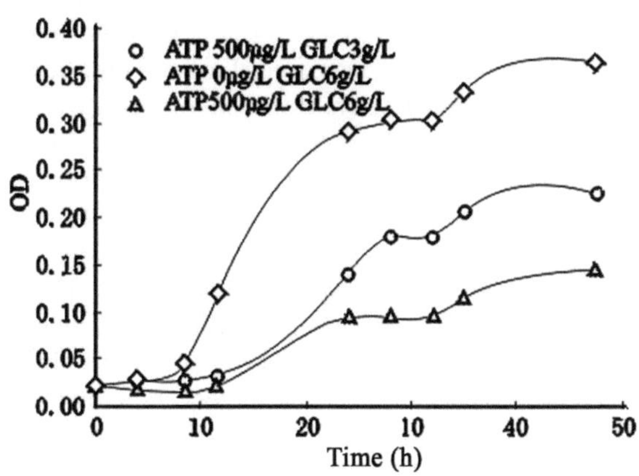

Fig. 5.21 Growth curve for *Pseudomonas* sp. in different concentrations and combinations of ATP and glucose. *OD*, optical density

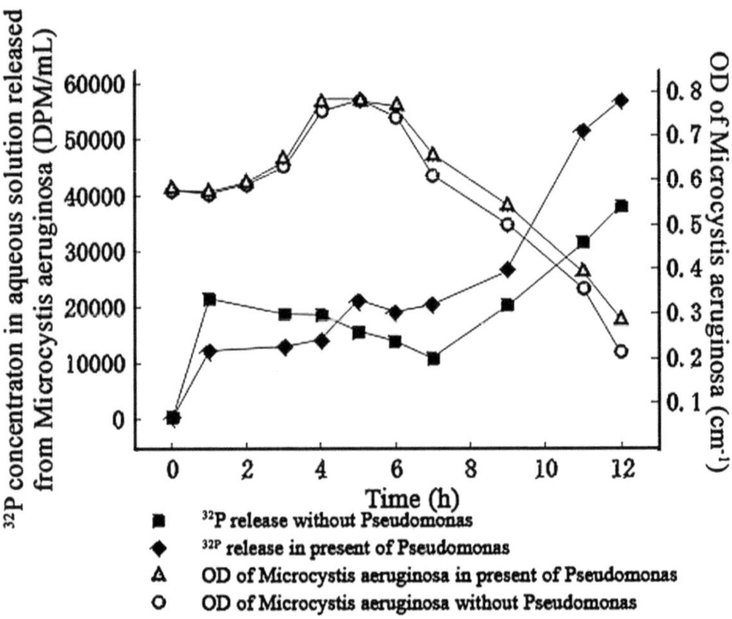

Fig. 5.22 Effect of attached *Pseudomonas* sp. X on release of ^{32}P and growth of *Microcystis*. DPM, dissolved particulate matter; *OD*, optical density

keeps the phosphorus concentration in *Microcystis* at a certain level and accelerates its growth. Thus, in this phase, the existence of adjacent *Pseudomonas* sp. X could enhance growth of *Microcystis*. In contrast, the presence of *Pseudomonas* sp. X in the late stable and declining phases accelerates phosphorus release from *Microcystis* and accelerates its decline.

Pseudomonas sp. X influences the distribution of phosphorus released from *Microcystis*. Studies show that after adding *Pseudomonas*, there is little fluctuation of phosphorus in the water (Fig. 5.23), a situation similar to that when *Pseudomonas* is not present (see Fig. 5.22). Meanwhile, phosphorus concentration in the *Pseudomonas* has increased. Furthermore, because *Pseudomonas* can absorb phosphorus, it can accelerate phosphorus release from *Microcystis*.

Microcystis provides nutrients for the attached bacteria. Therefore, growth of the attached bacteria is closely related to the condition of the *Microcystis* (Sommaruga & Robarts, 1997; Worm & Søndergaard, 1998; Brunberg, 1999). The growth of *Pseudomonas* was logarithmic, with no stable or decline phases. The maximum growth rate of *Pseudomonas* coincided with *Microcystis* entering the declining phase, when there were many dead cells and much organic matter was released (Fig. 5.24). The amount of phosphorus absorbed by *Pseudomonas* was determined by its biomass and the phosphorus concentration in the cell. *Pseudomonas* had long delay and log phases during the whole growth period of *Microcystis*. Only when *Microcystis* entered the stable phase of growth did the phosphorus content of *Pseudomonas* cells reach maximum.

Fig. 5.23 Distribution of ^{32}P released from *Microcystis* in the presence of *Pseudomonas*

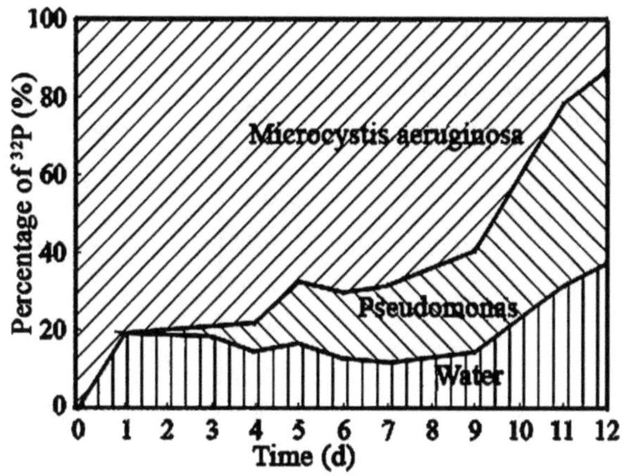

The decrease of synthesized protein and polysaccharide leads to bacterial invasion into the algal cell, subsequently causing algal autolysis (Caiola, 1991). Even when *Microcystis* and the bacteria were separated by a filter, the presence of the bacteria still could hasten the decline of the algae.

(4) Influence of extract from algae on growth of Pseudomonas sp. X.

Growth of *Pseudomonas* sp. X requires an environment provided with growth factors (Fig. 5.24). Furthermore, with increasing yeast extract concentration, optical density also increased in the culture medium (Fig. 5.25). When an extract from *Microcystis* was added to the culture medium, instead of the yeast extract, a similar situation pertained (Fig. 5.26). This result indicates that *Microcystis* probably provides growth factors for the attached *Pseudomonas* sp. X, in addition to nutrients, as already described.

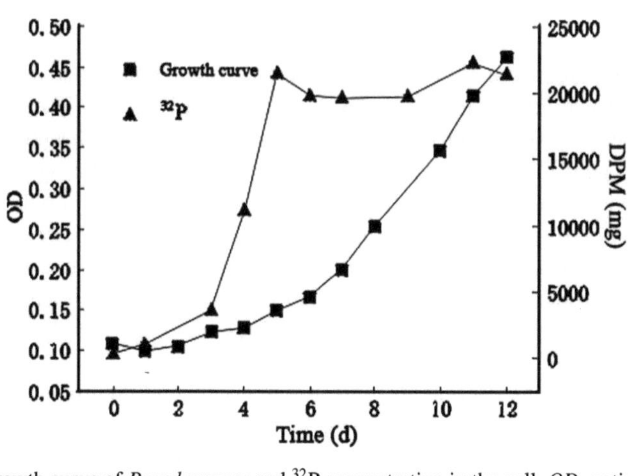

Fig. 5.24 Growth curve of *Pseudomonas* and ^{32}P concentration in the cell. *OD*, optical density

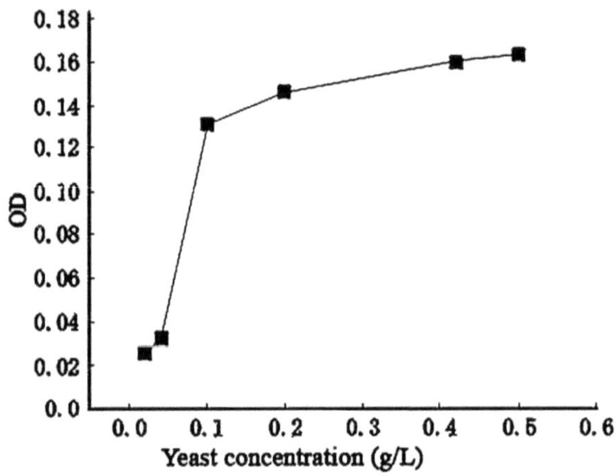

Fig. 5.25 Influence of yeast concentration on growth of *Pseudomonas*. *OD*, optical density

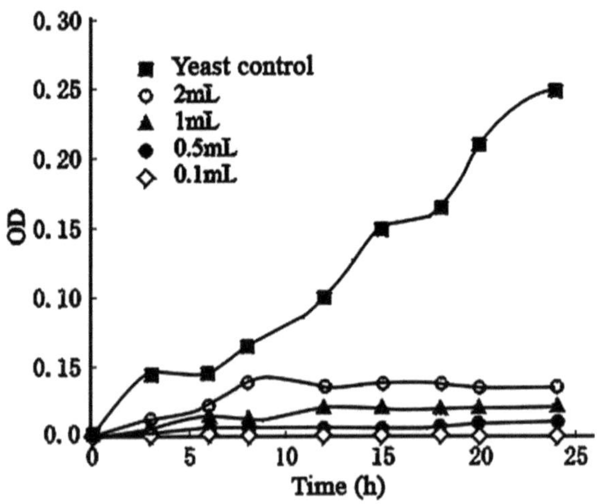

Fig. 5.26 Influence of extract from *Microcystis* on growth of *Pseudomonas* sp. *OD*, optical density

Extracts prepared from *Microcystis* in different growth phases had different influence on the growth of *Pseudomonas* sp. X (Fig. 5.27). Details of the algal density from which extract was prepared are shown in Table 5.4. The bacteria could not grow in malt agar (MA) culture medium without *Microcystis* because the growth factors were not present. The attached bacteria could only begin growth in the presence of *Microcystis* or material excreted from it. *Pseudomonas* grew well in the extract from *Microcystis* in the declining phase, followed by that from the stable phase and that from the log phase. From the log phase to the stable phase, numbers of algal cells and amount of excreted metabolites increased, providing increased carbon for growth of *Pseudomonas* sp. X.

When the algal cells decline, the exocytosed organic and inorganic components can change and may provide a superior medium for the attached bacteria (Brunberg, 1999); this may explain why *Pseudomonas* sp. X grows best in extract from the declining phase of *Microcystis* growth.

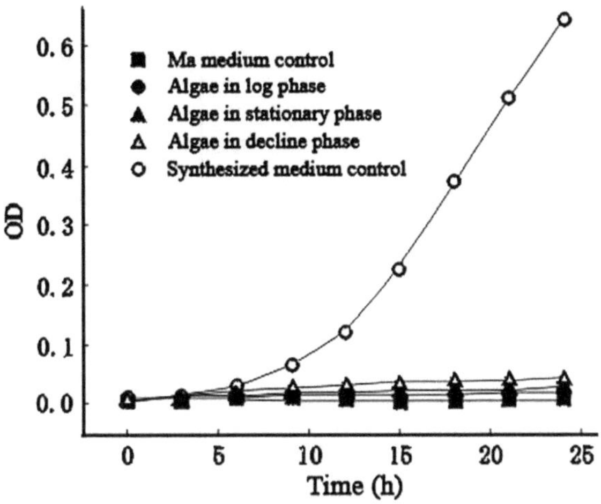

Fig. 5.27 Comparison of the growth of *Pseudomonas* sp.: influence of fluids extracted from *Microcystis* of different growth phases of *Pseudomonas* sp. *OD*, optical density

Table 5.4 Algal cell concentrations at the three different phases of *Microcystis* growth used for the extraction

	Log phase	Stable phase	Decline phase
Algal cell concentration (cell/mL)	5.8×10^6	2.9×10^7	1.6×10^7

Growth of *Pseudomonas* sp. X in *Microcystis* extracts was much lower than that in the culture medium, probably because of differences in growth factors and organic carbon sources. An adequacy of yeast extract as growth factor was added into the culture medium; however, preparation of the extracts from *Microcystis* involved treatment to remove bacteria, which probably also destroyed other material.

5.2.2.4 Metabolites of Microcystis Used by Pseudomonas sp. X

A large amount of small soluble organic molecules would be produced during *Microcystis* growth. It is helpful to use acetate as an example by which to understand the relationship between the attached *Pseudomonas* and their utilization of *Microcystis* metabolic products. Such studies may provide insights into biological control for the problem of nutrient enrichment in lakes.

Pseudomonas was cultured with sodium acetate instead of glucose as the carbon source, the most favorable concentration being about 200 mg/L. Low concentrations of acetate enhanced bacterial growth, whereas high concentrations initially restrained but then accelerated growth. In general, with increasing acetate concentration, the log phase of growth was longer, and the optical density when the stable period was reached was higher (Fig. 5.28). When *Pseudomonas* sp. X used acetate as its carbon source at 15°–30 °C, growth was inhibited with decreasing temperature, and the lag phase was longer (Fig. 5.29).

When the concentration of acetate exceeded 200 mg/L, the growth of *Pseudomonas* sp. X was inhibited, and at higher concentrations the inhibitory effect was

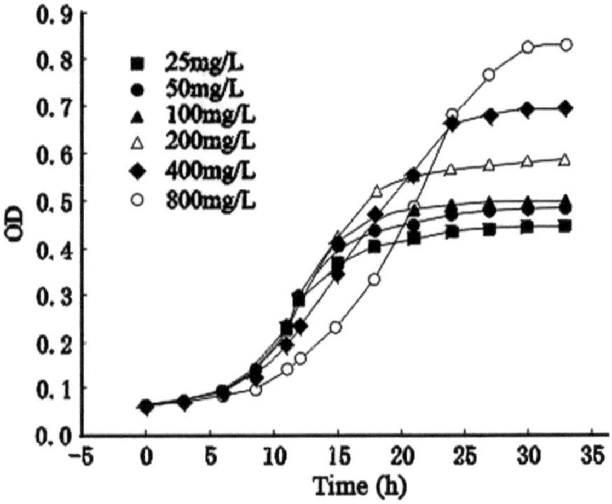

Fig. 5.28 Growth curve of *Pseudomonas* sp. X in different concentrations of sodium acetate. *OD*, optical density

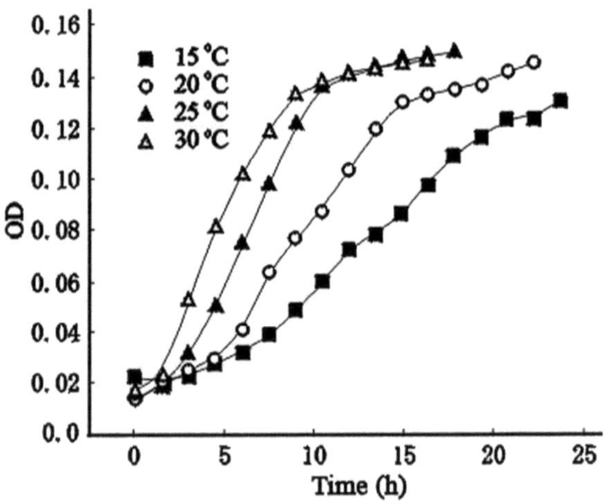

Fig. 5.29 Growth curve of *Pseudomonas* sp. X in sodium acetate at different temperatures. *OD*, optical density

stronger. In Lake Taihu, the acetate excreted by *Microcystis* is minimal, and most of the metabolic products also are concentrated in the adhesive sheath, which provides a good growth environment for the attached bacteria.

References

Alldredge, A. L., U. Passow & B. E. Logan, 1993. The abundance and significance of a class of large transparent organic particles in the ocean. Deep-Sea Research I 40: 1131–1140.

Azam, F., T. Fenchel, J. G. Field, J. S. Gray, L. A. Meyer-Reil & T. F. Thingstad, 1983. The ecological role of water-column microbes in the sea. Marine Ecology Progress Series 10: 257–263.

Becquevort, S., V. Rousseau & C. Lancelot, 1998. Major and comparable roles for free-living and attached bacteria in the degradation of *Phaeocystis* derived organic matter in Belgian coastal water of the North Sea. Aquatic Microbial Ecology 14: 39–48.

Berman, T., 1970. Alkaline phosphates and phosphorus availability in Lake Kinneret. Limnology and Oceanography 24: 541–547.

Biddanda, B. A., 1985. Microbial synthesis of macro particulate matter. Marine Ecology Progress Series 20: 241–251.

Boavida, M. J. & R. T. Heath, 1983. Are the phosphatases released by *Daphnia magna* components of its food? Limnology and Oceanography 29: 641–645.

Brunberg, A., 1999. Contribution of bacteria in the mucilage of *Microcystis* spp. to benthic and pelagic bacterial production in a hypereutrophic lake. FEMS Microbiology Ecology 29: 13–22.

Caiola, M., 1991. Bdellovibrio-like bacteria in *Microcystis aeruginosa*. Algological Studies 64: 369–376.

Carney, J. J. & J. J. Elser, 1990. The strength of zooplankton-phytoplankton coupling in relation to trophic state. In: Tilzer M. M. & C. Serruya (eds.), Ecology of large lakes. New York: Springer-Verlag, 615–631.

Cembella, A. D., N. J. Anita & P. J. Harrison, 1984. The utilization of inorganic and organic phosphorus compounds as nutrients by eukaryotic microalgae: a multidisciplinary perspective: part 2. CRC Critical Reviews in Microbiology 11: 13–81.

Chróst, R. J., 1986. Algal-bacterial metabolic coupling in the carbon and cycle in lake. In: Meguar, F. & M. Gantar (eds.), Perspective in microbial ecology. Ljubljana: Slovene Society for Microbiology, pp. 360–366.

Chróst, R. J., 1991. Environmental control of the synthesis and activity of aquatic microbial ectoenzymes. In: Chróst, R. J. (ed.), Microbial enzymes in aquatic environments. New York: Springer-Verlag, 29–59.

Chróst, R. J. & J. Overbeck, 1987. Kinetics of alkaline phosphatase activity and phosphorus availability for phytoplankton and bacterioplankton in Lake Plußsee (north German eutrophic lake). Microbial Ecology 13: 229–248.

Chróst, R. J., R. Wcislo & G. Z. Halemejko, 1986. Enzymatic decomposition of organic matter by bacteria in a eutrophic lake. Archiv für Hydrobiologie 107: 145–165.

Chróst, R. J., U. Münster, H. Rai, D. Albercht, P. K. Witzel & J. Overbeck, 1989. Photosynthetic production and exoenzymatic degradation of organic matter in euphotic zone of a eutrophic lake. Journal of Plankton Research 11: 223–242.

Dakhama, A., J. Noüe & M. C. Lavoie, 1993. Isolation and identification of antialgal substances produced by *Pseudomonas aeruginosa*. Journal of Applied Phycology 5: 297–306.

Ducklow, H. W., D. L. Kirchman, H. L. Quinby, C. A. Carlson & H. G. Dam, 1993. Stocks and dynamics of bacterioplankton carbon during the spring bloom in the eastern North Atlantic Ocean. Deep-Sea Research II 40: 245–263.

Gao, G., X. Y. Gao & B. Q. Qin, 2000. Experimental study on the PO_4^{3-} threshold of the alkaline phosphatase activity in Taihu Lake. Journal of Lake Sciences 12(4): 353–359 (In Chinese with English abstract).

Gu, Y. F., Y. Luo, W. Y. Ma, Z. Y. Zhou & H. J. Cai, 2000. Effects of temperature, organic carbon, nitrogen and phosphate on the growth of *Bacillus* sp. isolated from *Microcystis aeruginosa*. Chinese Journal of Applied and Environment Biology 6(1): 86–89 (In Chinese with English abstract).

Hantke, B., P. Fleischer, I. Domany, M. Koch, P. Pleß, M. Wiendl & A. Melzer, 1996. P-release from DOP by phosphatase activity in comparison to P excretion by zooplankton. Studies in hardwater lakes of different trophic level. Hydrobiologia 317: 151–162.

Havens, K. E., 2001. Complex analyses of plankton structure and function. The Scientific World Journal 1: 119–132.

Healey, F. P., 1973. Characteristics of phosphorus deficiency in *Anabaena*. Journal of Phycology 9: 383–394.

Healey, F. P., 1978. Physiological indicators of nutrient deficiency in algae. Mitteilungen der Internatinalen Vereinigung für Limnologie 21: 34–41.

Herndl, G. J., 1988. Ecology of amorphous aggregations (marine snow) in the Northern Adriatic Sea: II. Microbial density and activity in marine snow and its implication to overall pelagic processes. Marine Ecology Progress Series 48: 265–275.

Herndl, G. J., 1992. Marine snow in the Northern Adriatic Sea: possible causes and consequences for a shallow ecosystem. Marine Microbial Food Webs 6: 149–172.

Jamet, D., C. Amblard & J. Devaux, 1997. Seasonal changes in alkaline phosphatase activity of bacteria and microalgae in Lake Pavin (Massif Central, France). Hydrobiologia 347: 185–195.

Jones, J. G., 1972. Studies on freshwater bacteria: association with algae and alkaline phosphatase activity. The Journal of Ecology 60: 59–75.

Kuenzler, E. J. & J. P. Perras, 1965. Phosphatase of marine algae. Biological Bulletin 128: 271–284.

Lian, Y. W., Y. L. Wang, T. L. Zhen & H. S. Hong, 1999. Advance in the research on interaction between red tide algae and bacteria. Marine Sciences 1: 35–37 (In Chinese with English abstract).

Liu, L. L., Y. F. Gu, Y. Luo, W. Q. Ma, Z. Y. Zhou & H. J. Cai, 2000. On the growth and phosphorous metabolism of bacterium isolated from *Microcystis aeruginosa* in Taihu Lake. Journal of Lake Sciences 12(4): 373–378 (In Chinese with English abstract).

Looij, A. V. & B. Riemann, 1993. Measurements of bacterial production in coastal marine environments using leucien: application of a kinetic approach to correct for isotope dilution. Marine Ecology Progress Series 102: 97–104.

Overbeck, J., 1991. Early studies on ecto-and extracellular enzymes in aquatic environments. In: Chróst, R. J. (ed.), Microbial enzymes in aquatic environments. New York: Springer-Verlag, 1–5.

Petterson, K., 1980. Alkaline phosphatase activity and algal surplus phosphorus and phosphorus-deficiency indicators in Lake Erken. Archiv für Hydrobiologie 89: 54–87.

Pomeroy, L. R.,2001. Caught in the food web: complexity made simple? Scientia Marina 65(suppl 2): 31–40.

Porter, K. G. & Y. S. Feig, 1980. The use of DAPI for identifying and counting aquatic microflora. Limnology and Oceanography 25: 943–948.

Porter, K.G., H. Paerl, R. Hodson, M. Pace, J. Priscu, B. Riemann, D. Scavia & J. Stockner, 1988. Microbial interactions in lake food webs. In: Carpenter, S. R. (ed.), Complex interactions in lake communities. New York: Springer-Verlag, 209–227.

Posch, T. & H. Arndt, 1996. Uptake of sub-micrometer and micrometer-sized detrital particles by bacterivorous and ommivorous ciliates. Aquatic Microbial Ecology 10: 45–53.

Qin, X. M. & J. Z. Zou, 1997. Study on the effects of N, P, Fe-EDTA, Mn on the growth of a red tide dinoflagellate *Scripsiella trochoidea*. Oceanologia et Limnologia Sinica 28(6): 594–597 (In Chinese with English abstract).

Reichardt, W., 1971. Catalytic mobilization of phosphate in lake water and by Cyanophyta. Hydrobiologia 38: 377–394.

Reichardt, W., J. Overbeck & L. Steubing, 1967. Free dissolved enzymes in lake water. Nature (London) 216: 1345–1347.

Reim, R. L., M. S. Shane & R. E. Cannon, 1974. The characterization of a *Bacillus* capable of blue-green bactericidal activity. Canadian Journal of Microbiology 20: 981–986.

Rhee, G. Y., 1973. A continuous culture study of phosphate uptake, growth rate and polyphosphates in *Scenedesmus* sp. Journal of Phycology 9: 495–506.

Richardson, L. L. & K. D. Stolzenbach, 1995. Phytoplankton cell size and the development to microenvironments. FEMS Microbiology Ecology 16: 185–192.

Rivkin, R. B. & E. Swift, 1979. Diel and vertical patterns of alkaline phosphatase activity in the oceanic dinoflagellate *Pyrocystis noctiluca*. Limnology and Oceanography 34: 107–116.

Schindler, D. W., 1977. Evolution of phosphorus limitation in lakes. Science 195: 260–262.

Shanks, A. L. & D. Trent, 1979. Marine snow: microscale nutrient patches. Limnology and Oceanography 24: 850–854.

Sherr, E. B. & B. F. Sherr, 1987. High rates of consumption of bacteria by pelagic ciliates. Nature (London) 235: 710–711.

Sherr, E. B. & B. F. Sherr, 1988. Role of microbes in pelagic food web: a revised concept. Limnology and Oceanography 33: 225–1227.

Sherr, E. B. & B. F. Sherr, 2000. Marine microbes: an overview. In: Kirchman, D. L. (ed.), Microbial ecology of the oceans. New York: Wiley-Liss, 13–46.

Simon, M., H. P. Grossart, B. Schweitzer & H. Ploug, 2002. Microbial ecology of organic aggregates in aquatic ecosystems. Aquatic Microbial Ecology 28: 175–211.

Sommaruga, R. & R. D. Robarts, 1997. The significance of autotrophic and heterotrophic picoplankton in hypertrophic ecosystems. FEMS Microbiology Ecology 24: 187–200.

Sorokin, Y. I., 1999. Aquatic microbial ecology. Leiden: Backhuys Publishers.

Stewart, A. G. & R. G. Wetzel, 1982. Phytoplankton contribution to alkaline phosphatase activity. Archiv für Hydrobiologie 93: 265–271.

Stockner, J. G. & K. G. Porter, 1988. Microbial food webs in freshwater planktonic ecosystem. In Carpenter, S. R. (ed.), Complex interactions in lake communities. New York: Springer-Verlag, 69–97.

Wetzel, R. G., 2001. Limnology: lake and river ecosystem, 3rd edition. London: Academic Press.

Worm, J. & M. Søndergaard, 1998. Dynamics of heterotrophic bacteria attached to *Microcystis* spp. (Cyanobacteria). Aquatic Microbial Ecology 14: 19–28.

Wu, C. H., X. R. Wang & H. Sun, 1997. Establishment of models between the growth of *Selenastrum capricornutum* and several phosphorus fractions in the lake water. Environmental Chemistry 16(4): 341–347 (In Chinese with English abstract).

Wynne, D. & G. Y. Rhee, 1986. Changes in alkaline phosphatase activity and phosphate uptake in P-limited phytoplankton, induced by light intensity and spectral quality. Hydrobiologia 160: 173–178.

Wynne, D., B. Kaplan & T. Berman, 1991. Phosphorus activities in Lake Kinneret phytoplankton. In: Chróst, R. J. (ed.), Microbial enzymes in aquatic environments. New York: Springer-Verlag, 220–226.

Zhao, Y. J. & Y. D. Liu, 1996. Possible microbial control on the adverse impacts of algae: current information about the relationship between algae and microbes. Acta Hydrobiologica Sinica 20(2): 471–475 (In Chinese with English abstract).

Zhou, Y. Y. & X. Y. Zhou, 1997. Seasonal variation in kinetic parameters of alkaline phosphatase activity in a shallow Chinese freshwater lake (Donghu Lake). Water Research 31: 1232–1235.

Zhou, Z. Y., Y. Luo, W. Q. Ma & H. J. Cai, 1998. Identification and determination of growth curve of four bacterium isolated from Taihu Lake. Journal of Lake Sciences 10(4): 59–62 (In Chinese with English abstract).

Chapter 6
The Lake Ecosystem: Structure and Development

6.1 Characteristics of Chlorophyll *a* and Primary Production

Dingtian Yang, Yunlin Zhang, and Weimin Chen

Despite numerous investigations and the advent of net-pen aquaculture in Lake Taihu, there is still no comprehensive review of the structure and function of the ecosystem. Many investigations had been carried out since 1950 (the entire lake, or Meiliang Bay). Among these investigations, some large surveys have included (1) research on Wuli Bay by the Institute of Hydrobiology, Chinese Academy of Sciences in 1951 (Wu, 1962); (2) a hydrobiological study on Eastern Taihu Bay by the Biology Department of East China Normal University in 1959 (ECNU, 1959); (3) a comprehensive investigation by the Nanjing Institute of Geography and Limnology in the 1960s (Nanjing Institute of Geography and Limnology, 1965; (4) an environmental quality investigation and study by the Lake Taihu environment quality research group in 1981; (5) an aquatic biology resource investigation by the freshwater fishery center from 1981 to 1985; (6) an investigation of aquatic organisms in relation to agriculture and net-pen culture by the Nanjing Institute of Geography and Limnology in 1982 and 1984–1985; (7) an investigation by the Nanjing Institute of Geography and Limnology in 1987–1988; (8) investigations of aquatic organisms relative to water quality by the Chinese Academy of Freshwater Fishery Center and the Taihu Basin Authority, compiled by the Ministry of Water Resources Conservancy and others in "the 7th 5-year plan" and "the 8th 5-year plan" period; and (9) dynamic monitoring of the flora and fauna of Meiliang Bay in Lake Taihu, as well as laboratory studies of the Taihu Laboratory for Lake Ecosystem Research, conducted from 1991 to the present. This chapter provides a summary of the changes in the ecosystem structure of Lake Taihu.

Phytoplankton is a fundamental primary component of lake ecosystems, not only absorbing nutrients but also serving as food for zooplankton and large filter-feeding

Yang, D.T.
South China Sea Institute of Oceanology, Chinese Academy of Sciences, Guangzhou 510301, P. R. China
e-mail: dtyang@scsio.ac.cn

Qin, B. Q. (ed.), *Lake Taihu, China*,

animals. Phytoplankton plays an important role in the carbon (C), nitrogen (N), and phosphorus (P) biogeochemical cycles and in the energy exchange of a lake ecosystem. Furthermore, phytoplankton dynamics affects changes in the biological resources of the lake and in the structure and function of its subsystems. In addition, phytoplankton absorbs light through chlorophyll *a*, chlorophyll *b*, and other pigments, converting light energy to chemical energy.

6.1.1 Chlorophyll a (Chl a)

6.1.1.1 Monthly and Partial Variations of Chl a

Chlorophyll is the basic molecule that allows phytoplankton to carry out photosynthesis; it also provides an important index of primary production. The monthly monitoring data of the Taihu Laboratory for Lake Ecosystem Research from 1998 to 2001 showed in-lake concentrations of Chl a from 1.9 to 236 $\mu g/L$, with a mean of 22.2 $\mu g/L$. Chl a concentrations were highest in Meiliang Bay, a site with serious pollution problems, and decreased gradually towards the centre of the lake. Seasonally, Chl a began to increase in March, ranging from 0.6 to 27.7 $\mu g/L$ with a mean of 6.3 $\mu g/L$, and continued to rise with increasing water temperature until September or October. The lowest concentration was recorded in January–February, with a range of 1.74–11.1 $\mu g/L$ (mean, 5.7 $\mu g/L$) and differed seasonally in the surface and bottom layers. During the winter months, Chl a was higher in the warmer bottom layers than in surface water. During spring and summer, the highest Chl a concentration was recorded in the depth layer at 20–40 cm from sunlight and ultraviolet radiation.

Chl a concentration differs spatially. The highest concentration is usually observed in Wuli Bay. In Meiliang Bay, Chl a concentration gradually declined from the mouth of the Liangxi River towards the inner and outer regions of the bay. This spatial distribution was attributed mainly to the two following factors. (i) N and P concentrations at the river mouth and inner section of the bay were significantly higher than those of the outer bay as a result of the effect of the inflow river plume (Annals of Taihu Laboratory for Lake Ecosystem Research from 1998 to 2001). The Liangxi and Zhihugang Rivers from Wuxi and Wujin contain sufficient nutrients to stimulate phytoplankton growth. The 1991–1999 monthly monitoring data showed that total phosphorus was second in importance to water temperature only in controlling phytoplankton biomass in Meiliang Bay (Chen et al., 2003). (ii) Higher Chl a concentration is related to meteorological conditions in Meiliang Bay. Southern, southeastern, or southwestern winds prevail in Meiliang Bay in the summer months, because of the summer monsoon, and cause the algae in central and south parts to migrate into the north part of the lake, piling up in Meiliang Bay. For example, these three wind directions accounted for 91% of wind frequency in 2004 (Bai et al., 2005). In Eastern Taihu Bay, Chl a concentration in the water column was low because the extensive growth of macrophytes inhibited the phytoplankton by nutrient competition.

6.1.1.2 Vertical Distribution of Chl a During the Day

The vertical distribution of Chl a in the littoral zone differed with time of day. At 07:00, the highest concentration was in the surface water, at 09:00 it was at depth 0.2 m, at 11:00 at 0.5 m, and at 13:00 at 1 m. The highest Chl a concentration occurred at increasing depth with increasing light intensity, showing that the phytoplankton moves down to avoid excess light at the water surface; however, in the open lake, with its high turbidity, the highest Chl a concentration usually appeared at the 0.5 m depth from 09:00 to 15:00.

6.1.1.3 Annual Variation of Chl a Concentration

Chl a concentration can be used as an index of phytoplankton biomass and, generally, Chl a accounted for 0.1–0.5% of total biomass. From 1998 to 2000, the monthly patterns of Chl a at stations 0 and 8 (Fig. 6.1) showed higher concentrations from March to September than from October to February (Fig. 6.2). In general, two or three Chl a peaks were found per year, coinciding with phytoplankton blooms during spring–summer and summer–autumn. Higher Chl a concentrations occurred at station 0 than at other stations, because station 0 was located at the river mouth, where nutrient concentrations were high (Fig. 6.2). Although a smaller range of Chl a concentrations was observed at station 8 than at station 0, a distinct monthly pattern still was observed.

6.1.2 Phytoplankton Primary Production in Meiliang Bay

6.1.2.1 Measurement Method

Primary production (PP) in situ was measured by using either the dark and light bottle dissolved oxygen method (the Winkler method) or the radiolabeled carbon (^{14}C) uptake method. In Lake Taihu, the Winkler method was used routinely. The more sensitive ^{14}C method usually is used to measure PP in oceans, deep lakes, and oligotrophic lakes. Net PP (NPP) is equal to the difference of gross PP (GPP) and respiration by bacteria, zooplankton, macrophytes, and phytoplankton. In Meiliang Bay, most PP was by phytoplankton, as no macrophytes occur in the bay.

6.1.2.2 Daily Variation of PP in Meiliang Bay

In July 2001, measurements were made by the Winkler method to determine the daily variation of PP in the littoral zone near Taihu Laboratory for Lake Ecosystem Research (TLLER). The contribution of 2 h GPP at noon to the daily integrated GPP of the water column was larger than that in the morning and in the afternoon. In addition, the contribution of the morning GPP was generally larger than that in the afternoon.

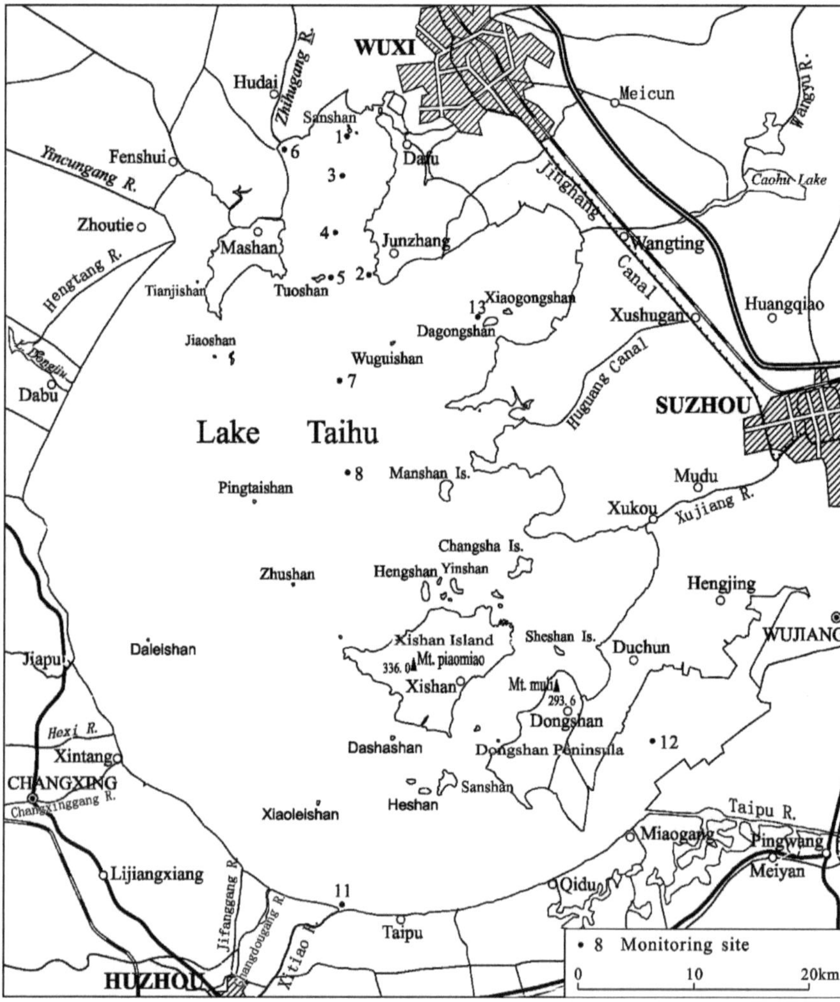

Fig. 6.1 Distribution of sampling stations

6.1.2.3 Monthly and Yearly Variation of PP in Meiliang Bay

Primary production varied among months. The monthly PP monitoring in Meiliang Bay from May 1998 to August 1999 showed that the monthly pattern of PP was similar to that of water temperature and Chl a concentration. The maximum PP in Meiliang Bay was noted in May [5.7 mg $O_2/(L \cdot d)$] and the minimum PP in December [0.6 mg$O_2/(L \cdot d)$].

PP in Meiliang Bay, measured as GPP and respiration (dissolved oxygen), showed great differences during sampling from October 1991 to February 1993 (Fig. 6.3). The pattern of PP during the year was similar to that of water temperature. The 1992 results showed that the highest GPP, 8.9 mg $O_2/(L \cdot d)$, was measured in

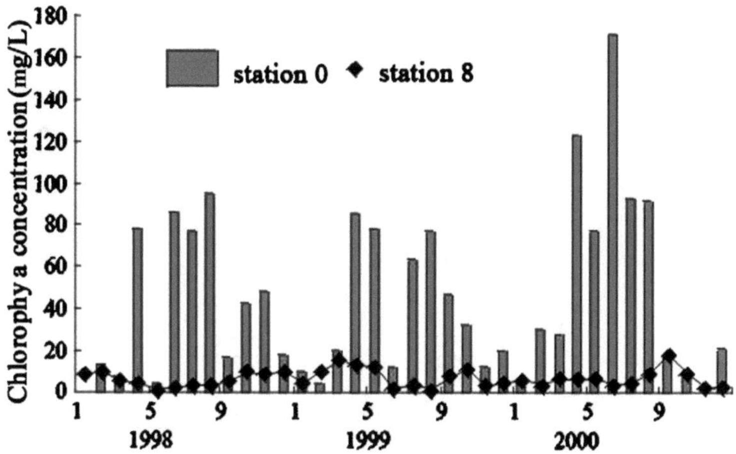

Fig. 6.2 Chlorophyll *a* (Chl a) concentration by month from 1998 to 2000 for two sampling stations in Lake Taihu

July, and the lowest PP, 0.6 mg $O_2/(L \cdot d)$, in January. The high PP in July accounted for about 20% of the annual PP in 1992.

6.1.2.4 Vertical Distribution of PP in Meiliang Bay

During July 26–28, 2001, water samples from the littoral zone near TLLER were collected to measure dissolved oxygen (DO). The concentration of DO at depth 0.2 m was higher than that at the surface at 08:00, 12:00, and 16:00, but the concentrations at different depths were almost the same at 19:00 (Fig. 6.4). Vertically, the highest NPP was at depth 0.2 m, and low values were recorded at depth 6 m depth. NPP significantly decreased at depths of more than 0.4 m (Fig. 6.5).

6.1.3 The Relationship Between Chl a and Primary Production

The mean NPP per Chl a concentration in May, October, and December 1988 and August 1999 was 0.022, 0.025, 0.019, and 0.024 mg/(mg · L · h), respectively. The analysis of approximately 30 groups of NPP per Chl a concentration found in different layers in Meiliang Bay from 1998 to 1999 showed that most (23 groups) had values between 0.012 and 0.028 mg/(mg · L · h) at depth 0.2 m. The mean value was 0.021 mg/(mg · L · h) (Table 6.1).

A formula could be obtained from these data:

$$\text{NPP} = 0.021\ C_{\text{Chl a}} \tag{6.1}$$

where NPP is the net primary production (mg $O_2/(L \cdot h)$) and $C_{\text{Chl a}}$ is Chl a concentration at depth 0.2 m (μg/L).

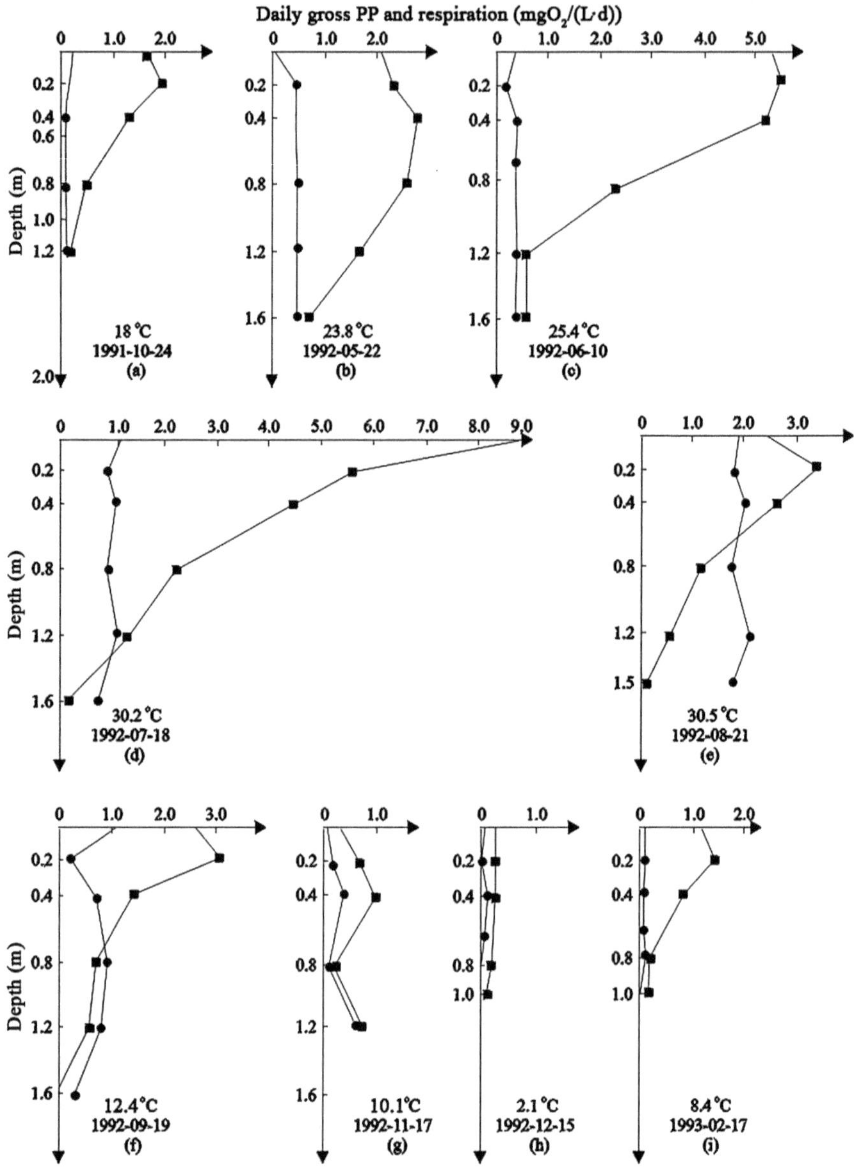

Fig. 6.3 Depth profiles of gross primary production (GPP) and respiration in Meiliang Bay in different seasons: GPP (■) and respiration (•)

6.1.4 Ecological Factors Affecting Primary Production

The principal ecological factors affecting PP are water temperature, phytoplankton biomass, underwater irradiance, total suspended matter (TSM), nutrients, and transparency.

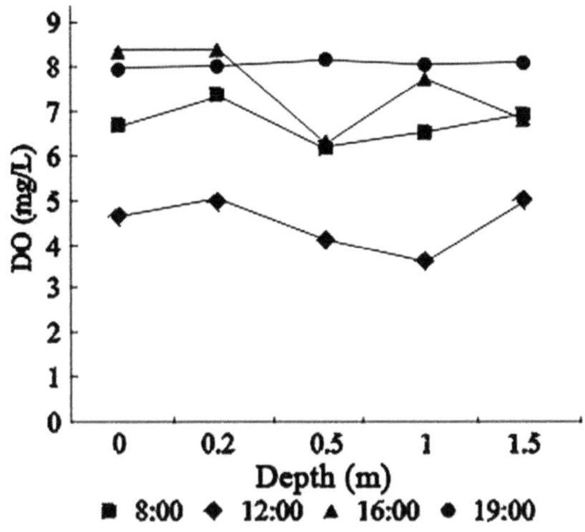

Fig. 6.4 Temporal changes in depth profiles of dissolved oxygen (*DO*) (diurnal)

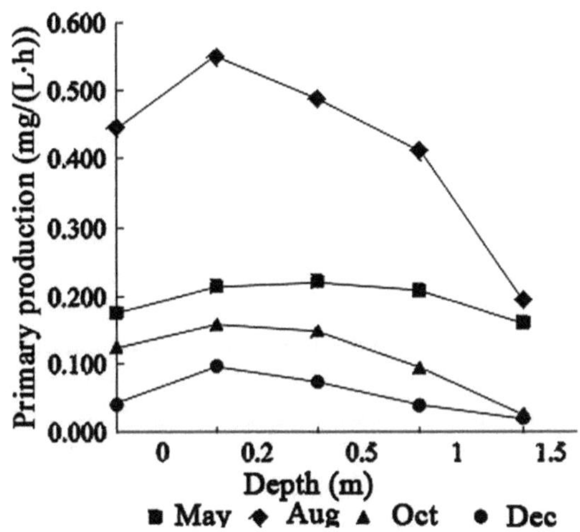

Fig. 6.5 Temporal changes in depth profiles of net primary production (NPP) (monthly)

6.1.4.1 Water Temperature

Phytoplankton usually showed rapid growth when temperature exceeded 10 °C. The optimum water temperature for growth of most algae was about 20 °C, at which algal enzymes were most active. In general, enzyme activity increases two- to fourfold per 10 °C. Between 10 °C and 25 °C, Chl a concentration increased with increasing

Table 6.1 The range and mean value of primary production (PP) in different water layers [in mg/(mg · L · h]

Depth/m	0	0.2	0.4	0.6	1.0
Range	0.008–0.024	0.012–0.028	0.011–0.025	0.011–0.021	0.00046–0.0227
Mean	0.016	0.021	0.018	0.016	0.0084

water temperature and then decreased at higher temperatures. The maximum concentration for daily GPP per Chl a occurred at 24 °C.

6.1.4.2 Underwater Irradiance

Euphotic depth is defined as the penetration depth of 1% of the surface irradiance. Below this depth, macrophyte growth is inhibited because of insufficient sunlight. At greater depths, phytoplankton NPP is generally negative. Euphotic depth can be calculated by the following formula:

$$D_{eu}(\lambda) = 4.605/K_d(\lambda) \tag{6.2}$$

where $D_{eu}(\lambda)$ and $K_d(\lambda)$ are the euphotic depth (m) and the diffuse attenuation coefficient (m^{-1}) at wavelength λ, respectively.

The euphotic depths of Lake Taihu were generally shallow, varying within a range of 1.6–1.8 m with an average of 1.5 m. In Meiliang Bay, the shallowness of euphotic depth was caused by high concentrations of TSM and phytoplankton. Wind-induced sediment resuspension, with mixing of water and substrate, often increased the concentrations of suspended matter, which caused high light attenuation, resulting in the shallow euphotic depth.

At the water surface, there was significant photoinhibition of phytoplankton photosynthesis because of the strong light intensity and short-wave radiation. Wavelengths of light between 250 and 260 nm strongly inhibit photosynthesis, and phytoplankton are inhibited from absorbing light from 670 to 680 nm. In addition, increasing water temperature also enhances photoinhibition.

Smith et al. (1980) reported that ultraviolet (390 nm) and visible light (430 nm) are each responsible for half the photoinhibition. The inhibition by ultraviolet radiation is much weaker if there was a high concentration of chromophoric dissolved organic matter (CDOM) in the water, which strongly absorbs ultraviolet radiation. In Lake Taihu, CDOM concentration was relatively high as a consequence of sewage input from rivers and algal blooms, which caused rapid attenuation of ultraviolet radiation. However, the highest photosynthesis and PP were generally found at depth 20 cm as a result of light inhibition in the surface water in summer during strong surface irradiance.

Data for underwater irradiance and GPP in the different water layers are shown in Table 6.2. During normal daily variation of underwater irradiance, the maximum PP layer moved from the surface in the morning to depth of 0.2 m at noon, and then gradually returned to the surface in the evening. Obviously, the light intensity at noon inhibited PP in the surface water.

The depth profiles of daily GPP and respiration in different seasons are shown in Fig. 6.3. The most productive layer for photosynthesis was generally found at depth 20–40 cm in Meiliang Bay. When surface photosynthetically available radiation (PAR) was weak, photosynthesis of the surface layer was markedly higher than that in other layers. In contrast, when surface PAR was strong, photosynthesis in the surface layer was less than that in deeper layers, and the most productive layer

Table 6.2 Underwater irradiance and gross primary production (GPP) in the different water layers on October 25, 1992

Depth	0 m		0.2 m		0.4 m		0.8 m	
Time	Irradiance	GPP	Irradiance	GPP	Irradiance	GPP	Irradiance	GPP
6:00–9:00	0.54	0.163	0.31	0.130	0.114	0.103	0.043	0.077
9:00–11:50	2.34	0.191	1.33	0.290	0.498	0.173	0.190	0.078
11:50–15:00	2.4	0.177	1.36	0.277	0.512	0.177	0.194	0.101
15:00–17:00	0.63	0.136	0.36	0.128	0.135	0.096	0.052	0.064
17:00–06:00	0		0		0	0		0

Irradiance, $E/(m^2 \cdot h)$; GPP, mg $O_2/(L \cdot h)$.

moved to depth 80–100 cm. Therefore, surface irradiance could significantly affect phytoplankton PP.

6.1.4.3 Phytoplankton Biomass

Phytoplankton biomass is the basic element of PP, and Chl a concentration is an index of phytoplankton biomass. A significant and positive linear relationship was found between GPP and Chl a concentration based on a 1-year investigation in Meiliang Bay:

$$GPP = -0.63 + 0.197C_{\text{Chl a}} \ (r = 0.90) \tag{6.3}$$

where GPP was daily GPP integrated through the water column ($mgO_2/(L \cdot d)$) and $C_{\text{Chl a}}$ was Chl a concentration ($\mu g/L$).

6.1.4.4 Nutrients

Large volumes of industrial wastewater, agricultural fertilizer, and sewage discharged into water elicit eutrophication. Monitoring results of TLLER from 1997 to 2002 showed that excessive phytoplankton growth was caused mainly by high concentrations of N and P. In Meiliang Bay, eutrophication has caused *Microcystis* blooms from May to October since the 1990s.

In addition, water transparency and turbidity mainly affected the transmission and distribution of light and further influenced light absorption by phytoplankton and PP (Zhang et al., 2004).

6.1.5 Simulated Water Current Experiment and Primary Production

From May 8, 1999 to June 22, 1999, a simulated water current experiment was conducted at TLLER to determine the effect of water dynamics on biological succession. The results showed that water currents affect photosynthetic rate by two

Table 6.3 The photosynthesis rate and the percentage contribution of each layer to the water column integrated photosynthesis in the simulation experiment [in mg/(μg · h)]

Depth	Still water current				Slow water current				Fast water current			
	May 20	%	May 22	%	June 4	%	June 6	%	June 19	%	June 21	%
0–0.2 m	5.02	15.5	4.86	30.88	3.64	25.28	4.41	27.38	4.93	46.41	4.06	57.02
0.2–0.4 m	5.73	17.8	3.63	23.04	3.31	23.03	3.88	24.13	3.51	33.04	1.98	27.87
0.4–0.6 m	6.36	19.7	2.18	13.82	2.67	18.54	2.91	18.10	1.35	12.73	0.63	8.79
0.6–1.0 m	12.37	38.3	4.06	25.81	4.00	27.81	4.03	25.06	0.79	7.41	0.43	6.00
1.0–1.2 m	2.82	8.74	1.02	6.45	0.77	5.34	0.86	5.34	0.04	0.40	0.02	0.32
0.0–1.2 m	32.30	100	15.75	100	14.38	100	16.10	100	10.62	100	7.11	100
Surface irradiance (μmol/(m^2 · s))	692.5		125.0		627.5		286.3		680.0		237.5	
GPP (mg O_2/(m^2·h))	0.70		0.25		1.54		0.94		1.12		0.84	
NPP (mg O_2/(m^2·h))	0.60		−0.16		0.92		−0.21		0.08		−0.77	
ΣPAR (μE/m^2)	6.98×10^6		1.04×10^6		3.75×10^6		1.86×10^6		2.41×10^6		7.06×10^6	

NPP, net primary productivity; ΣPAR, sum of photosynthetically available radiation.

opposing processes: (1) sediment disturbance by currents releases nutrients from the sediment and the increased nutrient concentrations promote growth of algae and increase PP; and (2) sediment resuspension increases the concentration of TSM, promotes light attenuation, and decreases PP.

The highest photosynthesis rate, 32.30 mg/(μg · h), was recorded in still water, on May 20, under high surface irradiance and low light attenuation (Table 6.3). In slow currents, the integrated photosynthesis rates in the water column for strong and weak surface irradiance were 14.38 mg/(μg · h) and 16.10 mg/(μg · h), respectively. In fast currents, the photosynthesis rates were 10.62 mg/(μg·h) and 7.11 mg/(μg·h) for strong and weak surface irradiance, respectively.

The highest integrated GPP in the water column was recorded in slow current with a mean concentration of TSM of 12.04 mg/L. The second highest GPP was recorded in fast water with a mean TSM concentration of 42.65 mg/L. The lowest GPP was recorded in a still current with an average TSM concentration of 2.01 mg/L.

The highest NPP, 0.92 $mgO_2/(m^2 \cdot h)$, was recorded in a slow current with high surface irradiance. However, the next highest NPP, 0.60 $mgO_2/(m^2 \cdot h)$, was recorded in a still current with high surface irradiance. At all three water speeds, NPP was always negative when irradiance was weak.

6.2 Phytoplankton Community Structure: Succession and Response to the Environment

Wanping Zhou and Weimin Chen

6.2.1 Phytoplankton Community Structure

6.2.1.1 Species Composition of Phytoplankton

There are a variety of phytoplankton species in Lake Taihu. These microscopic organisms are the primary producers in the lake, and as such, form the foundation of the food chain for aquatic animals. Therefore, phytoplankton plays a pivotal role in any lake ecosystem. However, because of the ever-increasing rate of eutrophication in Chinese lakes, some algae, especially Cyanophyta, develop in great quantities, causing harmful blooms and deterioration of water quality.

Analysis of phytoplankton from 1990 to 1996 showed a total of 8 phyla, 116 genera, and 239 species. There were 24 genera and 53 species of Cyanophyta, 2 genera and 3 species of Cryptophyta, 4 genera and 6 species of Pyrrophyta, 6 genera and 9 species of Chrysophyta, 3 genera and 4 species of Xanthophyta, 24 genera and 48 species of Bacillariophyta, 6 genera and 15 species of Euglenophyta, and 47 genera and 101 species of Chlorophyta (Appendix A).

A comparison of phytoplankton community composition between the 1960s and the 1990s showed that the greatest changes occurred in the Chlorophyta and Cyanophyta. Twenty genera of Chlorophyta recorded during the 1960s were no longer found in the 1990s, while 18 genera collected in the 1990s were not found in the 1960s. Among the Cyanophyta, 1 genus recorded in the 1960s was not found in the 1990s, and 9 genera collected in the 1990s were not recorded in the 1960s. In addition, except for Wuli Bay, Cryptophyta frequent in the 1990s were not found in the 1960s. Overall, the dominant species, as well as the most frequently reported species, were similar in both the 1960s and 1990s, but oligotrophic species in the 1990s were generally rare compared with the 1960s.

6.2.1.2 Variation in Phytoplankton Abundance

Both abundance and biomass of phytoplankton increased from 1991 to 1995 (Fig. 6.6).

The mean phytoplankton abundance and biomass in February, June, August, and October from 1990 to 1995 are shown in Fig. 6.7. Mean abundance in 1995 was 1.8 times greater than that in 1991. Similarly, mean biomass in 1995 had increased by 0.8 times the value in 1991.

Zhou, W.P.
State Key Laboratory of Lake Science and Environment, Nanjing Institute of Geography and Limnology, Chinese Academy of Sciences, 73 East Beijing Road, Nanjing 210008, P. R. China

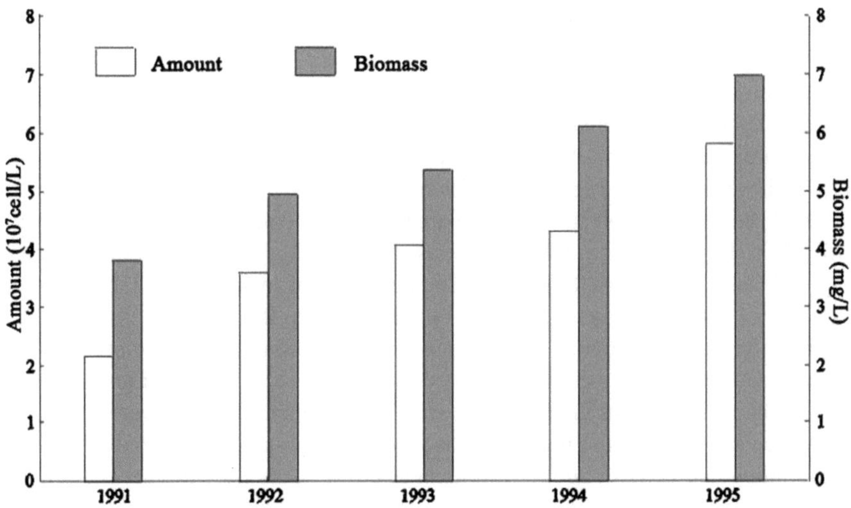

Fig. 6.6 Increase in annual mean phytoplankton abundance and biomass in Lake Taihu from 1991 to 1995

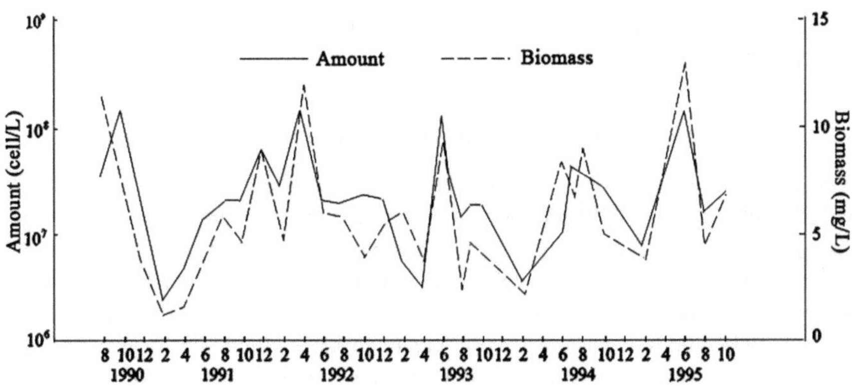

Fig. 6.7 Bimonthly variations of phytoplankton abundance and biomass in Lake Taihu from 1990 to 1995

The monthly variation of phytoplankton abundance and biomass is shown in Fig. 6.7. The maximum abundances appeared in June 1994 and June 1995, with values of 7.07×10^7 and 1.64×10^8 cell/L, respectively. The minimum abundance occurred in February in winter of both years, with only 5.54×10^6 and 8.69×10^6 cell/L, respectively. The maximum abundance was 18.9 times the minimum in 1994 and 1995. An anomaly of abundance of 7.95×10^7 cell/L appeared in December in 1992, which was 21.5 times the minimum in February in 1991. Although the maximum abundance occurred in different months from 1990 to 1995, the higher abundances generally fell in summer and lower abundances in winter. The monthly increase in biomass was similar to the increase in abundance.

6.2.2 Changes in the Phytoplankton from the 1950s to the 1990s

Lake Taihu is shallow, with many bays, and thus different microenvironments, each of which affects phytoplankton abundance and biomass. Based on the characteristics of each area, the lake can be divided into Wuli Bay, Meiliang Bay, Eastern Taihu Bay, and the open lake region. Phytoplankton species number and abundance in different areas and periods can be compared, based on data from the 1950s, 1960s, and 1990s. Phytoplankton abundance was generally higher in northern Meiliang Bay and lower in Eastern Taihu Bay. The hierarchy of mean annual biomass was Wuli Bay > Meiliang Bay > open lake > Eastern Taihu Bay.

The differences in phytoplankton abundance and biomass were determined by the dominant species. If the dominant species were the same in the compared areas, greater abundance meant greater biomass. If the dominant species were different, abundance and biomass did not correspond, because biomass was related to the size of the individual cells in addition to abundance. In Wuli Bay, biomass was high because of sufficient phytoplankton abundance and a high percentage of diatoms and Cryptophyta with generally larger cell size than other groups. In Meiliang Bay, abundance was high, but biomass was low, because its phytoplankton was composed mainly of Cyanophyta with small individual cell size. In Eastern Taihu Bay, the biomass was lower than in other areas as a result of low abundance, although many large diatom cells occurred in this area.

6.2.2.1 Changes of Phytoplankton in Wuli Bay from the 1950s to the 1990s

Wuli Bay, southwest of Wuxi City, is a bay of about 5.4 km^2 with an average depth of 2 m. Wuli Bay, shaped like an isomerous "V," is divided by the Baojie Bridge into East and West Wuli Bay. In the 1950s, macrophytes flourished here and the water was clear. In the early 1960s, water quality remained good, and macrophytes were composed mainly of *Nymphoides peltatum*, bitter herbs, and *Symphytum* sp. After the late 1960s, the bay was enclosed to create an aquaculture site for fish, and the macrophytes degenerated year by year. Following the disappearance of the macrophytes in 1980s and the development of aquaculture, the purification capacity of the water gradually decreased and water transparency was reduced. Meanwhile, city sewage was discharged into the Bay, causing further deterioration in water quality and worsening eutrophication. Under these conditions, the phytoplankton species and their abundance varied each year. Some species that usually live in clear water decreased and eventually disappeared. In contrast, the abundance of those algae that contributed to eutrophication increased continuously and formed blooms.

In 1951, Cryptophyta dominated the phytoplankton, accounting for 46% of total abundance. The abundance of the representative species *Cryptomonas ovata* and *Chroomonas nordstedtii* exceeded that of all other species combined. The abundance of *C. ovata* peaked in May at 7.1×10^5 cell/L. Abundance in winter and summer was high, but slightly less than that of Bacillariophyta and Cyanophyta. Bacillariophyta was the second most abundant group, accounting for 30% of total abundance. Abundance was high throughout the year, especially in winter.

Abundance of the dominant species, *Cyclotella stelligera*, peaked in March at 3.4×10^5 cell/L. The mean annual abundance of *C. stelligera* was 5.3×10^4 cell/L, accounting for 66% of the mean total abundance of Bacillariophyta. Cyanophyta was the third most abundant algal group, accounting for 9% of total phytoplankton abundance. Cyanophyta accounted for the highest percentage of total abundance in summer. The abundance of the dominant species, *Chroococcus limneticus* var. *carneus* and *Chroococcus limneticus* var. *subsasus*, peaked in summer at 2.7×10^5 cell/L. During the 1950s, although *Microcystis aeruginosa* and *Microcystis flos-aquae* were commonly found, they were present in low numbers. This observation contrasts sharply with the present situation, in which these species abound in some areas, and sometimes predominate.

During the 1970s and 1980s, Wuli Bay experienced eutrophication and frequent *Microcystis* blooms. In the late 1990s, there were increased nutrients, contamination, and hypereutrophication but no more large-scale *Microcystis* blooms. Although *Microcystis* could be found in Wuli Bay in the summer months, its abundance was much lower than in Meiliang Bay. From the 1990s, the dominant species in Wuli Bay changed from *Microcystis* to *Aphanizomenon* sp., *Chroomonas acuta, Cryptomonas erosa, Cryptomonas ovata, Cyclotella* sp., *Melosira* sp., *Navicula* sp., and *Oscillatoria* sp. In the hypereutrophic bay, the competitive ability of these algae was much stronger than that of *Microcystis* (Chen et al., 1997); however, the biomass and Chl a concentration in Wuli Bay were still quite high. Therefore, the high biomass of *Aphanizomenon* sp., *Melosira* sp., and *Oscillatoria* sp. should be reduced further.

The biomass and percentage of six different phytoplankton taxa in summer 1994 and 1995 are shown in Table 6.4. The dominant taxa, by both biomass and percentage, were Cryptophyta and Bacillariophyta; the Cyanophyta biomass was not large.

Significant ecoenvironmental changes were recorded in Wuli Bay in recent decades. From 1951 until the 1990s, indicator species for eutrophic and oligotrophic conditions were present in a mixture. From 1994 to 1995, some clear-water species such as *Ochromonas mutabilis* and *Dinobryon divergens* vanished. In 1951, some attached phytoplankton, such as *Cocconeis placentula*, were present in high numbers each season, but they are scarce today because of the disappearance of submerged vegetation.

Table 6.4 Biomass and the percentage of six different phytoplankton taxa in summer 1994 and 1995 in Wuli Bay

Year	Biomass (mg/L)					
	Cyanophyta	Cryptophyta	Pyrrophyta	Bacillariophyta	Euglenophyta	Chlorophyta
1994	3.98	16.8	0.14	10.98	0.73	1.93
1995	2.04	5.7	1.35	15.46	0.4	1.42
	Percentage (%)					
1994	11.5	48.6	0.40	31.8	2.1	5.6
1995	7.7	21.6	5.1	58.6	1.5	5.4

6.2.2.2 Changes of Phytoplankton in Open Lake Taihu from the 1950s to the 1990s

The open areas of Lake Taihu include the areas other than Wuli Bay, Meiliang Bay, and Eastern Taihu Bay. In 1960, numerous blue-green algal blooms were recorded. During the 1990s, the area affected by blooms increased, but with no significant change of the dominant species. The dominant Cyanophyta species throughout the year included *M. aeruginosa, M. flos-aquae*, and *M. wesenbergi* in the open areas of Lake Taihu, while *Chroococcus limneticuss* and *C. minutus* were highly abundant in some areas of the lake, and *Cyclotella* sp. and *Melosira granulata* were present throughout the year and became locally dominant. Among the Cryptophyta, *C. erosa, C. ovata*, and *Chroomonas acuta* were abundant throughout the year.

More than ten species of *Scenedesmus* sometimes occurred in abundance in open Lake Taihu and other lake areas. The species and their distrubutions were as follows: *S. abundans*, Eastern Taihu Bay, Gonghu Bay; *S. acuminatus*, Wuli Bay and near Dapukou; *S. arcuatus*, Wuli Bay, Meiliang Bay; *S. bijuga*, Wuli Bay and near Sanshan, Tuoshan, and Wuguishan; *S. brasilliensis*, Gonghu Bay; *S. cavinatus*, near Tuoshan; *S. denticulatus*, Wuli Bay and the lakes near Tuo Mountain and Dapukou; *S. dimorpha*, Wuli Bay and near Sanshan, Tuoshan, and Wuguishan; *S. guadricauda*, all of the lake; *S. javaensis*, near Daleishan; *S. oboiquus*, near Sanshan Island; *S. perforatus*, Gonghu Bay; and *S. platydiscus* in Meiliang Bay near Sanshan Island.

High phytoplankton abundance was recorded between 1990 and 1995 (Table 6.5), especially in 1990 when levels reached 1.21×10^8 cell/L. Blue-green algae blooms were responsible for the high phytoplankton abundance, with the bloom in 1990 lasting for several months.

6.2.2.3 Changes of Phytoplankton in Eastern Taihu Bay from the 1950s to the 1990s

In 1959, the dominant phytoplankton genera in Eastern Taihu Bay included *Cosmarium, Euastrum, Microcystis, Melosira, Synedra, Surirella*, and *Staurastrum*. In the 1990s, the dominant genera were *Chroomonas, Cyclotella, Cryptophyta*, and *Melosira*, and, sometimes, *Microcystis*. Between 1959–1960 and 1990–1995, both species composition and dominant species changed markedly. *Cosmarium* changed the most, with decreasing abundance. Some genera such as *Arthrodesmus, Desmidium, Euastrum, Hyalotheca, Onychonema*, and *Pleurotaenium* have not been found

Table 6.5 Phytoplankton abundance and biomass in the open areas of Lake Taihu from 1987 to 1995

	1987–1988 $n = 4$	1990 $n = 3$	1991 $n = 6$	1992 $n = 6$	1993 $n = 7$	1994 $n = 5$	1995 $n = 4$
Abundance ($\times 10^8$cell/L)	–	1.21	0.31	0.76	0.27	0.44	0.38
Biomass (mg/L)	6.65	5.28	4.01	6.21	3.18	4.18	4.15

Data are mean values from 3–7 samples per year, indicated as $n = x$ in each column.

Table 6.6 Phytoplankton abundance and biomass in Eastern Taihu Bay

	July 1959	July 1987	August 1990	July 1993	August 1993	July 1994	August 1995
Abundance (10^4cell/L)	1.74	–	378	505	317	363	469
Biomass (mg/L)	–	4.74	3.14	3.58	2.06	1.97	2.54

for many years. Furthermore, some genera apparently absent in 1959 and 1960, such as *Chroomonas* and some species of Cryptophyta, became common in the 1990s, sometimes with high abundance.

In contrast to other areas of the lake, Eastern Taihu Bay today is characterized by flourishing macrophytes, clear water, and high transparency reaching to the lake bottom. The macrophytes could compete for light and nutrients with phytoplankton, and some macrophytes could secrete substances that restrain phytoplankton growth. Therefore, phytoplankton abundance and biomass in Eastern Taihu Bay were lower than in other areas of Lake Taihu.

During the 1950s to the 1990s, the highest phytoplankton abundance in Eastern Taihu Bay was found from 1990 to 1995 (Table 6.6). This occurrence might have been caused by the migrating of blue-green algae blooms from western Lake Taihu and the southern area of the lake, driven by wind and waves.

6.2.2.4 Changes of Phytoplankton in Meiliang Bay from the 1950s to the 1990s

Meiliang Bay, in northern Lake Taihu near the suburbs of Wuxi City, has an area of 132 km^2 and a mean depth of 2.0 m. Meiliang Bay is a tourism destination and the principal water source for Wuxi City (see Fig. 1.2). Because of industrial and agricultural pollution in the past 20 years, a large amount of domestic sewage, wastewater, and agricultural irrigation water was discharged into the Bay from the Liangxi River and the Zhihugang River. Meiliang Bay currently is one of the most eutrophic regions in Lake Taihu, and algal blooms are often recorded from May to October. Therefore, the succession of the dominant species of phytoplankton in Meiliang Bay was given great attention.

A comprehensive survey in 1960 by the Nanjing Institute of Geography and Limnology Chinese Academy of Sciences, found large-scale algal blooms in Meiliang Bay. The algal species were mainly *M. aeruginosa* and *M. flos-aquae*.

The variation of phytoplankton biomass in Meiliang Bay in 1960, 1981, 1988, and from 1991 to 1998 is shown in Fig. 6.8. Phytoplankton biomass markedly increased from 1960 to 1981 to 1988, and from 1991 to 1998, which might be caused by the increasing total nitrogen (TN) concentration (Fig. 6.8).

The dominant genera in Meiliang Bay included *Anabaena, Aphanizomenon, Microcystis*, and *Oscillatoria* (Cyanophyta), *Cyclotella, Melosira*, and *Navicula* (Bacillariophyta), *Chroomonas* and *Cryptomonas* (Cryptophyta), and *Pediastrum* and *Scenedesmus* (Chlorophyta). *Microcystis* was predominant, with highest frequency and abundance; the main species were *M. aeruginosa, M. flos-aquae*, and *M. pulverea*. These species multiplied rapidly in the summer months, and when

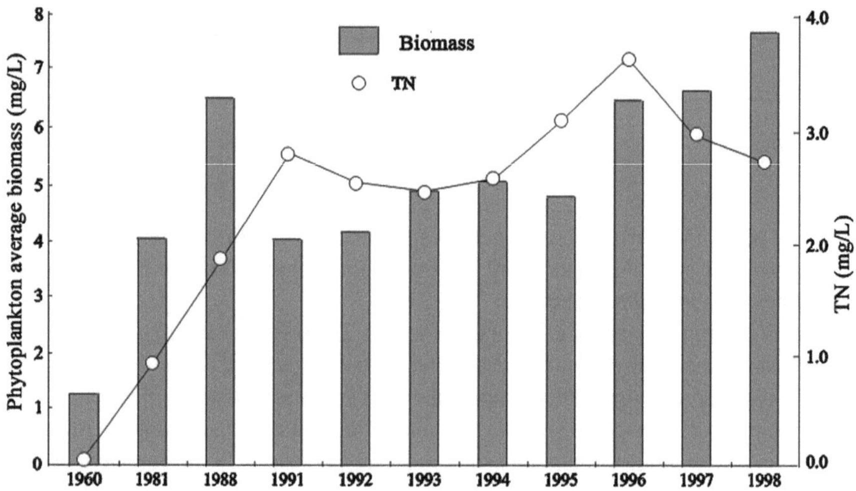

Fig. 6.8 Temporal changes in phytoplankton biomass and total nitrogen in Meiliang Bay, Lake Taihu, between 1960 and 1998. *TN*, total nitrogen (*circles*)

abundance was $> 10,000 \text{cell/mL}$, an algal bloom could form. *Microcystis* usually accounted for more than 99% of the total phytoplankton.

Monitoring data of 1992 and 1996–1999 showed a regular seasonal succession of the dominant taxa in Meiliang Bay. In spring, *Chroomonas, Cryptomonas, Pediastrum*, and *Scenedesmus*, along with *Microcystis*, often appeared massively in the inner part of Meiliang Bay. The composition and dominant genera of phytoplankton in the central and outer parts was similar to that of the inner part. However, the species with high biomass in the central and outer parts were fewer than those in the inner part. In the outer part of the Bay, the predominant species included *Melosira, Pediastrum*, and *Ulothrix*. In the summer, the quantity of the predominant species was lower than in spring. The predominant taxa included *Chroomonas, Cyclotella*, and *Microcystis* in the inner part; *Cryptomonas, Melosira*, and *Pediastrum* in the central part; and *Melosira* in the outer part. In the autumn, the quantity of the predominant species was less than in spring and summer. *Microcystis* predominated in the outer part with 100% frequency of occurrence. In contrast, it was not the predominant species in the inner and central parts, showing that *Microcystis* moved from the inner to the outer part of Meiliang Bay and to the open lake. For example, *Microcystis* blooms had expanded to the Pingtaishan area, in the center of Lake Taihu, in 2001. In winter, the number of dominant species with high biomass was much higher than that in summer and autumn. *Anabaena, Chroomonas*, and *Cryptomonas* were predominant in the inner part, and *Anabaena* and *Oscillatoria* were found in the central and outer parts. Peak abundances of *Microcystis, Chroomonasm* and *Cryptomonas*, and *Melosira* were recorded in July, April, and August, respectively.

Phytoplankton biomass in Meiliang Bay was characterized by annual variation and affected by factors such as wind speed and direction, nutrient concentration, monitoring sites, and sampling protocol. For example, in late July 1990, consecutive days of oppressive weather caused a large algal bloom, predominantly *Microcystis*, in Meiliang Bay. This effect was compounded by the prevailing southeast wind in Lake Taihu, which caused accumulation of the bloom that piled up to 0.5 m thick along the shore of Meiliang Bay. Phytoplankton abundance near a drinking water plant reached 1.32×10^9 cell/L, with a biomass of 10.82 mg/L. Subsequently, phytoplankton abundance was greatly reduced by a change in the wind direction. In this particular bloom, *Microcystis* accounted for 98% and 86% of the total abundance and biomass, respectively.

The seasonal variation pattern of phytoplankton biomass is shown in Fig. 6.9, with *Microcystis*, Cryptophyta, and diatoms with high biomass as examples. Seasonal changes in phytoplankton biomass showed that *Microcytis* predominated in the warmer seasons, while Cryptophyta and Bacillariophyta predominated in cold seasons (Fig. 6.9).

From the 1960s to the 1990s, the major seasonal changes in phytoplankton biomass were in spring, high abundance of *Cyclotella, Scenedesmus, Synedra*, Cryptophyta, and *Microcystis*; in summer, large-scale *Microcystis* blooms, along with *Cyclotella* and *Melosira granulata*; in autumn, *Microcystis* blooms persisted, with increased abundance of *Cryptomonas* and *M. granulata*; and in winter, greatly increased abundance of *Pediastrum, Scenedesmus*, and *Synedra*, sometimes with high abundance of *Cryptomonas*, but lacking an obvious dominant species.

The predominant genera in Wuli Bay and Meiliang Bay are compared in Table 6.7. The common characteristic of both locations was the occurrence of algal

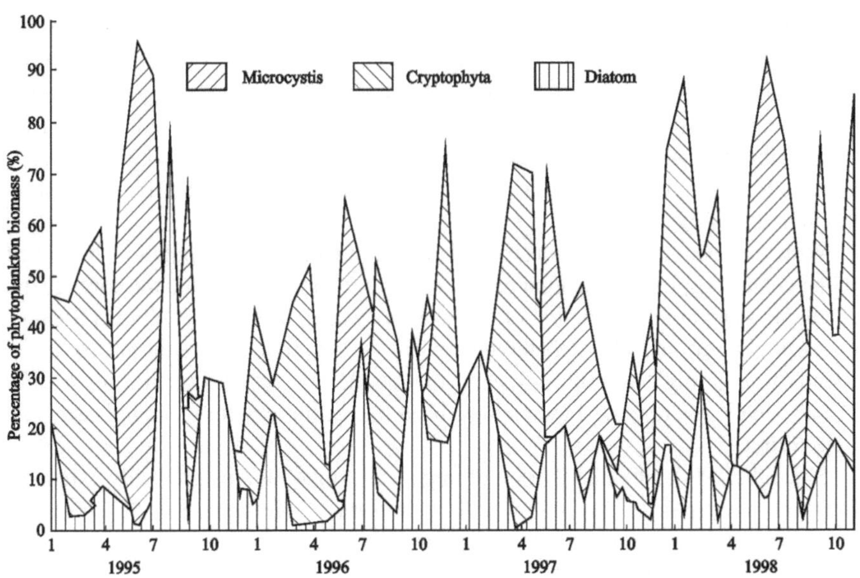

Fig. 6.9 Seasonal changes in predominant phytoplankton biomass in Meiliang Bay 1995–1998 (*1, 4, 7*, and *10* indicate months)

Table 6.7 Comparison of the predominant phytoplankton genera in Wuli Bay and Meiliang Bay

	1950s, Wuli Bay 1960s, Meiliang Bay	1980s	1990s
Wuli Bay	*Chroococcus*, Cryptophyta, *Cyclotella*	*Chroococcus*, Cryptophyta *Cyclotella*, *Microcystis*	*Chroococcus*, Cryptophyta *Cyclotella*, *Microcystis*
Meiliang Bay	*Anabaena Microcystis*	Cryptophyta, *Melosira*, *Microcystis*	Cryptophyta, *Melosira*, *Microcystis*

blooms. In Wuli Bay, these were usually *Cyclotella* blooms, whereas in Meiliang Bay the blooms were *Microcystis*. Within a given year, the abundance and biomass of the *Cyclotella* bloom in Wuli Bay and the *Microcystis* bloom in Meiliang Bay affected the whole phytoplankton succession process because of the duration and large scale of the blooms.

6.2.3 Response of the Phytoplankton to Changes in the Aquatic Environment

6.2.3.1 Microcystis Blooms in Lake Taihu

Microcystis blooms were recorded as early as 1959 (investigations of Eastern Taihu Bay by The East China Normal University) and 1965 (investigations of Lake Taihu by the Nanjing Institute of Geography & Limnology, Chinese Academy of Sciences, 1965). However, quantitative methods used in the early investigations differed from those used today, and thus past abundance cannot be compared directly with current data. In the early period (including 1959 and 1965), the aquatic environment of Lake Taihu was good, phytoplankton abundance was low, and *Microcystis* blooms appeared in only a few small strips; therefore, *Microcystis* was not detrimental to the function of the lake.

With the development of industry and agriculture, increased population, improved living conditions, increased input of polluted water into the lake, and the consequent change in structure and function of the lake ecosystem, *Microcystis* blooms became increasingly serious. Phytoplankton biomass in the area near Tuoshan (a small island) was as high as 600 mg/L, and of this, *Microcystis* accounted for 69.8% (Yang, 1996). *Microcystis* blooms have been one of the environmental disasters in Lake Taihu.

Since the 1990s, the distribution of large-scale algal blooms gradually extended from the inner part to the central and outer parts of Meiliang Bay. In 2000, the blooms reached Gonghu Bay, Dapukou, and the centre of the lake, but gradually decreased in the inner part of Meiliang Bay. Furthermore, blooms have become larger and more frequent over time. The occurrence of *Microcystis* blooms was related to increased nutrients and to other various factors in the lake. For example,

Zhu & Cai (1997) showed that different wind patterns significantly affected the horizontal and vertical distribution of *Microcystis* in the lake. When the wind speed was less than 2–3 m/s, the water surface was considered hydrodynamically as having no waves, and the *Microcystis* bloom on the surface drifted quickly to the shoreline, following the wind, and accumulated. However, if wind speed was faster than 2–3 m/s, waves would occur. Waves, wind disturbance, and average circulation mixed the bloom and distributed the algae evenly in the water column. In the algal bloom season, the movement of the *Microcystis* demonstrated that algal blooms were a hydrological process.

6.2.3.2 The Physiological and Ecological Characteristics of Microcystis and Algal Blooms

Eutrophication is accompanied by mass propagation of Cyanobacteria, formation and increase of algal blooms, and marked damage to the lake. The most frequent blooms occur in Meiliang Bay, an important drinking water resource for Wuxi City. As early as July 1990, *Microcystis* blooms were so extensive in Meiliang Bay that they blocked filters in the Meiyuan drinking water factory, resulting in a sharp drop in the quantity and quality of tap water available for Wuxi City. This problem not only affected the residents' demand for drinking water in the summer but also caused the partial or complete termination of production by 116 factories. The direct economic loss was as high as RMB1.3×10^8, equivalent to about US \$$1.6 \times 10^7$. Similar incidents have occurred frequently in recent years.

Because the blooms are mainly caused by *Microcystis*, it is important to understand their physiological and ecological characteristics. The *Microcystis* colonies, irregular in shape, and of variable size, are embedded in a transparent, thick, amylose pectin of polycarbohydrate. *Microcystis* cells, arranged so tightly that they cannot be separated easily, are round and contain air sacs. The cells are 4–5 μm in diameter and have a mean volume of about 85 μm^3. The diameter of a dividing *Microcystis* cell is about 7 μm. Numerous bacteria are observed in the pectin of the colony, and usually some protozoans, especially *Vorticella*, attach to the algae colonies. The quantity of bacteria increases with the number of cells per *Microcystis* colony.

The amount of cell division change with season, availability of nutrients, and other factors. The annual growth of *Microcystis* falls into the following three periods. During the first period, December–April, *Microcystis* is rarely found in the surface layer of the water, but is present in the sediments, both at the surface (0–3 cm deep sediments) and in the middle level (10–13 cm); these are cells that survived the previous winter (as confirmed by Wu et al., 2003). Although trace amounts of *Microcystis* could be found in the surface sediment all year round, from December to early April, it is highly abundant there. This period is characterized by high nutrient concentrations, low underwater light intensity, and low water temperature. Although sometimes a few *Microcystis* cells are found in the water column, cell division is not observed.

In the second period, April–September, *Microcystis* begin to multiply quickly, the multiplication rate being dependent on water temperature. Cell divisions are present in the water column, and abundance increases quickly. The highest rate of increase (r) measured in the laboratory was 1.27, and when rate of increase reaches 0.69, the amount of cells doubles every day. With favourable water temperature from 15° to 25 °C, and sufficient nitrogen and phosphorus, the frequency of increase rate ranging from 0.6 to 0.8 accounted for 55% (Fig. 6.10). The cells multiplied for 1 day or several days, and increased exponentially during 1–5 weeks.

During the third period, October–December, cell abundance decreases in October and November, but cell divisions can still be found, and algal blooms are still recorded. The photosynthetic efficiency of *Microcystis* was greatly influenced by light. Simulation experiments showed that, at a light intensity weaker than 500 $\mu E/(m^2 \cdot s)$, photosynthetic efficiency was stable, with the rate being proportional to the light intensity (linear relationship) (Fig. 6.11). At light intensities above about 750 $\mu E/(m^2 \cdot s)$, the rate of photosynthesis remained constant at 12 mg/(mg · h) (Fig. 6.11). The light intensity in Lake Taihu was about 500–1, 000 $\mu E/(m^2 \cdot s)$ in cloudy weather and 2, 000–2, 500 $\mu E/(m^2 \cdot s)$ in clear weather.

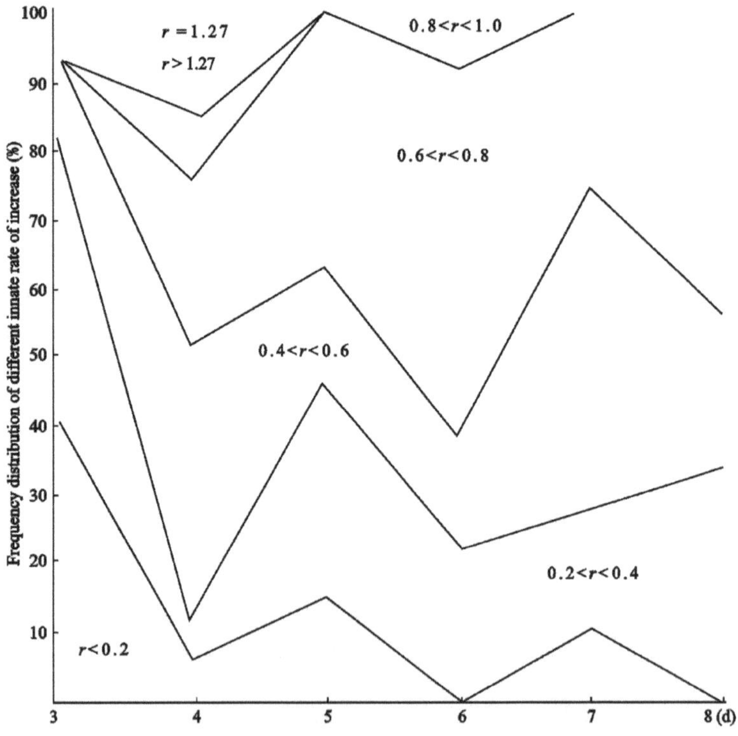

Fig. 6.10 Frequency distribution of the innate rate of increase of *Microcystis*

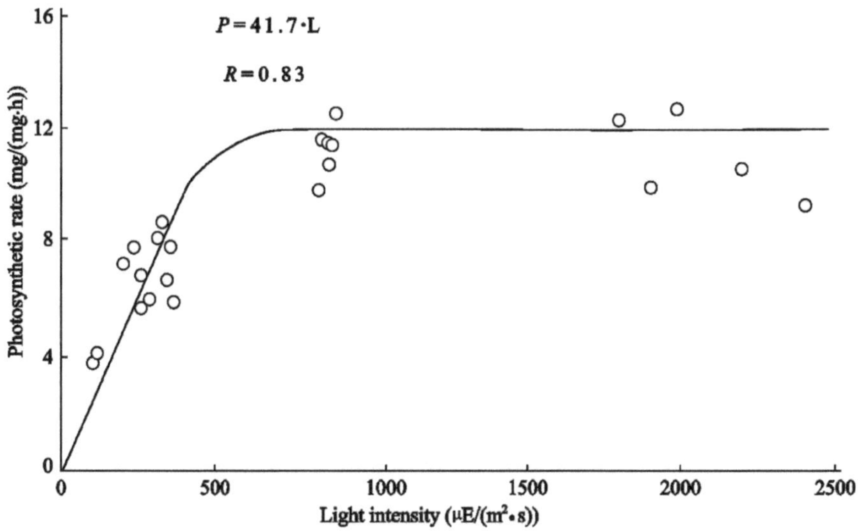

Fig. 6.11 Photosynthesis-Irradiance(P-I) curve for *Microcystis*

There was a significant vertical movement of *Microcystis* across a 24-h period. *Microcystis* floated to the surface at noon but moved down in the morning and afternoon. Furthermore, there were diurnal differences in the sizes of the colonies. Under laboratory conditions, colonies formed in daytime were small, about 7.47–8.95 μm in diameter, and their growth rate was $0.28\,h^{-1}$. In contrast, colonies formed at night were large, about 7.57–13.7 μm in diameter (Chen et al., 1998).

Water temperature significantly influenced *Microcystis* growth. Generally, when water temperature dropped below 13 °C, growth rate slowed. Experiments showed that when temperature was 10°–30 °C, the photosynthetic efficiency of *Microcystis* increased exponentially. When the temperature was above 35 °C, photosynthetic efficiency declined. Experiments showed that 25°–30 °C was the optimal growth temperature, and summer water temperature in the north of Lake Taihu was usually 30°–32 °C (Chen et al., 1998).

Experimental data also showed that 1 mg PO_4^{3-}-P could produce 384.6 μg Chl a, equivalent to 128.2 mg phytoplankton. Similarly, 1 mg total dissolved nitrogen (TDN) could produce 33.3 μg Chl a, equivalent to 11.1 mg phytoplankton. The experiment showed that when PO_4^{3-}-P concentration in the water column was low, alkaline phosphatase activity increased, causing increased decomposition and transformation of organic phosphorus, and increased use of PO_4^{3-}-P concentration by phytoplankton. In contrast, with increased PO_4^{3-}-P concentration, alkaline phosphatase activity was decreased, and PO_4^{3-}-P concentration gradually dropped. Thus, the alkaline phosphatase activity in the phytoplankton was influenced by the PO_4^{3-}-P concentration in the water, and that the threshold value was 0.02 mg/L. When PO_4^{3-}-P concentration was below this threshold, the alkaline phosphatase activity of phytoplankton in Lake Taihu (mainly *Microcystis*) increased markedly (Gao & Gao, 2000).

Formation of blooms was related to strength and direction of both wind and waves. Observation showed that with wind from the south and southwest, Chl a concentration produced by *Microcystis* could be five times higher outside the enclosure than that inside the enclosure. Thus, most of the *Microcystis* outside the enclosure might originate from phytoplankton driven by the wind. The aggregations driven by wind and waves caused the spatial differences of the phytoplankton (Cai and Chen, 1998).

Ingestion of *Microcytis* by crustacean zooplankton was investigated in a series of experiments using an E/Eone particle counter. The crustaceans, dominated by *Ceriodaphnia cornuta*, ingested *Microcystis* 8–15 μm diameter at an average feeding rate of 0.13–0.15 ml/(individual·h). Phytoplankton and bacteria were its main food, however. Some *Microcystis* also could be ingested by rotifers. Furthermore, *Microcystis* was ingested by fish, as well as degraded by bacteria. Nevertheless, because the consumption of *Microcystis* was significantly lower than its growth, it would be difficult to control *Microcystis* blooms using zooplankton.

6.2.3.3 N/P and the Response of Phytoplankton to the Environment

Since 1950, continued economic development around the lake has caused profound changes in the aquatic environment. Because of the rapid growth of industry, agriculture, and urbanization, a large amount of untreated wastewater has been discharged into the lake. The increase of nutrients significantly affected the development of eutrophication. A rapid eutrophication set in 1983–1984, accompanied by elevated concentrations of nitrogen (N) and phosphorus (P) in the water. A significant positive correlation was found between total nitrogen and Chl a concentration, based on the mean data from 1991 to 1995. The correlation coefficient was as high as 0.96, and the formula was as follows:

$$\mathrm{C}_{\mathrm{Chl}a} = -8.5C_{\mathrm{TN}} + 11.5 \qquad (6.4)$$

where $\mathrm{C}_{\mathrm{Chl\ a}}$ was chlorophyll a concentration (μg/L) and C_{TN} was the concentration of total nitrogen (mg/L).

A similar positive correlation was also found between total phosphorus and Chl a concentration.

Between the 1950s and 1980s, the average water level of Lake Taihu was 3.02 m. Since the 1990s, the average water level has increased to 3.16 m, some 12 cm higher than before. The seasonal variation and the average height of the water level directly influenced the growth of aquatic organisms in the lake and may even have caused the disappearance of submerged aquatic vegetation (SAV). In addition, the extensive area of aquaculture accelerated the deterioration and local extinction of SAV in the lake, especially in Wuli Bay. The extinction of SAV greatly decreased the roughness of the lake bed, which in turn directly influenced the concentration of TSM. The higher the concentration of TSM, the lower the transparency of the water. Increasing concentration of TSM caused by the disappearance of SAV caused decreasing water transparency in Meiliang Bay. The correlation between these two parameters is represented by the following formula:

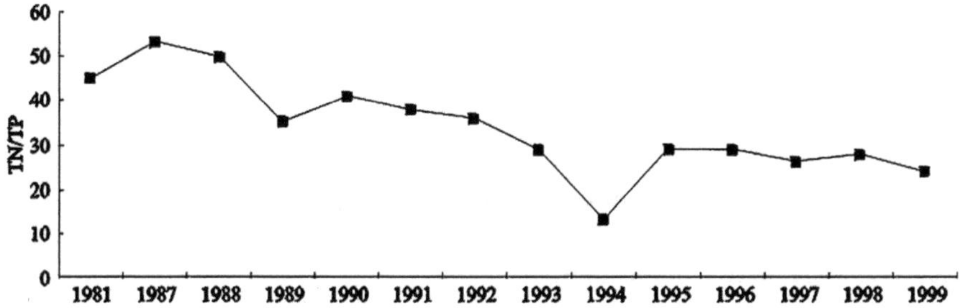

Fig. 6.12 Temporal changes in the ratio of total nitrogen/total phosphorus (TN:TP) in Lake Taihu, 1981–1999

$$ST = 3.28TSM^{0.56}(r^2 = 0.84, n = 94, P < 0.01) \quad (6.5)$$

where ST is transparency (m) and TSM is total suspended matter concentration (mg/L).

The abnormal increase of *Microcystis* was related to the nutrient condition of the water, especially total phosphorus (TP) concentration, and also to the ratio of TN/TP. Usually, the ratio of TN/TP in phytoplankton cells is 16:1. A ratio in the range of 10:1–15:1 is suitable for phytoplankton growth. When the ratio exceeds 15:1, phytoplankton growth is limited by phosphorus; in contrast, when the ratio is below 10, phytoplankton growth is limited by nitrogen.

Before 1980, the ratio was high. In 1981, it was 45:1, much higher than the theoretical value of 16:1. Subsequently, the ratio decreased, and in the late 1990s, it was below 30:1 (Fig. 6.12). The value of TN/TP was gradually becoming closer to the theoretical value, which was more suitable for phytoplankton growth. Therefore, eutrophication will remain a serious problem in Lake Taihu in future years.

6.3 Analysis of the Zooplankton

Weimin Chen

6.3.1 Zooplankton Composition and Abundance

6.3.1.1 Zooplankton Composition

The zooplankton is mainly composed of Protozoa, rotifers, and cladoceran and copepod crustaceans. Zooplankton is both the natural food of fish and an aquatic resource used by people. This community plays an important role in environmental sciences, especially monitoring, because the number of zooplankton species can be regarded as an index of the trophic level of lakes. Therefore, it is important to monitor the zooplankton as an integral part of research on environment and eutrophication.

The zooplankton was sampled on 29 occasions from 1990 to 1995, and 73 genera and 101 species were identified (Table 6.8). The quantity of zooplankton was, in descending order of number of genera and species, rotifers, Protozoa, cladocerans, and copepods; some species were abundant, others were rare (Appendix B).

Species of zooplankton vary considerably in their temperature tolerance. Some have a broad tolerance and are found in the lake throughout the year; others have narrow temperature tolerances, with distinct preferences for either a warm or cool environment, and thus are found only in certain seasons. Most of the zooplankton species in Lake Taihu (some two-thirds of the total) have a broad temperature tolerance and are found throughout the year.

Wu (1962) found 193 species of zooplankton in Wuli Bay in 1951. At that time, Wuli Bay was a small shallow bay, covered by floating vegetation that extended from the bank to the centre. This complex environment allowed for a much greater number of species than found today.

The main difference in the zooplankton community of the 1950s and 1960s as compared with the 1990s is the decrease in the number of protozoan species from 92 to 25. In the same time period, the number of rotifer species remained rather constant, while the numbers of cladoceran and copepod species decreased slightly. Wuli Bay was eutrophic in the 1950s and 1960s, while other areas, especially along the bank of western Taihu Bay, were mesotrophic with abundant macrophytes. In western Taihu Bay, in the first decade of the 21st century, there are few macrophytes remaining; the water column is eutrophic, and the composition of zooplankton has greatly changed, with a marked decrease in the number of species.

Chen, W.M.
State Key Laboratory of Lake Science and Environment, Nanjing Institute of Geography and Limnology, Chinese Academy of Sciences, 73 East Beijing Road, Nanjing 210008, P. R. China
e-mail: chenwm@niglas.ac.cn

Table 6.8 Zooplankton species composition by major taxa in Lake Taihu from 1990 to 1995

Year	Sample number	Protozoa		Rotifer		Cladoceran		Copepod		Total	
		Genus	Species	Genus	Species	Genus	Species	Genus	Species	Genus	Species
1990	3	8	11	13	23	7	8	6	7	34	49
1991	6	13	20	26	39	14	19	11	13	64	91
1992	6	11	18	26	38	13	18	9	11	59	85
1993	5	12	20	27	40	12	17	9	11	60	88
1994	5	10	18	26	40	12	17	8	9	56	84
1995	4	13	20	25	39	12	17	8	9	58	85
Total	29	18	25	29	44	14	19	11	13	73	101

In three main areas of Lake Taihu, the zooplankton composition was distinctive. Wuli Bay in the north had been a center of fish aquaculture for many years. Western Taihu Bay is the main body of the lake, with an extensive surface area. Eastern Taihu Bay is large, long, and narrow, with lush aquatic vegetation. The species of zooplankton, particularly the Protozoa and rotifers, reflected the different environmental conditions and nutritional status in the three areas of the lake.

Among the zooplankton in Wuli Bay, the dominant species were the Protozoa *Askenasia volvox, Didinium balbianii*, and *Vorticella aequilata*; the rotifers *Brachionus angularis, Brachionus calyciflorus, Polyarthra* spp., and *Rotaria rotatoria*; the cladocerans *Bosmina coregoni* and *Daphnia longispina*; and the copepod *Sinocalanus dorrii.*

In eastern Taihu Bay, the dominant species of zooplankton were the Protozoa *Difflugia globulosa* and *Tintinnopsis* spp., with *Vorticella aequilata* (February 1995), *Epistylis rotans*, and *Stentor polymorphus* (August 1995) common in Meiliang Bay and Zhushan Bay in the northwest of the lake; the rotifers *Keratella cochlearis, Keratella quadrata*, and *Polyarthra* spp. (throughout the year), and *Asplanchnopus multiceps, Lecane ungulata*, and *Hexarthra fennica* (seasonally abundant); the cladocerans *Ceriodaphnia cornuta* and *D. longispina*, with abundant *Diaphanosoma brachyurum, Diaphanosoma* sp., and *Moina micrura* in the northwest lake; and the copepods *Limnothora sinensis* and *Sinocalanus dorrii* (common all year), and *Cyclops vicinus, Microcyclops javanus*, and *Schmackeria inopinus* often collected.

In Eastern Taihu Bay, the dominant species were the Protozoa *D. globulosa, Tintinnopsis* spp., and *Trombidium viride*; rotifers (similar to Western Taihu Bay) including *K. cochlearis, K. quadrata*, and *Polyarthra* spp., and in aquacultural enclosures, *B. angularis* or *B. calyciflorus*; the cladocerans *B. coregoni, C. cornuta*, and *D. longispina*, and *Alona guttata, Alonella excisa, Chydorus gibbus, Chydorus sphaericus, Graptoleberis testudinaria*, and *Sida crystallina* generally found in areas with aquatic vegetation (mainly in Eastern Taihu Bay); and copepods *L. sinensis* and *S. dorrii.*

6.3.1.2 Zooplankton Abundance

The densities of zooplankton differ in the three main areas, and decrease from Wuli Bay to western Taihu Bay to Eastern Taihu Bay (Table 6.9). The densities of zooplankton also differ seasonally, increasing gradually from April to August and decreasing gradually from October to February (Table 6.9).

Zooplankton consists of heterotrophic organisms that feed on bacteria, algae, and organic detritus. Carnivorous species feed on other zooplankton and have a higher position in the food chain. Whether food is abundant or not is one of the key factors determining the fate of the zooplankton community, and thus the quantity of zooplankton can reflect the nutritional status of a water body. This concept is the theoretical foundation for estimating the state of eutrophication by the quantity of zooplankton, an approach commonly used in China and other countries.

Table 6.9 Regional and seasonal changes in zooplankton densities from 1990 to 2000

Lake region	Year	Sampling sites	Zooplankton abundance (individuals/L)						
			Feb.	Apr.	Jun.	Jul.	Aug.	Oct.	Dec.
Wuli Bay	1990	6	–	–	–	–	10,301	1,530	5,053
	1991	2	813	1,178	1,148	–	202	202	1,835
	1992	3	835	296	4,511	–	238	238	942
	1993	3	15,101	3,648	2,404	–	107	107	–
	1994	3	602	–	2,107	1,034	68	68	–
	1995	3	2,737	–	1,570	–	1,368	1,368	–
	1998		2,506	–	–	–	804	–	–
	1999		2,602	–	–	–	2,934	–	–
	2000		4,941	–	–	–	5,800	–	–
Western Taihu Bay	1990		–	–	–	–	4,776	1,351	761
	1991	27	80	734	409	–	1,184	214	94
	1992	27	216	94	739	–	3,296	307	14
	1993	27	156	232	415	–	345	68	–
	1994	27	50	–	594	407	246	53	–
	1995	27	541	–	133	–	354	217	–
	1997	27	510	558	1,255	1,833	763	601	777
	1998		2,988	786	321	878	1,490	711	2,024
	1999		637	1,024	1,279	1,153	3,164	2,185	2,713
	2000		1,018	888	1,028	1,524	3,294	806	1,251
Eastern Taihu Bay	1990	8	–	–	–	–	2,845	–	–
	1991	11	11	–	–	–	433	–	–
	1992	11	75	255	237	–	683	640	–
	1993	11	101	208	1,121	–	109	365	–
	1994	11	3	387	412	–	41	248	–
	1995	11	139	–	586	–	331	44	–
	1998		108	–	165	–	203	260	–
	1999		204	–	–	341	1,554	76	–
	2000		135	–	–	–	140	–	–

If no data were collected, this is indicated in the table by –.

The index system is as follows: oligotrophic, zooplankton $< 1,000$ individuals/L; mesotrophic, 1,000–3,000 individuals/L (premesotrophic 1,000–2,000 individuals/L, and late mesotrophic 2,000–3,000 individuals/L); and eutrophic, $> 3,000$ individuals/L. The frequency distribution of the quantity of zooplankton in each of the three areas of Lake Taihu for the 6-year period 1990–1995, classified according to the above index, is shown in Table 6.10. These data are then transformed into the percentage area of Lake Taihu with the different trophic levels in Table 6.11.

In Wuli Bay, the zooplankton reached late mesotrophic and eutrophic levels once or twice a year, with frequencies of 11% and 22%, respectively (see Table 6.10), the highest among the three areas of the lake. Therefore, Wuli Bay was a eutrophic water body from the start. Its trophic state is similar to that of Lake Donghu in Wuchang, Hubei, in which eutrophication is attributable to anthropogenic activity. Lake Donghu is a midsize shallow lake in the Yangtze River that in the 1950s became well known for its beautiful scenery and pure water. However, since the 1980s, the population has increased, industry and agriculture have developed,

Table 6.10 Frequency distribution of zooplankton densities (ind/L) in each year and area of Lake Taihu (1990–1995)

		1990	1991	1992	1993	1994	1995	Total	Percentage
Wuli Bay	Total number of sampling sites	18	18	18	16	14	12	95	
	Density < 1,000 (%)	27.8	50.0	50.0	37.5	71.4	25.0	44.2	43.1
	1,000–2,000 (%)	5.6	38.9	22.2	12.5	21.4	41.6	23.1	23.2
	2,000–3,000 (%)	11.1	11.1	11.1	12.5	7.2	16.7	11.6	11.6
	> 3,000 (%)	55.5		16.7	37.5		16.7	22.1	22.1
Western Taihu Bay	Total number of sampling sites	81	161	160	135	134	108	779	
	Density < 1,000 (%)	38.3	88.2	78.1	89.6	93.3	86.1	82.0	82.0
	1,000–2,000 (%)	33.3	5.0	13.1	8.9	5.2	9.3	10.7	10.7
	2,000–3,000 (%)	13.6	2.5	3.2			0.9	2.7	2.7
	> 3,000 (%)	14.8	4.3	5.6	1.5	1.5	3.7	4.6	4.6
Eastern Taihu Bay	Total number of sampling sites	24	64	64	54	55	44	305	
	Density < 1,000 (%)	37.5	92.2	90.6	96.3	90.9	97.7	88.9	88.8
	1,000–2,000 (%)	37.5	4.6	6.2		9.1		6.9	6.9
	2,000–3,000 (%)	4.2	1.6	1.6	3.7		2.3	2.0	2.0
	> 3,000 (%)	20.8	1.6	1.6				2.2	2.3

ind/L, individuals per liter.

Table 6.11 The percentage area of Western Taihu Bay with different trophic levels in 1991–1995

Year	Oligotrophic (< 1,000 ind/L)	Pre-mesotrophic (1,000–2,000 ind/L)	Late-mesotrophic (2,000–3,000 ind/L)	Eutrophic (> 3,000 ind/L)
1991	59.2	14.8	14.5	11.5
1992	26.8	27.7	12.0	33.5
1993	66.6	31.0		2.4
1994	80.3	17.3		2.4
1995	75.5	22.0	2.5	7.2

ind, individuals.

discharge of domestic sewage and industrial wastewater into the lake has increased, and aquaculture activities, particularly the cultivation of grass carp, have also increased. The result has been local extinction of the aquatic vegetation and a dramatic increase in algal biomass. Currently there are blue-green algae (Cyanophyta) blooms in summer, and the water is turbid and odoriferous, with accompanying changes in water chemistry and biology. Although the composition of the zooplankton has not changed significantly, its abundance has increased, particularly that of Protozoa and rotifers (Dalian Fisheries University, 1985). The situation described above for Lake Donghu also applies to Wuli Bay for the 1980s. The latter lake has been an area of aquaculture production since the 1970s and also has had fertilizer added in attempts to improve the nutritional quality of the water.

In the 1980s, increasing tourism lead to the establishment of numerous hotels around Wuli Bay, and large quantities of domestic sewage, including detergent, flowed into the lake, causing eutrophication. In Wuli Bay, the protozoan *Vorticella aequilata* greatly increased, reaching densities as high as 9,100 individuals/L in February 1993. The genus *Vorticella* is composed of species that feed on organic detritus; therefore, these species are commonly seen in the activated sludge of wastewater treatment plants. The appearance of large numbers of *Vorticella* in the lake indicated significant organic pollution. Many *B. calyciflorus* and *B. angularis* rotifers appeared in Wuli Bay in recent years, reaching densities of 700 individuals/L in April 1993 and February 1995. *Brachionus angularis* lives in natural pools, rearing ponds, and the inlets of shallow lakes and creeks containing abundant organic matter. It is considered to be of the β-medium degradation type (Huang, 2001). *Brachionus calyciflorus* and *B. angularis* have the same ecological behavior, and both of them belong to the β-medium degradation type (Huang, 2001). The appearance of these two rotifer species in Wuli Bay indicated the water had already been polluted by organic matter.

In contrast, in western Taihu Bay, the water quality was generally better than in Wuli Bay. In the former, the frequency of zooplankton abundance reaching late mesotrophic and eutrophic levels was 2.7% and 4.6%, respectively. In some areas of western Taihu Bay, such as Meiliang Bay, the trophic level was similar to that in Wuli Bay. In Meiliang Bay, zooplankton indicated late mesotrophic or eutrophic conditions once or twice a year. Along the shore of Meiliang Bay, the areas of great natural beauty became congested with rapidly growing tertiary industry, such as catering, that produced a large amount of domestic sewage and polluted the area. In August 1995, *Stentor polymorphus* was found at levels as high as 800 individuals/L in Meiliang Bay. This protozoan is commonly found in water polluted with restaurant waste, and it is now abundant in Meiliang Bay. The wastewater from Mashan is discharged into Zhushan Bay.

The fish farms of Meiliang Bay cause significant organic pollution, especially from uneaten fish feed and fish excreta. High densities of Protozoa, reaching 3,900 individuals/L in February 1995, are distinct indicators of this type of organic pollution. The monitoring points along the shore of western Taihu Bay and other bays reached late mesotrophic and eutrophic levels in 1991. Eutrophic levels were reached again in 1992 because of the influence of wastewater; however, because the pollution was erratic, nutrient conditions were erratic. In a wide area around Pingtai Mountain in western Taihu Bay, zooplankton levels usually indicated oligotrophic conditions, and premesotrophic levels were reached only occasionally.

In Eastern Taihu Bay, the water quality was consistently better than that in Wuli Bay and western Taihu Bay. Late mesotrophic or eutrophic conditions were reported in few years and in few sampling locations, with frequencies of 2.2% and 2.8%, respectively. This area is typified by lush aquatic vegetation, and is located in the lower reaches of Lake Taihu. Therefore, nutrients did not accumulate, because they were absorbed by the aquatic vegetation or drained out with the lake water through the river.

The rapid development of fish farming in recent years has been accompanied by worsening water quality (Yang et al., 1995). In monitoring locations in the aquaculture areas, zooplankton reached mesotrophic levels in 1991, and some locations were eutrophic in 1992. In addition, the mouth of Duchun was affected by land wastewater, and zooplankton reached premesotrophic and late mesotrophic levels in 1991, 1992, and 1993.

6.3.2 Succession of Dominant Species

The quantity of zooplankton in Lake Taihu generally peaks in August, and less frequently in April or June. The nature of the peak reflects the composition of the zooplankton community, which is dominated by Protozoa and rotifers, with cladocerans and copepods making up only 1–2% of the total. Peak times for reproduction of the dominant genera are April to June for *Polyarthra* and August for *Difflugia* and *Tintinnopsis* spp. The quantity of the dominant species determined the total amount of zooplankton.

Cladoceran density usually peaks in May. *Bosmina* can be found throughout the year and peaks in autumn; other cladocerans peak in the late summer to early autumn. The seasonal succession of cladocerans is shown in Fig. 6.13. The order of appearance (beginning in spring) usually is *Leptodora kindti-Diaphanosoma* spp. (mainly *D. sarsi*)-*C. cornuta-Moina* spp. (mainly *M. micrura* and *M. rectirostris*)-*Bosmina* spp. (mainly *B. coregoni*) and *Chydorus* spp. in late autumn; in winter, there are few cladocerans because of low water temperature.

Adult calanoids were dominated by *S. dorrii*, which occurred throughout the year with a peak in spring. *Cyclops* peaked in summer and autumn. The seasonal dynamics of the immature stages of copepods is uncertain because of the difficulties in distinguishing between calanoids and *Cyclops* in the nauplius and copepodid stages. Copepods were in a growth stage in February, with few individuals in each stage of development. The densities of calanoid adults were increasing and peaked in spring. Therefore, it could be concluded that the calanoids have one generation per year. Some small *Cyclops* living in warm water may have rapid and continuous generations.

6.3.3 Succession of Zooplankton and Response to Environmental Change

Both species composition and quantity of zooplankton have changed greatly since the 1960s. The overall trends are decreases in the number of large-sized species and in zooplankton quantity, especially for crustaceans. The succession of the zooplankton communities is influenced by abiotic and biotic factors. The abiotic factors include water temperature, transparency, suspended solids, and waves; the biotic factors are phytoplankton, bacteria, fish, and aquatic vegetation.

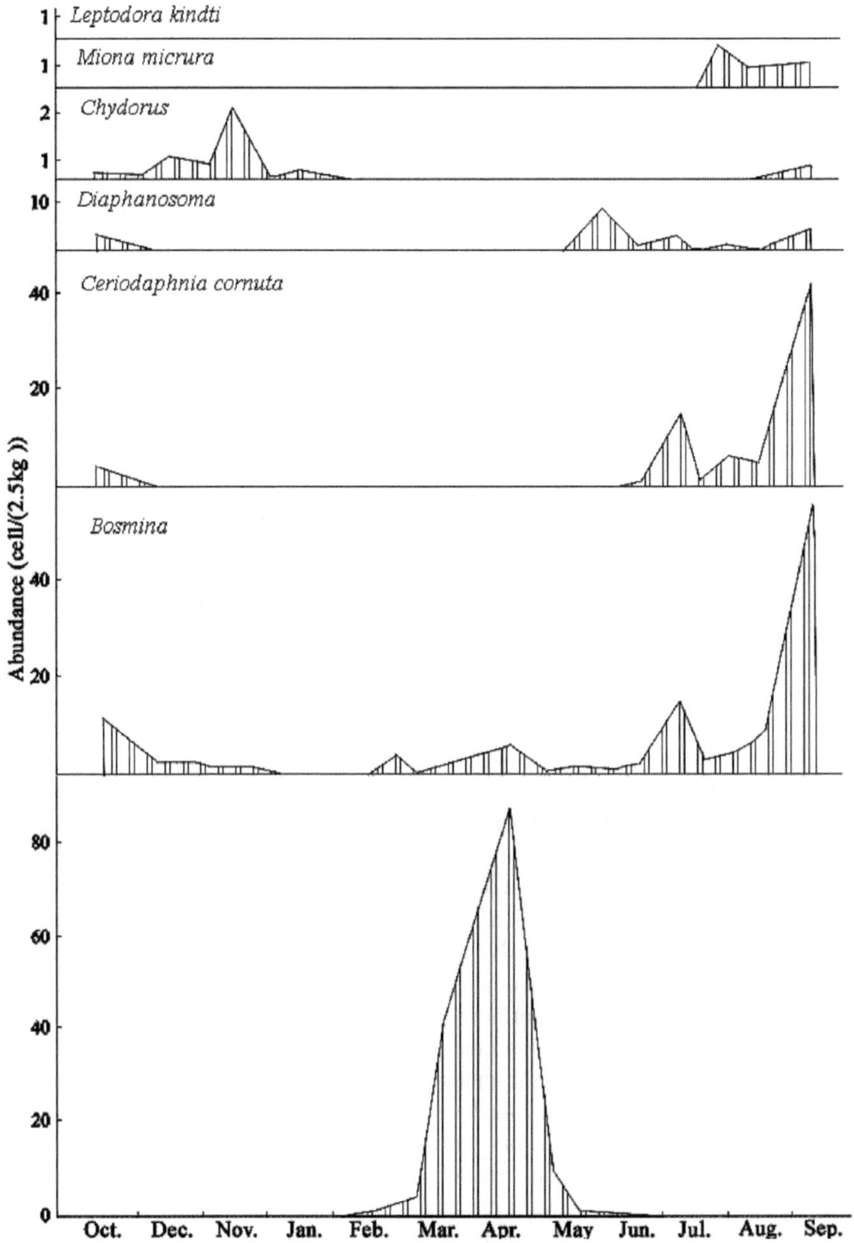

Fig. 6.13 Seasonal succession of cladocerans in Meiliang Bay from 1991 to 1992

Environmental change quickly affects reproduction of the zooplankton. Generally, Lake Taihu has an autumn flood period in August. For example, at the end of August 1994, typhoon No. 17 caused large-scale precipitation and dramatic cooling, resulting in a cataclysm of the hydrological regime, which affected reproduction of *Tintinnopsis* spp. and other Protozoa. Therefore, in 1994, the highest density of zooplankton in each area of Lake Taihu was not observed in August, but in June, when *Polyarthra* spp. predominated.

There is significant interannual variation in zooplankton quantity because of differences in the hydrological regime. Lake Taihu is a subtropical lake, affected by strong seasonal rains and typhoons. A floodgate has been set up at the mouth of the river, which allows the water to drain out during flood years and introduces water from the Yangtze River when the water level in the lake is low. The hydrological regimes are different each year and affect the quantity of zooplankton. Generally, the quantity of zooplankton is low in a flood year and larger in years with normal or low water flow. Such factors should be considered in evaluations of the water quality of Lake Taihu.

Solar radiation is the single most important ecological factor affecting the growth and reproduction of zooplankton. Lake Taihu is located close to the East China Sea and has a monsoon climate. The seasonal succession of climate is distinct. From spring to summer, solar radiation gradually becomes stronger, and water temperature rises; therefore, zooplankton reproduces rapidly in these two seasons. From autumn to winter, solar radiation gradually weakens and water temperature decreases; many zooplankters are dormant, with only a few thriving in cold conditions, and some eurythermal species still present in the lake.

Some unusual zooplankton peaks have been recorded in Lake Taihu. For example, abundant *Vorticella aequilata* in Wuli Bay in February 1993 and in the northwest of West Lake in February 1995 was responsible for the peak zooplankton quantities occurring in the winter of those years.

Phytoplankton and aquatic vegetation significantly influenced the succession and quantity of zooplankton. *Microcystis* blooms were continuously found in Meiliang Bay, so both the quantity and the biomass of oligotrichida and peritrichida increased up to 70% of the total biomass of Protozoa. The eutrophic water provided sufficient food for growth and reproduction of Protozoa. Therefore, the quantity and the biomass of Protozoa were obviously related to eutrophication (Cai, 1998). In addition, because of changes in water quality in Lake Taihu since 1999, curly pondweed was present in some areas of northern Meiliang Bay in spring, so *Chydorus* sp., which had not been seen for a long time, was found again in spring 2002.

The driving factors affecting the annual change of zooplankton quantity are water temperature, transparency, suspended solids, and waves, as already stated (Chen and Qin, 1998). These four factors are complementary to each other and are of particular importance in a large shallow lake such as Lake Taihu. The water was clear and stable when there were no waves, and the concentration of suspended solids was low. When the waves increased, transparency greatly decreased and the concentration of suspended solids markedly increased. Suspended solids affected not only the transparency of the water but also the release of sediment nutrients, which then in turn influenced the growth of phytoplankton and bacteria. During these processes, wave-driven water dynamics played a critical role.

A simulation experiment, conducted in a large laboratory water enclosure, confirmed these relationships (Chen et al., 2000). The experiment was carried out at the Taihu Laboratory for Lake Ecosystem Research (TLLER) from May 8 to June 24, 1999. The bottom of the enclosure was evenly covered with 10 cm sediment, to which Lake Taihu water was added to a depth of 1.2 m. Filtered (33-μm mesh met) water was added every other day to compensate for loss to evaporation. The air temperature was kept constant at about 25 °C. After each treatment was set up, sampling and measurement were carried out for 5 successive days. The experiment was divided into three water currents: in *still water*, the enclosure was filled with Lake Taihu water, covered, and kept still for 10 days, then monitored May 19–23; in *slow current*, water current velocity over the sediment surface was held at 0.124 m/s for 10 days, then with velocity held at 0.124 m/s, monitoring was done June 3–7; and in *fast current*, water velocity was increased and held at 0.319 m/s for 10 days, then monitored June 18–22.

In the still water, TSM concentration ranged from 0.73 to 4.07 mg/L, with a mean of 2.01 mg/L, and transparency was as high as 120 cm. In the slow current, TSM concentration ranged from 11.17 to 14.37 mg/L, with a mean of 12.04 mg/L, and transparency was 65 cm. In the fast current, TSM concentrations ranged from 40.60 to 47.67 mg/L, with a mean of 42.65 mg/L, and transparency was only 25 cm.

In the still water, there were many species of zooplankton; cladocerans were predominant with 87.9% of total abundance, followed by copepods, including *B. coregoni, C. cornuta, D. longispina, D. brachyurum*, and *Moina* spp. In the slow current, *C. cornuta* and *C. sphaericus* were found, and in the fast current, there were only *Harpacticus* sp. and *M. rosea.* Of particular importance was the high percentage of cladocerans in still water conditions. The cladocerans took unicellular algae, organic detritus, bacteria, and Protozoa as food. The high quantity of cladocerans made the quantity and biomass of the algae lower; therefore, abundant cladocerans have important ecological effects.

With increasing current speed from still water to slow current, the percentages of Protozoa and rotifers increased, while the percentages of copepods, and particularly cladocerans, decreased in density. Protozoa (4,820 individuals/L) and rotifers (1,240 individuals/L) occurred in the slow current. The growth of some Protozoa was supported by high levels of suspended solids. The dominant species of rotifers were *Brachionus* spp. and *Keratella* spp., whose preferred food sizes were 0.5–20 μm. When current speed increased from still to slow, the densities of cladocerans and copepods declined to 0.6 and 1.8 individuals/L, respectively, from lack of food. In a fast current, the mean densities of Protozoa and rotifers were 0 and 460 individuals/L whereas the densities for cladocerans and copepods were only 0.2 and 0.5 individuals/L, respectively.

It is evident that water dynamics plays an important role in the succession of zooplankton. In large shallow lakes, water dynamics move zooplankton from one area to another, and also reduce food availability for large zooplankton, such as cladocerans and copepods, by driving sediment resuspension and increasing the concentration of suspended solids. At the same time, increased suspended solids and nutrients caused by waves quickly promote growth of algae, Protozoa, and small rotifers with short life spans. Furthermore, an increase in algae was related to changes in zooplankton density. An increase of zooplankton quantity generally followed an increase of phytoplankton. At the same time, bacteria were the main food of zooplankton, so the amount of bacteria directly affected changes in zooplankton density.

A further factor affecting zooplankton dynamics is the fishery. In 2003, the output of Lake Taihu fishery was 25,000 t. *Coilia ectenes taihuensis* and whitebait composed 50–60% of the total catch. Both these fish species, along with others, feed on zooplankton. Thus, it is clear that numerous complex interacting factors resulted in the sharp decrease in zooplankton and marked increase in phytoplankton density.

6.4 Species Composition, Succession, and Environmental Response of Zoobenthos

Weimin Chen

Zoobenthos, the fauna living at the bottom of the lake, are an important component of the aquatic ecosystem. The zoobenthos of Lake Taihu includes Annelida, Arthropoda (including the larvae of aquatic Insecta), Coelenterata, Molluscoidea, Mollusca, Nemathelminthes, Platyhelminthes, and Porifera.

6.4.1 Community Structure

6.4.1.1 Species Composition

The zoobenthos is dominated by Annelida, Mollusca, and Insecta, with the remaining taxa rare. The Annelida in Lake Taihu belong to the Hirudinea, Oligochaeta, and Polychaeta. Although there was only one species of Polychaeta and several species of Hirudinea, there were numerous species of Oligochaeta. The Mollusca include abundant and widespread Gastropoda and Lamellibranchiata, and these taxa are economically important. Among the Arthropoda, there were few species of Crustacea, but these had large numbers of individuals and were of economic importance. In contrast, among the Insecta, Coleoptera, Diptera, Ephemeroptera, Hemiptera, Odonata, Plecoptera, and Trichoptera were present, with larvae of some species of Chironomidae (Diptera) being both numerous and widespread; some may serve as environmental indicators.

Zoobenthos have a large size, long life expectancy, and a small active area. These organisms respond acutely to environmental change and can provide an index for pollution monitoring and water quality evaluation. The study of the community structure of zoobenthos can be used to objectively analyze and evaluate the nutritional status of the lake.

There are 61 species of zoobenthos in Lake Taihu (Appendix C). Their community structure reflects the location and characteristics of the lake, located in the south of the Yangtze Delta, and linked to the sea by the Yangtze River. It is shallow, with a rather stable water level, which fluctuates only within approximately 2 m; generally, it cannot dry up. The basin of the lake is flat, and includes several large bays. Some estuarine species also occur in Lake Taihu, for example, *Nephthys oligebranchia, Novaculima chinensis*, and *Neorhynchoplax introversus*. During 6 years in the 1990s, 43 taxa were found, including 1 species of Polychaeta, 6 species of

Chen, W.M.
State Key Laboratory of Lake Science and Environment, Nanjing Institute of Geography and Limnology, Chinese Academy of Sciences, 73 East Beijing Road, Nanjing 210008, P. R. China
e-mail: chenwm@niglas.ac.cn

Oligochaeta, 15 species of Gastropoda, 13 species of Lamellibranchia, 3 species of Crustacea, and 5 species of aquatic Insecta and their larvae (Appendix C).

6.4.1.2 Density and Biomass of Zoobenthos

The zoobenthos increased markedly in both density and biomass between 1960 and 1988 (Table 6.12).

There were marked differences in the density and biomass of zoobenthos among the three different areas of the lake in the 1990s (Table 6.13). In the Wuli Bay area, with fish farming and eutrophic water, zoobenthos was dense, but had low biomass because the predominant species were small aquatic Oligochaeta worms and larvae of Chironomidae. In Eastern Taihu Bay, with abundant submerged macrophytes, the zoobenthos had low density and high biomass, consisting principally of large snails and lamellibranchs. In western Taihu, with few submerged macrophytes, the zoobenthos density and biomass were intermediate between those in the other two areas.

6.4.2 Distribution of Zoobenthos

The zoobenthos demonstrated a diversity of morphology, behavior, feeding strategies, and habitat preferences. Aspects of the ecology of the main zoobenthos species are now described.

Nephthys oligebranchia (Polychaeta) is 20 mm long, bears parapodia on each segment of its body that allow it to swim, and is omnivorous. Its preferred habitat is the soft organic-rich mud of East Lake Taihu. There are three common species of Tubificidae (Oligochaeta) in Lake Taihu; *Branchiura sowerbyi* and *Tubifex cathayensis* are large, up to 150 mm long, while another, *Limnodrilus hoffmeisteri*, is only 35–65 mm long. The tubificids dig the anterior part of their body into the mud, extend the posterior part in the water for respiration, and feed upon the organic material in the mud. *Limnodrilus hoffmeisteri* was found in east Lake Taihu, Meiliang Bay, and Zhushan Bay, along with some deltas in the mouth of the lake. The river flow velocity decreases upon entering the lake, and sand and organic substances accumulate there; thus, the deltas become soft substrate, organic-rich areas, suitable for *L. hoffmeisteri*.

Most of the Naididae species (Oligochaeta) in Lake Taihu are less than 20 mm long, except *Branchiodrilus hortensis*, which reaches 50 mm in length. The finger-like gills and long hair-like bristles on the sides of the cephalic gills enlarge the surface area of the body and increase buoyancy; thus, these species can crawl in the bottom of the lake, and float in the water column. All the oligochaetes live among the macrophytes. *Dero digitata* hides itself in a secreted velum hanging on the vegetation. *Slavina appendiculata* cannot float, but crawls among the macrophytes. The mode of feeding still is unknown for several species of Naididae.

Hirudinea, parasitic or semiparasitic zoobenthos, have not been well investigated in Lake Taihu. Some species such as *Glossiphonia complanata* and *Hemiclepsis*

Table 6.12 Temporal changes in density and biomass of zoobenthos between 1960 and 1988

Year	Molluscs		Annelids and insects		Total		Sampling sites and frequency
	Density (ind/m^2)	Biomass (g/m^2)	Density (ind/m^2)	Biomass (g/m^2)	Density (ind/m^2)	Biomass (g/m^2)	
1960	48.00	43.21	23.50	0.89	71.50	44.1	170 sites in July
1980–1981	110.62	44.50	27.12	0.30	137.73	44.80	110 sites once per season
1987–1988	153.75	77.12	32.25	1.47	180.00	78.59	39 sites, four times (Mar., Aug., Oct., Apr.)

ind/m^2, individuals per square meter.

Table 6.13 Density and biomass of zoobenthos in three areas of Lake Taihu, 1990–1995

Lake region		1990			1991						1992		1995
		Aug.	Oct.	Dec.	Feb.	Apr.	Jun.	Aug.	Oct.	Dec.	Apr.	Oct.	Oct.
Wuli Bay	Density (ind/m^2)	2033	1033	2400	1675	97	204	711	230	1445	1478	620	2047
	Biomass (g/m^2)	11.3	26.1	28.8	26.2	19.6	13.2	17.5	7.1	23.1	25.0	3.2	31.6
Western Lake	Density (ind/m^2)	286	296	253	109	131	294	689	369	244	116	351	252
Taihu	Biomass (g/m^2)	72.0	69.3	105.2	26.9	33.8	50.4	98.5	60.4	62.84	27.8	57.7	29.5
Eastern Taihu	Density (ind/m^2)	220	418	378	340	194	152	183	232	273	185	277	400
Bay	Biomass (g/m^2)	150.5	162.7	224.6	99.4	150.6	92.8	132.5	100.4	110.4	140.9	149.8	47.3

ind, individuals.

kasimiana are known to parasitize aquatic animals such as clams, terrapins, and frogs; *Hirudo nippinia* and *Witmania pigra* parasitize livestock and humans. After taking their blood meal from the host, the leeches detach, and fall to the bottom where they are found among the macrophytes, under stones, and especially in the wharf area, where livestock and people gather.

Cipangopaludina cathayensis and *Cipangopaludina chinensis* are large snails, with shell heights of 10–60 mm and widths of 35–40 mm. *Angulagia polyzenata* and three *Bellamya* species are medium-sized snails, with shell heights of 25–27 mm and shell widths of 15–25 mm. Other snails, such as species of Lymnaeidae, Melaniidae, and Planorbidae, are small. The mouth of the snail has an oral ball, bearing the radula (hard, teeth-like structures). The radula enables the snail to obtain nutrients by scraping the tissues of macrophytes and epiphytes, and thus the distribution of snails in Lake Taihu is closely related to that of the macrophytes. Snails are therefore mainly distributed along the bank where the macrophytes grow, and in the harbor in Eastern Taihu Bay and other bays. The large- and medium-sized snails crawl among the macrophytes, and the small species are found among the leaves. *Cipangopaludina cathayensis* and *C. chinensis* are mainly found in the harbor where the submerged macrophytes grow. *Angulagia polyzenata* and *Bellamya aeruginosa* are found near the shoreline where the water is quiet. *Bellamya quadrata, Plarafossurulus sinensis*, and *Semisulcospira cancellata* live among submerged macrophytes. *Alocinma longicornis, Polypylis hemisphaerula*, and *Radix* species inhabit the back of the leaves.

Lamellibranchia is represented by ten species and two subspecies, Corbiculidae by two species, and Mytilidae and Splenacae each by one species. Except for *Limnoperna lacustris*, these mollusks submerge their anterior ends in the mud; water enters the mantle cavity via the siphon, gases are exchanged, and phytoplankton, small plankton, and organic debris are filtered out to provide nutrition. *Limnoperna lacustris* is a small mollusk, 8–30 mm long, found adhering to posts, poles, stones, and the shells of other mollusk species. *Anodonta woodiana, Cristaria plicata*, and *Hyriopsis cumingii* are large, with shells 180–290 mm long, 70–170 mm high, and 45–100 mm wide. The Mollusca in Lake Taihu have been overfished in an effort to satisfy the pearl industry, and as a consequence, large individuals are now rare. *Novaculima chinensis* (Solecurtidae) can reach 46 mm long, 16 mm high, and 10 mm wide; however, such large individuals have not been collected recently. *Corbicula fluminea* is small, with adults reaching 40 mm long, 37 mm high, and 20 mm wide; *Sphaerium lacustre* is smaller, with adults reaching 11 mm long, 9 mm high, and 4 mm wide.

Although all the lamellibranchs in Lake Taihu, except *L. lacustris*, share a common mode of life and feeding strategy, they have different geographic distributions. Some species of *Anodonta* spp. are found in the bays where the water is quiet and the substrate is soft and full of organic substances, such as in Eastern Taihu Bay. In contrast, members of the Unioninae and Corbiculidae prefer flowing currents and sandy substrates and are found mainly in harbors where the rivers flow into the lake.

Among the crustacean arthropods, water lice and gammarids are small, 10–20 mm long. These crustaceans are often found in qualitative samples but rarely appear in quantitative samples; they are mainly distributed in the bays of Lake Taihu. The common shrimp *Polaemon modestus*, 30–40 mm long, is found in western Lake Taihu. This species prefers weak light and undergoes a diurnal migration, staying in the bottom of the lake in the daytime and swimming to the upper level of the water at night. When there is moonlight, the shrimps congregate. One of the other main shrimp species in the lake is *Macrobrachium nipponensis*; at 60–90 mm long, this species is markedly larger than *P. modestus*. Most of *M. nipponensis* are found where macrophytes grow abundantly, mainly in Eastern Taihu Bay. *Caridina* sp. and *Palamortes sinensis* are small shrimps, 20–30 mm long, found among the macrophytes in Eastern Taihu Bay, other bays, and the shore zone. All the shrimps in Lake Taihu have the same mode of life and feeding activity. They live on the bottom of the lake, crawl by pleiopods on the cephalothorax, and swim by appendages on the abdomen. They actively ingest food such as algae, macrophyte debris, small zoobenthos, and mosquito larvae; most of their food is plant material.

Crabs are migratory crustaceans that mate and spawn where the river meets the sea. The eggs attach to the appendages of the female crab, and incubate there. The juveniles grow in the lake, and can attain a weight of 200 g in 1 year. *Neothnchoplex introversus* breeds in the mouth of the river, and this small crab enters Lake Taihu with the tide. After sexual maturation in late autumn, the new adult returns to the river to mate. Crabs burrow in the lake, and actively ingest a diversity of foods such as leaves, stems, seeds, *Corbicula* mollusks, and small fish.

The Chironomidae of Lake Taihu have not been studied in detail, although it is known that some species are widely distributed in the lake. In general studies of zoobenthos, the larvae of *Chironomus* spp. and *Tanypus* spp. have been collected. There are two kinds of *Chironomus* spp., one being 20 mm long, the other 30 mm long. Both have a bright red body and a chocolate-brown head; they appear mostly in summer and autumn. Some *Tanypus* species are about 10 mm long, the body wall is clear and yellowish, and the head is chocolate-brown. *Tanypus* species appear mainly in spring and summer and are distributed where macrophytes grow and along the bank of the lake. The chironomid larva secretes silk from the salivary gland to cover itself, and the nest is covered by soft, organic-rich mud. Chironomids are omnivorous, feeding on debris and small organisms; and usually are found where aquatic earthworms are found.

The zoobenthos in Lake Taihu can be divided into grazers, such as the snails, and omnivores, such as the lamellibranchs, which feed upon debris and algae. Zoobenthos are low in the trophic pyramid, belong to a short food chain, and play an important role in the cycling of material in the aquatic ecosystem of Lake Taihu. Zoobenthos comprise many species and abundant standing crops in Lake Taihu, and most species can be used as fishery resources. When they are caught and removed from the ecosystem, the nutrition load will decrease, slowing eutrophication.

Lake Taihu can be divided into four regions reflecting the habitat and distribution of zoobenthos (Fig. 6.14): (1) Eastern Taihu Bay, Xuekou Bay, and Wangting Bay, with abundant submerged and emergent aquatic plants, where gastropods predominate and oligochaetes are rare; (2) the centre of western Lake Taihu, with sandy sediment, and low levels of organic material ($< 1\%$), total nitrogen ($< 1\%$), and total phosphorus ($< 0.06\%$); because of the poor environment, zoobenthos is scarce, and the predominant species are *C. fluminea* and *Stenothyra globra*; (3)

Fig. 6.14 Distribution of the predominant species of zoobenthos in Lake Taihu. *1*, Chironomidae larvae and Oligochaeta; *2, Corbicula fluminea, Nephthys oligobranchia*, and *Stenothyra glabra; 3, Corbicula fluminea* and *Stenothyra glabra; 4*, Bivalvia

the bank of western Lake Taihu and Meiliang Bay, with abundant sediment, and organic material ($> 1\%$), where the zoobenthos is mainly composed of *S. globra, C. fluminea*, and *N. ligebranchia*, and *B. sowerbyi* is present at the entrance of the lake; (4) Wuli Bay and the mouth of the Liangxi River, which has poor water quality as a result of serious organic pollution, and where the zoobenthos mainly is composed of Oligochaeta (*Limnodrilus claparedianus* and *L. hoffmeisteri*) and chironomid larvae. In the Liangxi River, the water is very badly polluted and dissolved oxygen at the river bottom is almost zero; as a consequence, fish, shrimps, and large benthic invertebrates are locally extinct.

6.4.3 Response of Zoobenthos to Environmental Changes

6.4.3.1 Changes in Species Composition

Comprehensive investigations of the zoobenthos of Lake Taihu in the early 1960s and early and late 1980s showed that the density gradually increased. For Mollusca (mainly *Bellamya* spp. and *C. fluminea*), the density at the beginning of 1980s was 130% higher than in the early 1960s; in the late 1980s, the density was 39% higher than in the early 1980s. Biomass also had increased; at the start of the 1980s, it was 3% higher than in the 1960s, and in the late 1980s, biomass was 42% higher than in the early 1980s. For Annelidea (mainly *L. hoffmeisteri*) and aquatic insects (mainly Tendipedidae larvae), density in the early 1980s was 15% higher than in the early 1960s; in the late 1980s, density was 16% higher than in the early 1980s. In contrast, biomass at the beginning of the 1980s was 67% lower than that at the start of the 1960s, however, in the late 1980s biomass was almost 400% higher than in the early 1980s.

From the early 1960s to the late 1980s, the increase of zoobenthos in Lake Taihu was mainly affected by human activity. The mollusks *Bellamya* spp.and *C. fluminea* are the main fishery resources, being used as food for the human population and also as food for the fish raised in the fish farms. From the 1960s to the 1970s, significant expansion of the fish breeding activities in Lake Taihu increased the need for food for these fish, and snails and clams were harvested in large quantity. Over the years, harvesting methods have also placed further demands on the mollusk populations. Instead of the previous net-harvesting, dredge pumps now are used to collect the mud with snails and crabs. Catch efficiency was greatly increased, and the resources were largely destroyed. During 1988–1996, there were more than 400 ships catching snails each year; not until 1999 had their number decreased to 100. Annual harvesting reaches approximately 10,000 t, with the main target being *C. fluminea*. Because *Bellamya* spp. live in the macrophyte areas, they are difficult to harvest (Cao et al., 2000).

Based on the monetary value, the fishery for *C. fluminea* is second in importance to that for silverfish in Lake Taihu. A study of *C. fluminea* resources from 1997 to 1998 showed that the stocks in Lake Taihu comprised approximately 4.7×10^4 t and that about 5.8×10^3 t was harvested annually. This resource is unevenly distributed, with 72% of the lake area having low-density populations of only 10 individuals/m^2. Small *C. fluminea* predominate, with those $<$ 1.5cm constituting just over 50% of the population, and those $>$ 1.8cm long equalling only 14%. The mollusks reach sexual maturity when shell length is approximately 11 mm. Continual unchecked harvesting may destroy this resource, so the Fishery Resource Management Department now regulates the mesh diameter of the dredge pump, thus selectively catching *C. fluminea* by size, and regions of the lake are protected for breeding. Because *C. fluminea* has a very high reproductive potential, and matures in 1 year, the resource can be restored quickly; this is the main reason why the density of Mollusca increased in the early 1980s. Biomass and density of Mollusca do not necessarily increase concurrently. During the population reestablishment of *C. fluminea*, young individuals constituted a large percentage. The average individual weight was 0.9 g at the beginning of the 1960s, but it declined to 0.4 g in 1980s. Although the density more than doubled, the biomass did not increase obviously. In the late 1980s, the average individual weight was 0.5 g, although the percent increase in density decreased (from 130% in the early 1980s to 39%); biomass increased from 3% in the early 1980s to 42% in the late 1980s.

The molluscan resources of Lake Taihu went through three stages. The first stage was a steady increase of the population in the 1970s, with good environmental conditions suitable for growth and reproduction of *C. fluminea*. Also during this period, a large area of Lake Taihu was enclosed for fish farms (from 1969 to 1974, the enclosed area was only 1.3 km^2); however, this did not have a marked impact on the snail resources. In the second stage (the 1980s), the commercial harvesting methods and intensity did not have a marked negative impact on the snail and clam populations. In 1980, commercial harvesters began to use ships, and by 1984 the resources had decreased. As a consequence, measures to protect *C. fluminea* were introduced; fishing was stopped for half a year and the protected areas were enlarged. The resource quickly improved. During the third stage, the *C. fluminea* resources gradually decreased. The quantity of *C. fluminea* harvested for export had risen since 1990, and there were more than 400 ships working in the lake. Overfishing led to a decline in the stock, and although the number of ships was much reduced in 1997, the *C. fluminea* stock had declined to nearly 75% of levels 5 years earlier (Cao et al., 2000).

Discharge of wastewater and aquaculture directly influences the accumulation of organic material. For the former, Xiaomeikou Bay, in the south of Lake Taihu, provides a typical example. Although *Limnodrilus hoffmeisteri* had not been collected in this area in the 1960s, it had become common by the early 1980s. The reason is thought to be the establishment of a paper mill on the upper stream, discharging a large volume of organic waste into the lake and providing suitable conditions for *L. hoffmeisteri*. Aquaculture developed in the early 1980s, with the net-pen fish farms being concentrated in the large bays such as Eastern Taihu Bay and Zhushan Bay. Much bait was thrown into the lake, and the uneaten portions and the waste from the fish (principally faeces) accumulated at the bottom of the lake, forming a habitat for worms and chironomid larvae. The density of worms in the farms was 300–400 individuals/m^2, six to eight times higher than that in the uncultivated areas (Wu, 1991).

The rich community of zoobenthos reflects the location and characteristics of Lake Taihu, situated in the south of the Yangtze Delta, connected with the river close to the sea. The lake has a north-subtropical climate with abundant rain, which provides it with abundant water. Although it is shallow, the water level is stable, with an average depth range of 1.26–1.76 m across many years (Sun & Huang, 1993), and the lake usually does not dry up. The bottom is flat, and there are several large bays. The species of zoobenthos that are typically found near the mouth of rivers, such as *Neothnchoplex introversus, Nephthys oligebranchia*, and *Novaculima chinensis*, are also found in Lake Taihu, because the lake communicates with the river and the sea. In contrast, these species are not found in Lake Boyang or Lake Dongting, large shallow lakes in the middle reaches of the Yangtze River; these lakes do communicate with the Yangtze River but are not close to the sea.

Lake Taihu lies in an area that is densely populated and has well-developed agriculture and industry. It has a long history of exploitation, and anthropogenic impacts on the structure, quantity, and distribution of zoobenthos have become even greater in recent decades. For example, engineering activities and enclosing tideland for cultivation directly and indirectly reduced the available habitat for zoobenthos. Overfishing has resulted in decreased biodiversity of zoobenthos. With increasing development of agriculture and industry in the drainage area, ever-increasing nutrients flow into the lake, worsening eutrophication. As a consequence, the biomass of some pollution-resistant species such as *L. hoffmeisteri* and chironomid larvae has increased.

Estimation of the quantities of active zoobenthos (such as swimming shrimps and burrowing crabs) by common sampling methods is more difficult than for less mobile species. Because both shrimps and crabs are of economic importance, data on fisheries catch (supplied by the Fisheries Department) can be used to indicate biomass. *Palaemon modestus* is dominant in the yield of shrimp, constituting about 52% of the total shrimp harvest. The yield of *Macrobrachium nipponensis* ranked second to that of *Palaemon modestus*, composing about 28%; *Palamoretes* sp. and *Caridina* sp. represent about 19.5%. The total annual shrimp harvest has been 500–700 t, which is 5–10% of the total harvest of the fisheries. The annual shrimp harvest fluctuates greatly, with the maximum being 1,155 t in 1969. The annual variation in the harvest has been influenced greatly by hydrometeorological conditions and fishing intensity in recent years. The maximum annual harvest of crab was 750 t; however, floodgates currently prevent the young crabs from returning to the lake, and there is no more harvest.

6.4.3.2 Use of Zoobenthos as Indicators of the Nutrient Status of the Lake

Benthic invertebrates are key components of the lake ecosystem, and their community characteristics are comprehensive indicators of water quality, especially of nutrient status. For example, the numbers and proportions of Mollusca, Oligochaeta, and aquatic Insecta larvae are considered an index of nutrient status; lakes dominated by Mollusca indicate oligotrophic conditions; lakes with similar numbers of Mollusca, Oligochaeta, and midge larvae are mesotrophic; lakes with numbers of Oligochaeta and midge larvae exceeding that of Mollusca are eutrophic. Zoobenthos and water chemistry data collected in 1990–1995 (Table 6.14) showed that the environments of Lake Taihu could be divided into three types. In the hypertrophic area, specifically Wuli Bay, polluted by organic substances, Oligochaeta and Chironomidae larvae predominate absolutely. In the eutrophic area, western Lake Taihu, Mollusca, Oligochaeta, and Chironomidae larvae dominate the zoobenthos. In the mesotrophic area of the lake, Eastern Taihu Bay, Mollusca are predominant. The nutrient status of some areas changed greatly from 1990 to 1995. In Zhushan Bay in 1990, snails and *C. fluminea* were the predominant zoobenthos, whereas in 1995, aquatic Oligochaeta and Mollusca were dominant and the water body was eutrophic. In the Liangxi River in 1995, no large benthic invertebrates were collected, showing that water quality was bad. In Meiliang Bay, water quality has worsened gradually with increase in the eutrophic level, and aquatic Oligochaeta and chironomidae larvae dominate the zoobenthos.

Table 6.14 Density of four taxa of zoobenthos in three areas of Lake Taihu, 1990–1995 (individuals/m^2)

Lake area	Aug. 1990			Aug. 1991		
	Wuli Bay	Western Lake Taihu	Eastern Taihu Bay	Wuli Bay	Western Lake Taihu	Eastern Taihu Bay
Oligochaeta	70	43	12	550	27	10
Mollusca Gastropoda	0	106	180	0	77	142
Mollusca Bivalvia	0	124	18	0	590	27
Chironomidae larvae	1936	13	10	161	5	4
	Oct. 1992			Oct. 1995		
Oligochaeta	500	40	20	460	26	65
Mollusca Gastropoda	0	76	252	0	56	327
Mollusca Bivalvia	0	200	80	0	160	6
Chironomidae larvae	120	10	5	1580	10	3

6.5 Fish Composition and Fisheries

Weimin Chen and Qinglong Wu

6.5.1 Fish Population Structure

Fishes of freshwater ecosystems are of enormous economic importance. In Lake Taihu, the diversity of freshwater fishes has decreased in recent decades. Investigations on the fishes carried out between 1959 and 1985 by the Nanjing Institute of Geography and Limnology Chinese Academy of Sciences (1965), the Department of Biology, East China Normal University (1959), and the Freshwater Fisheries Research Center, Chinese Academy of Fishery Science recorded a total of 106 species (Appendix D). "The History of Fishery in Lake Taihu" states that the 1981 investigation found 72 of the 106 species. Deng et al. (1997) reported that a total of 40 species of fish and shrimp were collected in 169 sampling occasions from February 1993 to January 1995. Fish collections by Zhu (2004) in the intersection between the east and west portions of Lake Taihu, as well as the appraisal for fish specimens, yielded a total of only 47 species. The fish species recorded as locally extinct in "Fish Name List in Lake Taihu" include *Acipenser sinensis, Macrura reetesii, Mugil cephalus, Mugil soiuy, Tracbiidermus fasciatu, Cynoglossus gracilis, Cynoglossus purpureomculatu*, and *Cynoglossus trigmintnus*. At present, there are an estimated 50 species of fishes in Lake Taihu (Appendix D).

The composition of the fish community is characteristic of the temperate zone. The fish fauna in Lake Taihu is a combination of species from the Indian plain and from Chinese rivers and streams. This community is consistent with that in the middle and lower reaches of the Yangtze River. Most fish species belong to the family *Cyprinidae* (carps), and the survival of these wild fish in changing environments is the result of a long-term evolutionary process.

There are three ecological groups of fishes in Lake Taihu. The main fish are nonmigratory and include species such as *Cyprinus carpio, Carassius auratus, Parabramis* sp., *Megalobrama* sp., *Erythroculter* sp., *Culter erythropterus, Reganisalanx brachyrostratis, Protosalanx hyalocranius, Neosalanx tanghahkeii taihuensis, Neosalanx oligodontis*, and *Coilia ectenes taihuensis*. The smaller second group includes fish that migrate between the river and the lake, such as *Coilia ectenes* and *Anguilla japonica*. The third group is found in Lake Taihu only because of the artificial restocking programs, and represents semimigratory fishes that travel between rivers and the lake, such as *Mylopharyngodon piceus, Ctenopharyngodon idellus, Hypohthalmichthys molitrix*, and *Parasilurus asotus*.

Chen, W.M.
State Key Laboratory of Lake Science and Environment, Nanjing Institute of Geography and Limnology, Chinese Academy of Sciences, 73 East Beijing Road, Nanjing 210008, P. R. China
e-mail: chenwm@niglas.ac.cn

6.5.2 Changes in the Composition of Fish Catches

6.5.2.1 Composition of Fish Catches

The main commercial fishes include *Coilia brachygnathus, Coilia ectenes taihuensis, Neosalanx tanghahkeii taihuensis, Aristichthys nobilis, Hypohthalmichthys molitrix, Cyprinus carpio, Ctenopharyngodon idellus, Protosalanx hyalocranius, Protosalanx hyalocranius, Parabramis pekinensis, Megalobrama amblycephala, Megalobrama terminalis, Erythroculter ilishaeformis, Erythroculter mongolicus, Hemibarbus maculatus, Siniperca chuatsi, Mylopharyngodon piceus, Carassius auratus, Parasilurus asotus, Ophiocephalus argus, Pseudobagrus fulvidraco, Odontobutis obscure, Hemirhamphus intermedius*, and *Elopichthys batnbusa*. The total catch and the proportions of the main commercial species for 1952–2000 are given in Table 6.15.

The natural environment of Lake Taihu is reflected in the diversity of its fish species. As a shallow lake, Lake Taihu has a long retention time. The sediments have high organic matter contents. Moreover, there are macrophytes along the lakeshore of both Eastern Taihu Bay and western Lake Taihu. Therefore, there is abundant benthos that serves as a food source of fish such as *Mylopharyngodon piceus, Cyprinus carpio, Hemibarbus maculatus, Pseudobagrus fulvidraco*, and *Odontobutis obscure*. Other species, such as *Carassius auratus, Ctenopharyngodon idellus*, and *Squaliobarbus curriculus* feed on the lake-bottom algae, detritus, and plant materials. The food chain of these fish is quite short, and they act as primary consumers. It is estimated that these fishes account for about 20% of the annual fish harvest. Species that feed on zooplankton, such as *Coilia ectenes taihuensis* and *Reganisalanx brachyrostratis*, are the predominant fish, accounting for about 50% of the fishery harvest. Zooplankton serve also as a source of food for *Parasilurus asotus* and the young

Table 6.15 The total fish yield and the percentages of catch for different species in Lake Taihu

Year	1952	1963	1973	1983	1993	2000
Total fish yield (10^4 kg)	406.3	855.23	1,151.39	1,531.62	1,407.24	2,520.67
Coilia spp.	15.8%	55.2%	57.4%	46.2%	24.8%	50.6%
Ice fishes	12.9%	4.1%	8.0%	5.6%	12.5%	6.4%
Shrimps	15.1%	6.7%	9.6%	10.0%	3.6%	7.1%
Hypohthalmichthys molitrix and *Aristichthys nobilis*	15.7%		7.5%	8.6%		6.4%
Cyprinus carpio and *Carassius auratus*	18.2%		7.8%	6.5%		14.7%
Mylopharyngodon piceus and *Ctenopharyngodon idellus*	2.8%		0.8%	1.6%		1.7%
Culter Basilewsky	5.5%		3.5%	2.0%	6.6%	0.7%
Others	13.9%		5.5%	19.6%		12.5%
Amount (kg/hm^2)	17.25	36.00	48.00	63.75	58.50	105.00

Data from Lake Taihu Fishery Administrative Committee of Jiangsu Province.

of some carnivorous fish; therefore, the zooplankton faces strong predation pressure, an important reason for eutrophication. There are many species of carnivorous fish such as *Culter erythropterus, Erythroculter dabryi, Erythroculter monogolicus, Elopichthys batnbusa, Siniperca chuatsi*, and *Siniperca schorzeri*. Although they represent only a relatively small percentage of the total fish harvest, they are important in fish interrelationships as they have a high position in the food chain.

As a consequence of natural changes in the environment and the disturbances from economic activities, especially the building of massive dams on the Yangtze River and enclosing of tidelands for cultivation, the quantity of migratory, semimigratory fish, and nonmigratory fish that spawn in the shore zone has been reduced significantly. Moreover, most fishes in the lake tend to be young, which has lead to the predominance of two species, *Coilia ectenes taihuensis* and *Reganisalanx brachyrostratis*, as the dominant fish species and share of the fishery at present.

6.5.2.2 Development of the Fish Community

Changes in the fish catches reflect changes in structure of the fish community. From 1952 to 1963, Lake Taihu experienced a rapid growth in fish catches. The increase in annual harvest was rapid, with the highest fishing catches surpassing 1×10^4 t per year. During this period, *C. ectenes taihuensis* represented 15.6–45.7% of the total catch, and *H. molitrix, P. asotus, C. carpio, C. auratus*, and *Parabramis* sp. accounted for 33.35–40.78% of the total catch. In addition, the migratory and semimigratory fishes also were represented in the total catches. From the end of 1958 to the beginning of the 1960s, hydraulic engineering construction of the floodgate to build the dam separating the river and the lake caused obstruction of the return channel for migratory and semimigratory fish, which gradually affected fishery production. The production of some migratory fish, such as *M. reetesii* and *C. ectenes*, was also reduced gradually. 1964–1973 was a relatively stable period for fisheries, and the annual average catches surpassed 1×10^4 t.

Dams were constructed successively on the canal that leads to Lake Taihu, and hindered or blocked the young of some migratory and semimigratory fishes from entering the lake, causing a great decline in aquatic resources. The *C. ectenes taihuensis* catch of small- to medium-size fish in the lake increased to 46.2–64.3% of the total catches during the period after 1973. The decreasing catches of large-size fish with high commercial value, such as *H. molitrix, P. asotus, M. piceus*, and *C. idellus*, were offset by the increasing catch of *C. ectenes taihuensis*, causing overall catch to increase. During this time, the number of motorized sailing vessels increased. The establishment of fishery production management provided a commitment to protect the natural fish populations, and resulted in the release of fish fry into the lake, which was important for increasing the fish catches.

From 1975 to 1983, the fish catch rose again. The fish catches per year surpassed 1×10^4 t, with an average of 1.2×10^4 t. The catches of *C. ectenes taihuenisis* reached peaks of 47–63% of the total catches in the lake, similar to catches from the 1960s. Although the quantity had changed slightly, the catch of large-size commercial fish (*H. molitrix, P. asotus, M. piceus, C. idellus*, and *Parabramis* sp.) was still somewhat low, and only accounted for 16% of total catches. During this period, enclosure of tideland for cultivation took place more rapidly than ever before. The areas of tideland enclosed for cultivation were 6, 017 hm^2 since 1949, 923 hm^2 in the 1950s, 6773 hm^2 in the 1960s, 8126 hm^2 in the 1970s, and 105 hm^2 in the 1980s (Dou et al., 1988). The water in these enclosed areas was originally shallow and rich in macrophytes. With the water surface reduced, the fish spawning area also sharply declined. The quantity of both young and adult fish in the lake decreased each year. *Erythroculter mongolicus*, which feeds on *C. ectenes taihuenisis*, decreased in number along with its food source. The reduction in macrophytes caused a corresponding reduction in spawning sites, eggs, and hatching juvenile fish because the nursery habitats were reduced. At the same time, fishing capacity and intensity were increasing. The combined result was a young and small-size fish stock. The numbers of *C. ectenes taihuensis* and *R. brachyrostratis* increased rapidly at the same time.

In 1984–1993, fish catches rose again and average annual catch was 1.5×10^4 t. The catch of *C. ectenes taihuensis* represented 33.5–54.7% of total fish catches, with the percentage slightly lower than in the 1970s. In 1984, the people's government of Jiangsu Province issued temporary restrictions on fishing. The increase in the number of fry, coupled with improvements in the quality of the released fry, led to improved fish production. Black carp and grass carp represented 20% of total fish catches, Chinese carp and crucian carp 30%, silver carp 33%, and *P. asotus* 17%. The survival rate, increment rate, and the returned catching rate increased in the lake. The catch of ice fishes surpassed 1,000 t/a, and in 1985, 1986, and 1990, it exceeded 2,000 t, greatly enhancing commercial revenue.

In 1984, a pen culture experiment at the agriculture experimental station in Eastern Taihu Bay, was successful, with a yield of 100 kg/hm^2. Aquaculture paved a new way for the country to explore development of experimental fisheries. Initially aquaculture had low output, low utilization, and low benefit; however, this situation changed. After long-term extensive use of net-pens (net-enclosed areas for fish cultivation), the high investment, high delivery of feed, and high output negatively affected the water quality and the environment.

The period 1994–2000 represented another stage of total fish catches. Annual average catches were about 2.0×10^4 t, which represented a yield per unit area of about 105 kg/hm^2 (not including aquaculture). The percentage of *C. ectenes taihuensis* in total fish catches remained at quite high levels. When *C. ectenes taihuensis* reached a length of 130 mm, it began to eat *R. brachyrostratis*. As most *C. ectenes taihuensis* were harvested above 130 mm, this was one of the impacts on the *R. brachyrostratis* resources. According to Tang (1987), there is a functional relationship between *C. ectenes taihuensis* and ice fishes. The increase of *R. brachyrostratis* coincided with the decrease of *C. ectenes taihuensis*. In 1985 and 1986, the catch of ice fishes surpassed 2000 t and was related to decreased catches of *C. ectenes taihuensis*. In the early 1990s, with the changes in the environment and the increase of fishing intensity, catches of ice fishes dropped each year from 1500 to 500 or 600 t during 1995, 1996, and 1997, respectively. The catch of ice fishes was restored to more than a thousand tons only after government restrictions on fishing took effect. Although the catches of *H. molitrix* and *P. asotus* increased along with the size and amount of restocking, the overall catches were similar to those in the 1970s and 1980s, but were only 15.7% of the catches in the 1950s. In the period 1994–2000, the catch of *M. piceu* and *C. idellus* was similar to that of *H. molitrix* and *P. asotus*. The catch of *C. carpio, C. auratus*, and *Parabramis* sp. increased along with the quantity of restocked juvenile fish. The catch of these fishes reached 14.7% in 2000, surpassing catches in 1970–1980, but the catch was still low compared with 18.2% in 1952.

6.5.3 Development of Hatchery Fishery and Its Effect on the Environment of the Lake

6.5.3.1 Development of the Hatchery Fishery

During the 1950s, massive construction gates and dams were built, which blocked the channel of fish migration. The massive reclamation of land by enclosing and filling the lake bottom during the following decade destroyed some of the fish spawning areas, which led to a sudden drop in fisheries resources. In 1964, a fishery management committee was established. They released hatchery-raised juveniles of a variety of species, including *Parabramis* sp. and the river crab (Table 6.16). Overall, the fishery industry in Lake Taihu mainly reflects the measures of artificial restocking.

6.5.3.2 Development of Aquaculture and Its Impacts on the Environment

Aquaculture of Lake Taihu is divided into fish pond culture in reclamation areas and net-pen culture (cultivation enclosed by nets in the lake), net-box culture, enclosure culture and pen culture, and three-net culture, but in present Lake Taihu, the "three-net" cultivation in the lake is extremely limited.

The pond culture fishery in reclaimed shoreline areas of the lake was established around the lake's perimeter. From 1976 to 1986, its development was rapid until early 1990 when the area of cultivation basically stabilized at about 3300 hm^2. The catches were about 25,000 t. The main fish were species that feed on grasses. *Ctenopharyngodon idellu* and *Parabramis* sp. accounted for about 80% of the total output (Table 6.17). The other fishes mainly included *Hypohthalmichthys molitrix, Parasilurus asotus, Carassius auratus*, and *Cyprinus carpio*. Fishes in pond culture are fed using macrophytes from Lake Taihu, *Sorghum sudanensa* and *Lopium perenise* planted on the pond dam and ridges between ponds, as well as plants such as wheat, and pellet fish food. The average harvests of the higher macrophytes were more than 500,000 t and of graminaceous grass nearly 200,000 t, and 1.18 kg food was used to produce 1 kg fish in the 1990s during the intermediate stages of the pond culture (Li, 1995).

During 1970–1980, a large-scale enclosure cultivation experiment was performed in Gonghu Bay by the Lake Taihu Fishery Administrative Committee of Jangsu Province, but this ultimately failed. In 1982, Nanjing Institute of Geography

Table 6.16 Annual numbers of fish fry released by artificial restocking in Lake Taihu (in 10^3 tail)[1]

Year	Silver carp	Big head	Black carp	Grass carp	*Megalobrama amblycephala*	*Cyprinus carpio*	*Carassius auratus*	Total
1957	60.00	60.00						120.00
1966	140.60	3.30	32.00					175.90
1967	48.00	67.00	25.00					140.00
1968		11.80						11.80
1969	72.00	52.00						124.00
1970	93.00	103.00	29.00		25.00			250.00
1971	56.00	67.00	52.00		12.00			187.00
1972	179.00	101.00	8.00		7.00			295.00
1973	109.40	111.80	14.30		277.40			512.90
1974		2.90	30.48		323.33	0.92		357.64
1975	31.68	391.93	18.86		396.52	117.32		982.82
1976	182.00		33.00		507.00	13.00		735.00
1977	211.42	324.43	60.35	23.03	416.00	18.44		1,053.67
1978	287.81	273.04	54.89	8.29	216.44	18.81		859.28
1979	266.06	361.28	73.61	15.08	307.94	2.82		1,026.79
1980	246.35	171.06	35.76	38.99	112.03	42.68		646.87
1981	314.30	232.76	17.82	115.60	297.31	61.22	6.86	1,047.12
1982	263.26	124.54	11.95	228.01	50.03	30.38		708.17
1983	253.18	269.80	30.39	105.31	40.71	119.58	8.24	827.21
1984	270.80	177.80	10.40	46.00	122.60	106.20		733.80
1985	240.24	337.93	0.48	60.13	117.68	50.84	11.33	818.63
1986	420.50	143.00	162.40	36.50	109.00	148.60		1,020.68
1987	123.55	50.09	172.57	46.60	472.75	77.05	2.11	944.71
1988	3.49	934.68	2.79	7.53	46.93	2.23	2.98	1,000.63
1989	213.03	138.23	56.49	19.04	123.64	246.23	58.62	1,355.27
1990	38.73	274.28	90.97	0.95	107.39	39.10	14.70	566.12
1991	98.14	579.12	72.79	41.38	282.39	11.06	12.41	1,098.34
1992	280.12	642.77	15.61	46.04	114.60	25.83	75.40	1,200.37
1993	28.11	1103.97	118.29	120.85	191.72	11.88	0.70	1,575.52

[1] Taihu Fishery Administrative Committee, 1995. The Management of Lake Taihu Fishery in 30 Years (1964–1994).

Table 6.17 Fishes species in pond culture and catches of the reclamation area in Eastern Taihu Bay

Year	Pond area (hm^2)	Total fish catch (t)	Catch of *Ctenopharyngodon idellu* (t)	Catch of *Parabramis* sp. (t)	Catch of other fishes (t)
1976	552.67	2,061	585.324	566.775	908.901
1977	675.53	2,767	774.76	702.818	1,289.422
1978	797.33	3,064	937.584	845.664	1,280.752
1979	910.87	3,616	1142.656	954.624	1,518.72
1980	982.00	3,745	1295.77	1,052.345	1,396.885
1981	1,034.00	4,013	1,444.68	1,099.562	1,468.758
1982	1,085.73	4,412	1742.74	1,292.716	1,376.544
1983	1,371.33	5,364	2,059.776	1,534.104	1,770.12
1984	1,733.33	7,775	3,187.75	2,192.55	2,394.7
1985	2,411.00	12,435	5,396.79	3,270.405	3,767.805
1986	3,418.27	17,788	7,044.048	5,087.368	5,656.584
1987	3,659.80	20,932	9,712.448	5,923.756	5,295.796
1988	3,447.27	22,534	1,0861.39	6,377.122	5,295.49
1989	3,529.87	23,015	12,059.86	5,477.57	5,477.57
1990	3,419.33	23,125	1,3320	5,041.25	4,763.75
1991	3,452.67	24,488	14,178.55	5,730.192	4,579.256
1992	3,306.00	24,205	13,724.24	6,317.505	4,163.26

and Limnology, Chinese Academy of Sciences, conducted an experiment in aquatic agriculture net-pen culture with an area of 0.15 hm^2 in Eastern Taihu Bay. In 1983, it was expanded to about 2.67 hm^2. Before 1990, it was limited to northwest from Jiaobai port to the Jishan port and the end of northeast. The cultivated species were *Ctenopharyngodon idellus* and *Parabramis* sp. After 10 years of technical exploration and slow development, by 1993 net-pen culture had spread to the entire lake, and cultivation, in addition to *Ctenopharyngodon idellus* and *Parabramis* sp., included the crab in nearly 40% of the area. After 1993, the net-pen aquaculture technique rapidly developed. In 2000, net-pen culture encircled the entire lake, with a cultivation area of $2,833\,hm^2$, accounting for 21% of the total area of Eastern Taihu Bay. The cultivation harvest was 3,317 t, with a value of nearly a hundred million RMB. The catches and value of the crab culture represented more than 80% of the net-pen culture and the total catches. At present, except for ship channels, all other areas of Eastern Taihu Bay are nearly completely covered by net-pen facilities.

The main impact on the environment by net-pen culture was an upsurge in nutrient levels, causing an increase in phytoplankton, and deposition of pollution aggregates (bait remnants and fish excreta) on the surface layer of the sediment (Wu, 2001). The high number of fish reduced the abundance of water grasses and shellfish. The rate of water flow was reduced 27–40% in the net-pen culture region. The large quantity of bait also caused a marked increase in the amounts of nitrogen and phosphorus, causing turbidity to rise and transparency to drop. In the net-pen culture yield of $750–7,500\,kg/hm^2$, the quantity of N and P in the water was equal to 64.96% and 64.81%, respectively, of the input. During the intermediate stages of net-pen culture during 1990, the production of 1 t fish resulted in the discharge of N (141.25 kg) and P (14.14 kg) into the lake (Yang et al., 1995).

This influence on the environment from the crab net-pen cultivation still persists in Eastern Taihu Bay. It discharged 571.8 kg N and 71.57 kg P into the lake per ton of river crab produced (Yang et al., 1995). The average data during the intermediate stages of the 1990s in Eastern Taihu Bay showed that the net-pen culture could produce 2.8% N and 0.1% P of total pollution to Eastern Taihu bay.

With the rapid expansion of the cultivation area and increase in catches, the ecological environment changed and water quality dropped. Especially in 1999, the average water quality already had reached class V of the national standard for surface water. But, the process of eutrophication continued. At present, Eastern Taihu Bay is already at the level of rich nutrition condition (Table 6.18) (Wu et al., 2000). Great changes in the biotic community in Eastern Taihu Bay have taken place. In the 1980s, diatoms and green algae were the dominant species; at the beginning of the following decade, green algae became predominant, and at present the dominant species are blue-green algae and green algae. *Microcystis* sp. became dominant in the high-temperature season, and the number of phytoplankton species was reduced. Phytoplankton abundance conspicuously increased from 2.02×10^6 cell/L in the 1980s and then to the present level of 4.66×10^6 cell/L. Zooplankton biomass of Eastern Taihu Bay dropped suddenly after 1980, with levels of only 0.534 mg/L in 1997. Compared with 1980, Protozoa increased, and the percentages of Cladocera and Copepoda decreased rapidly. The pressure of fish and crab aquaculture caused the food resources to rapidly decline. At the same time, the level of pollution increased, especially deposition of organic materials on the bottom. Some aquatic Oligochaeta resistant to pollution increased; the density and biomass of insect larvae also increased (Wu et al., 2001). The pollution caused by net-pen culture resulted in the rapid growth of aquatic vegetation. Expanded growth of *Zizania caduuciflora* around the net-pens and by *Zizania caduuciflora* along outside the floats occurred. The flowing water environment was suitable for the growth of *Vallisneria natans* community while *Potamogeton malaisnus* vanished (Wu et al., 2000).

A comparison of the data on sediments showed that the bottom pollution increased gradually (see Fig. 1.4). Especially in the last decade, TN and total organic carbon (TOC) accumulation in the surface layer of sediments in Eastern Taihu Bay increased significantly. Although external pollution existed, the utilization ratio of the bait was low, and the uneaten fish food sank to the lake bottom. Additionally,

Table 6.18 Eutrophication trends in water quality in Eastern Taihu Bay

Year	TN (mg/L)	NH_4-N (mg/L)	TP (mg/L)	COD_{Mn} (mg/L)	Chl a (μg/L)
1980	0.65	0.05	0.030	2.87	
1990–1991	1.01	0.14	0.043	5.50	3.5
1993–1994	1.04	0.20	0.075	5.57	4.3
1997	1.39	0.19	0.031	5.51	4.4
1998	0.83	0.21	0.037	4.59	3.1
1999	1.63	0.25	0.095	4.56	6.4

the net-pens also hampered the harvest of the macrophyte, *Zizania caduuciflora*, which resulted in increased plant remnants and sedimentation (Li et al., 2001).

At present, the composition of fish catches in Lake Taihu is not reasonable. It is necessary to study the current capacity and its ecological effects, to study the piscivorous fishes such as *Siniperca chuatsi*, to control the population of *Coilia ectenes taihuensis*, and to use the zooplankton to control the phytoplankton. At the same time, it is necessary to regulate aquaculture and prohibit all net-pen culture systems within the lake to restore water quality.

6.6 Structure and Succession of the Aquatic Vegetation

Weimin Chen

6.6.1 Species Composition of the Aquatic Vegetation

6.6.1.1 Macrophyte Species

In 1960, 66 macrophyte species were recorded, belonging to 48 genera and 29 families. In 1996, 66 macrophyte species were recorded, belonging to 51 genera and 30 families; however, there had been a shift in species composition during the 36 years between the surveys, as indicated by the observation that 43 species appeared in both periods but 23 species had been replaced by 23 newly appearing species since 1960 (Appendix E).

6.6.1.2 Spatial Distribution of the Macrophyte Species

Macrophyte species can be classified as emergent, floating-leaved, and submerged. Emergent vegetation is restricted to a zone less than 0.8 m deep, whereas floating-leaved vegetation is in part growing at depths to 1.2 m, and submerged aquatic vegetation is abundant to a water depth of 2.6 m.

These macrophytes are mostly present in Eastern Taihu Bay. Since 1960, macrophyte coverage of the bay bottom has been high, amounting to nearly 100%, in the Bay. At present, the dominant species are *Potamogeton maackianus, Potamogeton malaianus, Vallisneria natans, Phragmites communis, Zizania latifolia, Nymphoides peltatum*, and *Stipa tenacissima* associated with *Potamogeton crispus* and *Phoradendron flavescens*, which constitute the main communities of submerged macrophytes. *P. maackianus* is the predominant macrophyte species. *Potamogeton maackianus* covers approximately 40% of the lake bottom that has submerged aquatic vegetation, and mainly grows in the western and northern coastal areas of Eastern Taihu Bay, as well as in the pelagic region. *Potamogeton malaianus* is the dominant pelagic species. *Vallisneria natans* mainly occurs in the northwestern region of Eastern Taihu Bay. *Phragmites communis* is most abundant in the littoral zone of the northeastern part of the lake, while *Z. catifolia* is most abundant along the southeast coast. *Nymphoides peltatum* and *Trapa incisa* are abundant in the boundary region between dense stands of *Z. catifolia* and the outflow of the lake. *Stipa tenacissima* covers only a small area in the pelagic region of Eastern Taihu Bay. Other species, such as *Utricularia aurea, Najas minor*, and *Nymphaea tetragona*, are less abundant. *Elodea nuttalli* is a neophytic species growing in nearly every part of the Bay. Overall in Eastern Taihu Bay, almost the whole bay has been

Chen, W.M.
State Key Laboratory of Lake Science and Environment, Nanjing Institute of Geography and Limnology, Chinese Academy of Sciences, 73 East Beijing Road, Nanjing 210008, P. R. China
e-mail: chenwm@niglas.ac.cn

covered with aquatic vegetation, and the dry weight of above-lake-bottom biomass reaches 60×10^4 t, indicating dense aquatic vegetation.

Four bays extend from Meiliang Bay to the Dongshan Peninsula (Meiliang Bay, Gonghu Bay, Xukou Bay, and Eastern Taihu Bay). Generally, the aquatic vegetation is less dense in the north, such as Meiliang Bay, than in the south of the lake, such as Eastern Taihu Bay. Spatial distribution of submerged aquatic vegetation is sparse in the north and west, and dense in the south and east, where the dominant species is *P. malaianus*. This spatial distribution pattern may be associated with the summer monsoon, which occurs during the macrophyte growing season (May–October); wave effects are weak in the southeast part but strong in the northwest part of the lake. In addition, because of shallow depth in the southeast part, it is favourable to macrophyte growth.

At present, the coverage of aquatic vegetation on the northern coast of Gonghu Bay is low, approximately 1–10%. There, *P. communis* and *V. natanus* are the predominant species. In the southeastern part of the bay between Dagongshan Island and the mouth of Wangyuhe River, dense macrophyte stands extend along the shore and cover more than 40% of the lake bottom. About 70% of the lake bottom has macrophyte coverage greater than 90%, corresponding to an aquatic vegetation biomass of $10\,kg/m^2$.

West of the mouth of Eastern Taihu Bay and south of the Dongshan Peninsula, the aquatic vegetation density is heterogeneous, ranging between 5% and 90% coverage over an area of $100\,km^2$. The degree of coverage gradually increases from west to east. The dominant species were *P. malaianus* and *Hydrilla verticillata*. Dense submerged vegetation is found over an area of $60\,km^2$, extending from the southwest of the Dongshan Peninsula to the southern coast of the lake. Total macrophyte abundance and biomass there were as high as in Eastern Taihu Bay and had increased considerably in the 1990s.

The littoral zone of the western and the northern parts of the lake is sparsely covered by emergent vegetation, mainly *P. communis*. Submerged species such as *P. malaianus* are abundant at two sites close to Daqiangang, whereas the abundance of *P. communis* is low. In some places, the aquatic vegetation can extend to 2 km distant from shore. A survey performed in August 1992 in Zhushan Bay showed dense aquatic vegetation. The vegetated area accounted for approximately 60% of the total area of the bay, and in some places, the degree of coverage reached 80%, with the dominant species *P. maackianus* and *V. natanus*. During the 1990s, the density of the aquatic vegetation significantly decreased, which can be attributed to decreasing water quality and increasing eutrophication.

6.6.2 Succession of Aquatic Vegetation in Eastern Taihu Bay and Western Lake Taihu

6.6.2.1 Spatial Distribution of Aquatic Vegetation in 1960

(1) Eastern Taihu Bay: A total of 66 macrophyte species were found in the 1960 survey of Eastern Taihu Bay. The predominant species were *Z. latifolia, P. communis,*

P. malaianus, and *V. natanus*, with *P. malaianus* being most abundant. Among survey transects, differences in species composition and above-lake-bottom biomass were investigated. Generally, the biomass of the aquatic vegetation decreased from the littoral to the pelagic zone. The total biomass of aquatic vegetation, not including *P. communis*, was estimated to be 8×10^4 t, accounting for 80% of the total biomass of the aquatic vegetation in Lake Taihu.

(2) Coast of western Lake Taihu: In western Taihu, the macrophyte composition was less diverse, and the density of the aquatic vegetation was lower than in Eastern Taihu Bay. The dominant species in western Lake Taihu was *P. communis*, abundant along nearly the entire shore, and constituting a belt of emerged vegetation. In addition, there were stands of *Z. latifolia, Polygonum lapathifolium, Polygonum hydropiper, Sagittaria sagittifolia, Sagittaria pygmaea, Eleocharis tuberosa, Juncus setchuensis* var. *effusus*, and *Leersia japonica*. Some stands of floating-leaved vegetation (*Nymphoides peltatum*) and floating vegetation (*Salvinia natans*) were also found in this region. Several stands of floating-leaved (*N. peltatum*) and submerged species (*P. malaianus*) were found within 800 m beyond the *P. communis* belt. In western Taihu, the biomass of the aquatic vegetation amounted to only about 20% of the total biomass for the whole lake, even though the area of western Lake Taihu contains more than 95% of the total lake area.

(3) Islands in western Lake Taihu: There was no aquatic vegetation in the pelagic region of western Lake Taihu. In addition to some stands of *P. communis*, there was only sparse vegetation (*P. malaianus*) near the island.

(4) Tributaries of Lake Taihu: Lake Taihu has several tributaries. Along the banks of small rivers, *P. communi, Leersia japonica*, and *N. peltatum* were abundant species. In the central part of small rivers, *Najas minor, Hydrilla verticillata, V. natanus*, and *P. malaianus* were present. In medium- to large-size rivers, stands of *P. communi* were found at the river banks, and aquatic vegetation was rarely found in the central part. The survey carried out in 1997 estimated a total biomass of 36×10^4 t, of which *P. communis* constituted 22.5×10^4 t and submerged macrophytes 13.5×10^4 t. The macrophyte distribution is shown in Fig. 6.15.

6.6.2.2 Succession of the Aquatic Vegetation in Eastern Taihu Bay

In the 1960s and 1970s, human activities, such as enclosing tideland for cultivation and building aquaculture enclosures, caused a decrease in the density of aquatic vegetation in Lake Taihu. The 1997 survey showed that macrophytes were mainly abundant in Eastern Taihu Bay, the southern littoral zone, and the region south of Xishan Island. The area with 1% coverage of aquatic vegetation in Lake Taihu was estimated to be 450 km^2 and accounted for 19% of the total lake area. The area with 5% coverage of aquatic vegetation was estimated at 350 km^2 and accounted for 15% of total lake area. The temporal variation in the composition of the aquatic vegetation is given in Table 6.19. The numbers of all families, genera, and species of ferns, dicotyledons, and monocotyledons increased from 1960s to 1990s (Table 6.19). Species numbers of submerged aquatic vegetation and emergent

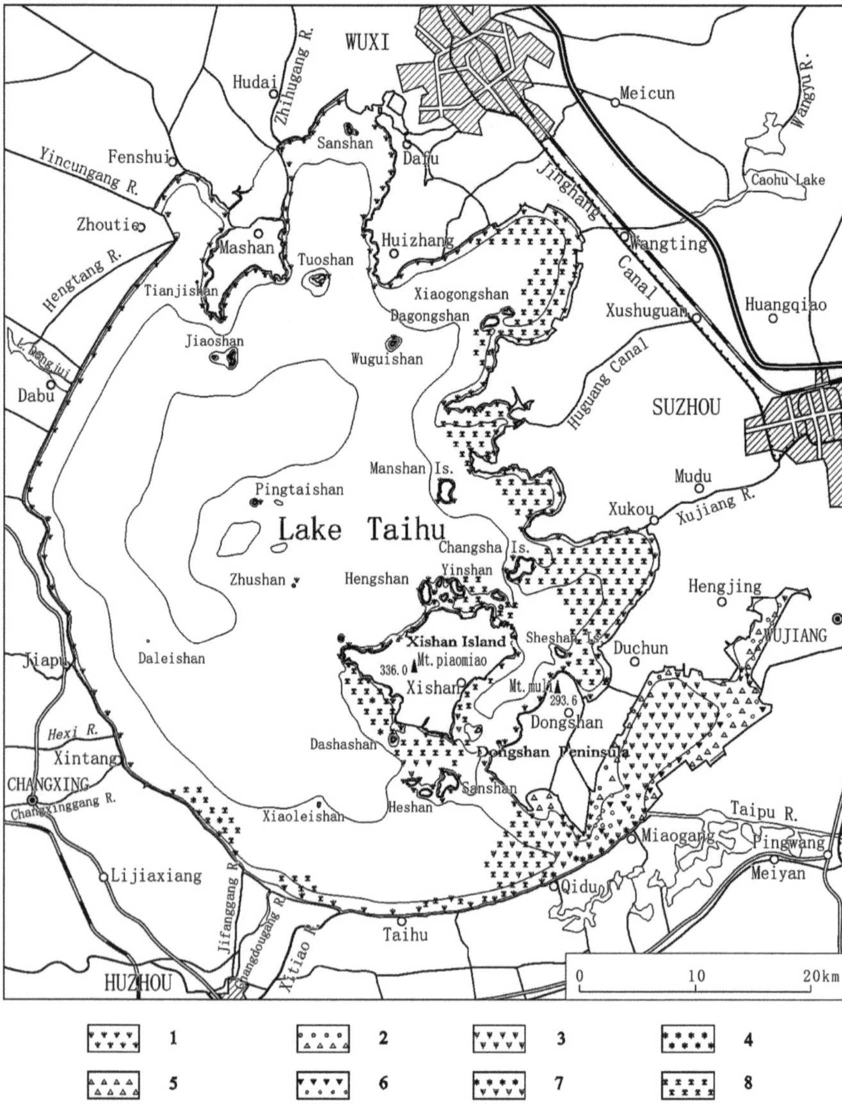

Fig. 6.15 Spatial distribution of aquatic vegetation in 1997 in Lake Taihu (Taihu Basin Authority. Compiled by Ministry of Water Resources & Nanjing Institute of Geography and Limnology, Chinese Academy of Sciences, 2000). *1, Phragmites communis*; *2, Zizania latifolia* and floating-leaved vegetation; *3, Potamogeton maackianus; 4, Vallisneria natanus; 5, Z. latifolia; 6, Nymphoides peltatum* and *Trapa incisa; 7, P. maackianus and V. natanus*; 8. *Potamogeton malaianus*

vegetation have increased, but species numbers of floating-leaved and floating vegetation have decreased, since the 1960s (Table 6.19).

In Eastern Taihu Bay, the aquatic vegetation covered an area of 126 km^2, amounting to 95% of the total area of the bay. The *P. maackianus* group had the greatest area covered and greatest biomass (Table 6.20). Submerged aquatic vegetation covered

Table 6.19 Temporal changes in composition (numbers of taxa) of aquatic vegetation in Eastern Taihu Bay (1960–1996)

Year	Fern			Dicotyledons			Monocotyledon			Species according to ecological type			
	F	G	S	F	G	S	F	G	S	SAV	FLV	FV	EV
1960	3	4	4	14	17	24	10	26	38	12	8	11	35
1981	3	4	4	14	20	31	12	23	31				
1996	4	4	4	17	22	31	13	30	39	16	1	9	38

F, family; G, genus; S, species; SAV, submerged aquatic vegetation; FLV, floating-leaved vegetation; FV, floating vegetation; EV, emergent vegetation.

74 km^2, accounting for 59% of the total vegetated area and 56% of the total area of Eastern Taihu Bay. *P. maackianus* was abundant in deep waters. The *P. maackianus* group, the *H. verticillata* and *P. maackianus* group, and the *P. malaianus* and *H. verticillata* group were the predominant macrophyte communities in Eastern Taihu Bay. Within the *P. maackianus* group, the following species were common: *N. peltatum, E. nuttalli, P. crispus, S. natans, V. natanus, P. malaianus*, and *P. crispus*.

Emergent vegetation covered an area of 46 km^2 in Eastern Taihu Bay, corresponding to 36% of the total vegetated area and 35% of the total area of the bay. Emergent vegetation was most dense in shallow waters along the lake shore. The main communities were (1) *P. communis*; (2) *Z. latifolia*; and (3) *Z. latifolia* and *Nelumbo nucifera*. Abundant species were *N. peltatum, E. nuttalli, P. crispus*, and *S. natans*.

Temporal-spatial distribution patterns of the aquatic vegetation in Eastern Taihu Bay are given in Table 6.20. In 1960, the area of *P. communis* comprised 77 km^2 in Eastern Taihu Bay. Thereafter, the wetlands covered with stands of *P. communis* were enclosed and developed into farmland, which caused the disappearance of *P. communis*.

Table 6.20 Areas of main aquatic vegetation communities in East Lake Taihu in 1960, 1981, and 1996 (Zhang et al., 1998)

Vegetation type	Community	Area (km^2)		
		1960	1981	1996
Emergent vegetation	*Phragmites communis*	77.0	8.0	8.4
	Zizania latifolia and *Nymphoides nucifera*	15.0	42.7	37.1
Floating-leaved vegetation	*Nymphoides indica*	0.0	0.0	3.6
	Nymphoides peltatum	0.0	0.0	3.1
Submerged vegetation	*Potamogeton maackianus* and *Elodea nuttalli*	0.0	7.3	66.4
	Potamogeton malaianus, Hydrilla verticillata, and *Vallisneria natanus*	160.0	74.0	7.4
Total		252.0	132.0	126.0

In the mid-1960s, *Z. latifolia* was introduced into Eastern Taihu Bay to support fisheries as fish food, and subsequently this species spread from the shore to the pelagic region. In 1960, *Z. latifolia* only covered 6% of the total vegetated area, but later this percentage increased to 32% in 1980 and 29% in 1996. *Z. latifolia* is a typical perennial plant of marsh areas and develops an extended root system rapidly growing into soil depths of 0.65 m. In 1996, the biomass of *Z. latifolia* reached $22,630 \times 10^4$ t, accounting for 32% of total biomass of aquatic vegetation in Eastern Taihu Bay. Residual stems and leaves of *Z. latifolia* decomposed, and the remaining particulate organic matter constituted an organic sediment in the profundal region. Furthermore, stands of *Z. latifolia* decreased resuspension and increased sedimentation by reducing the energy of wind-induced waves. Thus, sedimentation rates again increased. The thickness of the accumulated sediments in stands of *Z. latifolia* was estimated to be 0.4–1.2 m, and in some areas it reached 2.0 m.

The area covered with stands of submerged macrophyte associations decreased by 49% in 1981, and by 54% in 1996, as compared with the 1960s. In addition, a significant change was found in its species composition. In 1960, *P. malaianus, H. verticillata*, and *V. natanus* accounted for 65% of the total area covered with submerged vegetation. Large-scale use of net-pen enclosures for fish farming started in 1984, and used some dominant macrophytes, including *P. malaianus* and the neophyte *H. verticillata*, as fish food. *P. maackianus* and *E. nuttalli* gradually became the dominant species in Eastern Taihu Bay. In 1960, *P. maackianus* was rare in Eastern Taihu Bay, but in 1996, *P. maackianus* and *E. nuttalli* accounted for 52.7% of the area covered with submerged macrophytes. Around 1980, *P. malaianus, H. verticillata*, and *V. natanus* completely disappeared from the pelagic region of Eastern Taihu Bay, and only a few stands of these species remained in the zone connecting the bay with the open water area of Lake Taihu.

6.6.2.3 Characteristics of the Aquatic Vegetation in Western Lake Taihu

The area covered by aquatic vegetation in western Lake Taihu was less than in eastern Lake Taihu, and in the former, the biomass was low. Strong wind-induced waves in concert with eutrophication and increasing phytoplankton biomass increased turbidity and deeply altered the environmental conditions for macrophyte growth in western Lake Taihu. Since the 1960s, transparency greatly decreased, having adverse effects on the underwater light regime and inhibiting submerged macrophyte growth. The change in the underwater light regime, wave energy, and sedimentation, plus regulation of the water level by dams and floodgates, resulted in a rapid decrease and local extinction of the aquatic vegetation. At present, stands of *P. communis* exist only in a small bay in western Lake Taihu. A few specimens of *P. malaianus* and *N. peltatum* remain in stands of *P. communis*, which are restricted to the littoral zone. Zhushan Bay, which had large areas of submerged macrophytes, now has only small stands of *P. communis* remaining. In addition, the diversity of the macrophytes decreased.

The aquatic vegetation purifies the water by filtering seston and reducing resuspension. Therefore, restoration of the aquatic vegetation was considered to be an important measure to counteract eutrophication. Areas of weak dynamic action, such as bays, could be selected for the reestablishment of aquatic vegetation. The sedimentary environment and transparency could be improved by selecting one or two macrophyte species as pioneers, and further construction measures could sustain the growth of macrophyte stands. Once the water environment was suitable for sustaining aquatic vegetation, the macrophytes could grow and reproduce. In 1995, in the littoral zone near Taihu Laboratory for Lake Ecosystem Research (TLLER), little aquatic vegetation was found, except for a stand of *P. communis*. After this experimental area was enclosed in 1995, the macrophyte abundance and biomass gradually increased in the enclosed areas by natural restoration (Yang et al., 2002). This increase happened without human intervention.

6.6.3 Effects of the Environment on the Succession of the Aquatic Vegetation

Since 1960, the composition and biomass of the aquatic vegetation have experienced considerable change. The factors inducing this change were water depth, sediments, transparency, wind-induced waves, siltation, and human activities. The human activities included enclosing tideland for cultivation, enclosure cultures, crab culture, and further manipulations. In addition, release of pollutants and eutrophication had significant effects on the growth and distribution of water plants.

The normal succession of the aquatic vegetation resulting from environmental factors was affected by the change from natural conditions to an anthropogenically altered state. Human activities can be subdivided into direct and indirect actions. Direct action is characterized by aquaculture production, whereas indirect action is a function of drainage and eutrophication, which again is caused by urbanization and agricultural land use.

Much of the water of Eastern Taihu Bay rapidly changed into a terrestrial ecosystem because of the construction of wetland enclosures for cultivation. The area of Eastern Taihu Bay has been reduced by as much as $120\,km^2$, equal to 47% of the original bay area, in the past 40 years. The disappearance of the *P. communis* group coincided with a new succession of floating-leaved and submerged vegetation. Since 1980, large enclosure cultures have interrupted the natural succession processes. Enclosure cultures reduced currents and wind-induced waves, which accelerated the transition to a swamp. In turn, the increasing area of stands of *Z. latifolia* in this swamp changed the lake ecosystem structure. Between 1980 and 1996, only a slight expansion of the *Z. latifolia* group occurred, but this process still continues.

Between 1980 and 1996, the lake bottom was raised by as much as 0.15–0.45 m by increased sedimentation of fine material. In the 1950s, the surface sediment was a sandy silt. The organic matter was completely oxidized. The phosphorus and organic matter in sediment accounted for 0.023% and 0.073% of the dry weight, respectively. At the same time, mean annual phytoplankton density was only 26.7×10^4 cell/L. In 1992, the sediment in Wuli Bay had become fertile silt, with an intense fishy odor. The pH value of the sediment was neutral, the redox potential was less than -100 mV, and organic matter oxidation was low (Li, 1996a). These sediment characteristics caused the disappearance of the aquatic vegetation in Wuli Bay.

In eastern Wuli Bay, fisheries have a long history. In the early 1960s, *Ctenopharyngodon idellus* was bred in Wuli Bay. This species consumed the aquatic vegetation and profoundly changed the aquatic vegetation. *Silurus asotus* and *Parasilurus asotus* were the main fish species reared, and their population amounted to about 80% of the total fish population. Other fish species in Wuli Bay were *Mylopharyngodon piceus, Cyprinus carpio*, and *Megalobrama amblycephala*. The area-weighted annual yield amounted to 600–750 kg/km^2.

In 1951, before aquaculture started in Wuli Bay, macrophytes, such as *P. communis, T. incisa, H. dubia, V. natanus, P. crispus*, and *Z. latifolia*, were abundant. Construction of an artificial dyke along 90% of the coast caused the aquatic vegetation and wetlands to disappear (Li, 1996b). Similarly, the open water was almost completely covered by macrophytes, including *V. natanus, P. crispus*, and *M. spicatum*, in 1951; however, the rapid development of aquaculture and urbanization in Wuxi lead to gradual disappearance of this vegetation. This change, in turn, resulted in a decrease of the self-purification ability of the water, associated with increased suspended solids and changes in the underwater light regime. At present, the concentration of suspended solids has increased to 27–28 mg/L, and transparency is down to 45 cm.

In the late 1960s and early 1970s, aquatic vegetation was abundant in the north of Meiliang Bay. Starting with the enclosure near Mashan and the rapid development of industry and agriculture around the lake, this vegetation too gradually disappeared. At present, some stands of *P. communis* and scarce aquatic vegetation exist only in the coastal areas. It is difficult to restore aquatic vegetation, especially submerged vegetation, in western Lake Taihu because of the relatively great water depth and low transparency caused by sediment resuspension and frequent algal blooms. In addition, high pollutant concentrations and strong wind-induced waves inhibit the growth of submerged aquatic vegetation. It would take a long time to restore the original lake ecosystem structure and function, even if pollutant emissions could be regulated.

References

Bai, X. H., W. P. Hu, Z. X. Hu & X. H. Li, 2005. Importation of wind-driven drift of mat-like algae bloom into Meiliang Bay of Taihu Lake in 2004 summer. Environmental Science 26: 57–60 (In Chinese with English abstract).

Cai, H. J., 1998. The response of ciliated protozoans to eutrophication in Meiliang Bay of Taihu Lake. Journal of Lake Sciences 10(3): 43–47 (In Chinese with English abstract).

Cai, H. J. & W. M. Chen, 1998. The influence of the immigration and decomposition of *Microcystis* for the water environment of Taihu Lake. In: Cai, Q. M. (ed), Study on the environment and ecosystem of Lake Taihu (I). Beijing: China Meteorology Press, 149–157 (In Chinese with English abstract).

Cao, W. M., G. Zhou, J. M. Sheng, J. C., Ge, C. D., Zhu & L. K. Wu, 2000. The present situation and evolution of the *Corbicula fluminea* resources in Tai Lake. Journal of Nanjing Forestry University 24(suppl): 125–128 (In Chinese with English abstract).

Chen, W. M. & B. Q. Qin, 1998. The zooplankton distribution during winter and early spring and its environmental effects in Meiliang Bay, Taihu Lake. Journal of Lake Sciences 10: 10–16 (In Chinese with English abstract).

Chen, W. M., Y. W. Chen, X. Y. Gao & I. Yoshida, 1997. Eutrophication of Lake Taihu and its control. Agricultural Engineering Journal 6: 109–120.

Chen, W. M., Y. W. Chen & X. Y. Gao, 1998. Growth dynamics of *Microcystis* spp. in Lake Taihu. Journal of Lake Sciences 10(suppl): 347–356.

Chen, W. M., Y. W. Chen, B. Q. Qin, X. Y. Gao, G. Gao, J. Ji & Q. J. Xu, 2000. Experimental study on the biological community succession caused by water flow. Journal of Lake Sciences 12: 343–352 (In Chinese with English abstract).

Chen, Y., C. Fan, K. Teubner & M. Dokulil, 2003. Changes of nutrients and phytoplankton chlorophyll-*a* in a large shallow lake, Taihu, China: an 8-year investigation. Hydrobiologia 506–509: 273–279.

Dalian Fisheries University, 1985. Aquatic biology. Beijing: China Agricultural Press (In Chinese).

Deng, S. M., Z. J. Zang, H. X. Zhang & S. F. Li, 1997. A study on fish community structure in the open water zone of Taihu Lake. Journal of Fisheries of China 21(2): 134–142.

Dou, H. S., W. H. Ma, S. Z. Zhang & J. H. Deng, 1988. Environmental impacts of lake reclamation for cultivation in Taihu area. Acta Scientiae Circumstantiae 8(1): 1–9.

East China Normal University (ECNU), 1959. Investigation report of hydrobiology in East Lake Taihu (In Chinese).

Gao, G. & X. Y. Gao, 2000. Experimental study on the PO_4^{3-}-P threshold of the alkaline phosphatase activity in Taihu Lake. Journal of Lake Sciences 12(4): 353–358 (In Chinese with English abstract).

Huang, Y. Y., 2001. Pollution ecology of inland waters: principles and applications. Beijing: Science Press.

Li W. C., 1995. Fish ecological environment quality and control measurement in Lake Taihu. The project of "the Eighth Five-Year Plan" of the country (85-01-03-02).

Li, W. C., 1996a. On the adaptiveness of aquatic macrophyte to the sediment in Wuli Lake. Journal of Lake Sciences 8(suppl): 30–36 (In Chinese with English abstract).

Li, W. C., 1996b. Biological and environment succession in Wuli Bay of Taihu Lake along with the eutrophication progress. Journal of Lake Sciences 8(suppl): 37–45 (In Chinese with English abstract).

Li, W. C., K. N. Chen, Q. L.Wu & J. Z. Pan, 2001. Experimental studies on decomposition process of aquatic plant material from East Taihu Lake. Journal of Lake Sciences 13(4): 331–336 (In Chinese with English abstract).

Nanjing Institute of Geography, Academia Sinica, 1965. Investigation of Lake Taihu. Beijing: Science Press, 1–84 (In Chinese).

Smith, R. C., K. S. Backer, O. Holm-Hansen & R. Olson, 1980. Photo inhibition of photosynthesis in natural waters. Photochemistry and Photobiology 31: 585–592.

Sun, S. C. & Y. P. Huang, 1993. Lake Taihu. Beijing: Ocean Press (In Chinese with English abstract).

Taihu Basin Authority: Compiled by Ministry of Water Resources & Nanjing Institute of Geography and Limnology, Chinese Academy of Sciences, 2000. Atlas of ecology and environment of the Taihu Lake. Beijing: Science Press (In Chinese).

Tang, Y., 1987. On the population dynamics of lake anchovy in Taihu Lake and its rational exploitation. Journal of Fisheries of China 11(1): 61–73.

Wu, Q. L., 1991. Benthon inhabited in Eastern Taihu Lake and its pen fish farming area and preliminary assessment and division of the environment quality. Transactions of Oceanology and Limnology 4: 64–72 (In Chinese with English abstract).

Wu, Q. L., 2001. On the sustainable development of fishery in East Taihu Lake. Journal of Lake Sciences 13(4): 337–344 (In Chinese with English abstract).

Wu, Q. L., Y. H. Hu, W. C. Li, K. N. Chen & J. Z. Pan, 2000. Tendency of swampiness of East Taihu Lake and its causes. Acta Scientiae Circumstantiae 20(3): 275–279 (In Chinese with English abstract).

Wu, Q. L., K. N. Chen,Y. H. Hu & W. C. Li, 2001. Impacts of pen crab farming on environment in East Taihu Lake. Agro-Environmental Protection 432–434 (In Chinese with English abstract).

Wu, S. C., Y. W. Chen & G. Gao, 2003. Examination of alive algal cells in the sediments in Meiliang Bay of Taihu Lake during winter period. Journal of Lake Sciences 15(4): 339–344 (In Chinese with English abstract).

Wu, X. W., 1962. Limnology investigation of Wuli Lake in 1951. Hydrobiology Quarterly 1: 63–113.

Yang, Q. X., 1996. Algal bloom in Taihu Lake and its control. Journal of Lake Sciences 8(1): 67–73 (In Chinese with English abstract).

Yang, Q. X., W. C. Li, L. Yu & J. Wei, 1995. Pen-fish-farming development in East Taihu Lake and its effects on lake environment. Journal of Lake Sciences 7(3): 256–262 (In Chinese with English abstract).

Yang, L. Y., H. T. Liang, W. P. Hu, J. Ji, W. M. Chen, P. M. Pu & Q. M. Cai, 2002. The study on natural restoration of aquatic vegetation in the northern lakeside, Lake Taihu. Journal of Lake Sciences 14(1): 60–66 (In Chinese with English abstract).

Zhang, S. Z., G. X. Wang & P. M. Pu, 1998. The influence of algae-type eutrophication OH aquatic higher plants and hydrophytic vegetational reconstruction in Taihu Lake. Journal of Plant Resources and Environment 7(4): 52–57 (In Chinese with English abstract).

Zhang, Y. L., B. Q. Qin, W. M. Chen, G. Gao & Y. W. Chen, 2004. Experimental study on underwater light intensity and primary productivity caused by variation of total suspended matter. Advance in Water Science 15(5): 615–620 (In Chinese with English abstract).

Zhu, S. Q., 2004. Ichthyological survey of Lake Taihu during 2002–2003. Journal of Lake Sciences 16(2): 120–124 (In Chinese with English abstract).

Zhu, Y. C. & Q. M. Cai, 1997. The dynamic research of the influence of wind field on the migration of algae in Taihu Lake. Journal of Lake Sciences 9(2): 152–158 (In Chinese with English abstract).

Appendices

A Species of Phytoplankton in Lake Taihu

Bacillariophyta

Achnanthes lanceolata Breb.
Amphora ovalis Kütz.
Asterionella formosa Hass.
Attheya zachatiasi Brun.
Cocconeis pediculus Ehr.
Cocconeis placentula Ehr.
Cyclotella bodanica Eul.
Cyclotella comta Kütz
Cyclotella meneghiniana Kütz.
Cyclotella stelligera Cl. et Grun.
Cymatopleura solea (Breb.) W. Smith
Cymbella affinis Kütz
Cymbella parva (W. Smith) Cleve
Cymbella tumida (Greg.) Cl.
Cymbella ventricosa Kütz.
Diatoma vulgare Bory
Epithemia turgida (Ehr.) Kütz.
Epithemia zenra (Ehr.) Kütz.
Fragilaria capucina Desm.
Fragilaria crotonensis Kitton.
Gomphonema acuminatum Ehr.
Gomphonema parvulum (Kütz.) Grun.
Gyrosigma acuminatum (Kütz.) Rabh.
Gyrosigma kutzingii (Grun.) Cl.
Melosira granulata (Ehr.)
Melosira granulata var. *angustissima* f. *spiralis* Hust.
Melosira granulata var. *angustissima* Mull.
Melosira islandica Mull.

Melosira varians Ag.
Navicula exigua (Greg.) Mull.
Navicula placentula (Ehr.) Grun.
Navicula rhynchocephala Kütz.
Navicula simplex Krassk.
Nitzschia recta Hantsch
Pinnularia gibba Ehr.
Pinnularia viridis (Nitzsch.) Ehr.
Rhizosolenia longiseta Zach
Rhopalodia gibba (Ehr.) O. Mull.
Stauroneis anceps Ehr.
Stephano discus astraea (Ehr.) Grun.
Surirella biseriata Breb.
Surirella capronii Breb.
Surirella robusta Ehr.
Surirella robusta var. *splendida* (Ehr.) V.H.
Synedra acus Kütz.
Synedra ulna (Nitzsch.) Ehr.
Synedra ulna var. *biceps* (Kütz.) Schoenf.
Tabellaria fenestrata (Lyngb.) Kütz.

Chlorophyta

Acanthosphaera zachariasi Lemm.
Actinastrum hantzschii Lag.
Ankistrodesmus acicularis (A. Br.) Korsch.
Ankistrodesmus falcatus (Cord.) Ralfs
Asterococcus limneticus G.M. Smith
Characium rostratum Reinsch.
Chlamydomonas debaryana Gor.
Chlamydomonas ovalis Pasch.
Chlamydomonas simplex Pasch.
Chlamydomonas snowiae Printz
Chlorella pyrenoidosa Chick
Chlorella vulgaris Beij.
Chodatella cilliata (Lag.) Lemm.
Chodatella subsalsa Lemm.
Cladophora crispata (Roth.) Kütz.
Cladophora fracta (Dillw.) Kütz.
Closterium kutzingii Breb.
Closterium moniliferum (Bory.)
Closterium venus Kütz.
Closteropsis longissima Lemm.
Coelastrum microporum Nag.

Coelastrum reticulatum (Dang.) Senn.
Cosmarium angulosum Breb.
Cosmarium anisochondrum Nordst.
Cosmarium circulare Reinsch.
Cosmarium quadrum Lund.
Cosmarium subtumidum Nordst.
Crucigenia apiculata (Lemm.) Schm.
Crucigenia lauterbornei Schm.
Crucigenia quadrata Morr.
Crucigenia tetrapedia (Kirch.) W. et G.S. West
Dictyosphaerium pulchellum Wood
Echinosphaerella limnetica G.M. Smith
Errerella bornhemiensis Conr.
Eudorina elegans Ehr.
Franceia ovalis (Franc.) Lemm.
Geminella minor (Nag.) Hansg.
Geminella interrupta Turp.
Golenkinia paucispina W. et G.S. West
Golenkinia radiata Chod.
Gonatozygon monotaenium De Bary
Gonium formosum Pasch.
Gonium sociale (Duj.) Warm.
Kirchneriella lunaris (Kirch.) Moeb.
Kirchneriella obesa (West) Schm.
Micractinium pusillum Fres.
Mougeotia parvula Hass.
Nephrocytium agardhianum Nag.
Oocystis borgei Snow
Oocystis elliptica W. West
Oocystis lacustris Chod.
Oocystis parva W. et G.S. West
Pandorina morum (Mull) Bory
Pediastrum biradiatum Mey.
Pediastrum boryanum (Turp.) Men.
Pediastrum duplex Mey.
Pediastrum simolex (Mey.) Lemm.
Pediastrum simolex var.
duodenarium (Bail.) Rab.
Pediastrum tetras (Ehr.) Ralfs
Planktosphaeria gelotinosa G.M. Smith
Pleodorina californica Shaw
Pteromonas angulosa Lemm.
Quadrigula chodatii (Tan-Ful.) G.M. Smith
Scenedesmus arcuatus Lemm.
Scenedesmus abundans (Kirch.) Chod.

Scenedesmus acuminatus (Lag.) Chod.
Scenedesmus arcuatus (Chod.) Smith
Scenedesmus bijuga (Turp.) Lag.
Scenedesmus brasiliensis Bohl.
Scenedesmus cavinatus (Lemm.) Chod.
Scenedesmus denticulatus Lag.
Scenedesmus dimorphus (Turp.) Kütz.
Scenedesmus javaensis Chod.
Scenedesmus obiquus (Turp.) Kütz.
Scenedesmus perforatus Lemm.
Scenedesmus platydiscus (G.M. Smith) Chod.
Scenedesmus quadricauda (Turp.) Breb.
Schroederia robusta Korsch.
Schroederia setigera Lemm.
Schroederia spiralis (Printz.) Korsch.
Selenastrum gracile Reinsch.
Selenastrum minutum (Nag.) Coll.
Selenastrum westii G.M. Smith.
Sphaerocystis schroeteri Chod.
Sphaerozosma granulatum Roy et Biss.
Spirogyra communis (Hass.) Kütz.
Spondylosium planum (Woll.) W. et G.S. West
Staurastrum indentatum W. et G.S. West
Staurastrum paradoxum Mey.
Staurastrum polymorphum Breb.
Tetraedron caudatum (Cord.) Hansg.
Tetraedron trigonum (Nag.) Hansg.
Tetraedron trilobulatum (Reinsch.) Hansg.
Tetraedron tumidulum (Reinsch.) Hansg.
Tetrastrum hastiferum (Arn.) Korsch.
Tetrastrum staurogeniaeforme (Schr.) Lemm.
Ulothrix subtilissima G.S. West
Ulothrix tenerrima (Kütz.) Kütz.
Volvox aureus Ehr.
Volvox globator (L.) Ehr.
Westella botryoides (W. West) De Wild.

Chrysophyta

Chromulina ovalis Klebs
Chrysamoeba radians Klebs
Dinobryon bavaricum Imh.
Dinobryon cylindricum Imh.
Dinobryon divergens Imh.
Dinobryon sertularia Ehr.

Mallomonas caudata Iwan.
Ochromonas mutabilis Klebs
Synura uvella Ehr.

Cryptophyta

Chroomonas acuta Uterm.
Cryptomonas erosa Ehr.
Cryptomonas ovata Ehr.

Cyanophyta

Anabaena circinalis Rab
Anabaena flos-aquae (Lyngb.) Breb.
Anabaena oscillariodes Bory
Anabaena spiroides Kleb.
Anabaenopsis arnoldii Aptek.
Aphanizomenon flos-aquae (L.) Ralfs.
Aphanocapsa elachista W. et G.S. West
Aphanocapsa pulchra (Kütz.) Rab
Aphanothece elathrata W. et G.S. West
Aphanothece stagnina (Spr.) A. Br.
Calothrix stagnalis Gom.
Chroococcus limneticus Lemm
Chroococcus minor (Kütz) Nag.
Chroococcus minutus (Kütz) Nag.
Coelosphaerium aerugineum Lemm.
Coelosphaerium dubium Grun.
Coelosphaerium kuetzingianum Nag.
Dactulococcopsis acicularis Lemm.
Dactulococcopsis rhaphidioides Hansg.
Gloeocapsa kutzingiana Nag.
Gloeocapsa magma (Breb.) Holl.
Gloeocapsa punetata Nag.
Gloeothece linearis Nag.
Gloeotrichia echinulata (G.E. Smith) P. Richter
Gloeotrichia natans Rab.
Gomphosphaeria lacustris Chod.
Lyngbya circumcreta G.S. West
Lyngbya majuscula Harvey
Marssoniella elegans Lemm.
Merismopedia convoluta Breb.
Merismopedia elegans A.Br.
Merismopedia glauca (Ehr.) Nag.
Merismopedia minima G. Beck

Microcystis aeruginosa Kütz.
Microcystis flos-aquae Kirch.
Microcystis incerta Lemm.
Microcystis marginata (Menegh.) Kütz.
Microcystis pseudofilamentosa Crow.
Microcystis pulverea (Wood) Forti
Microcystis weserbergii Lomarek
Oscillatoria amphibia Ag.
Oscillatoria brevis (Kütz.) Gom.
Oscillatoria lacustris (Kütz.) Geitler.
Oscillatoria limnetica Lemm.
Oscillatoria princeps Vauch.
Oscillatoria tenuis Ag.
Raphidiopsis sinensis Jao
Rhabdoderma lineare Schm.
Rivularia dura Roth.
Spirulina major Kütz.
Spirulina prunceus W. et G.S. West
Synechococcus parvus Mig.
Synechocystis aquatilis Sauv.

Euglenophyta

Colacium vesiculosum Ehr.
Euglena acus Ehr.
Euglena caudata Hubn.
Euglena ehrenbergii Klebs.
Euglena oxyuris Schmar.
Euglena viridis Ehr.
Lepocinclis ovum (Ehr.) Lemm.
Phacus cadatus Hubn.
Phacus onyx Pochm.
Phacus pleuronectes (O.F. Müller: O.F.M.) Duj.
Phacus pyrum (Ehr.) Stein.
Strombomonas schauinslandii (Lemm.) Defl.
Trachelomonas armata (Ehr.) Stein
Trachelomonas oblonga Lemm.
Trachelomonas volvocina Ehr.

Pyrrophyta

Ceratium hirundinella (O.F. Müller) Schr.
Glenodinium gymnodinium Pen.

Glenodinium pulvisculus (Ehr.) Stein
Gymnodinium aeruginosum Stein
Peridinium bipes Stein
Peridinium volzii Lemm.

Xanthophyta

Botryococcus braunii Kütz.
Ophiocytium capitatum Woll.
Ophiocytium lagerheimii Lemm
Tribonema minus (Will.) Haz.

B Species of Zooplankton in Lake Taihu

Protozoa

Acanthocytis aculeata Hertwing and Lesser
Askenasia volvox Claparède and Lachmann
Actinosphaerium eichhorni Ehrenberg
Arcella vulgaris Ehrenberg
Carchesium polypinum Linné
Centropyxis aculeata Stein.
Curcurbitella sp.
Didinium balbianii nanum Kahl.
Didinium balbianii Fabre-Domergue
Difflugia acuminata Ehrenberg
Difflugia globulosa Dujardin.
Difflugia pyriformis Perty
Difflugia urceolata Carter
Dileptus sp.
Epistylis rotans Svec.
Epistylis urceolata Stiller
Euglypha tuberculata Dujardin
Litonotus lamella Schewiakoff
Paramecium caudatum Ehre.
Phryganella sp.
Spathidium spathula Müller-Woodruff et Spencer
Stentor polymorphus (Müller)
Strobilidium gyrans (Stokes)
Strobilidium velox Faure-Fr.
Strombidium viride Stein.
Tetrahymena sp.

Tintinnopsis sinensis Nie.
Tintinnopsis wangi Nie.
Tintinnidium entzii Chiang.
Tintinnopsis conicus Chiang.
Vaginicola sp.
Vorticella aequilata Kahl.
Vorticella marginata Stiller
Zoothamnium arbuscula Ehrenberg

Rotifera

Anuraeopsis fissa Gosse
Ascomorpha ecaudis Perty
Ascomorpha saltans Bartsch
Asplancha priodonta Gosse
Asplanchnopus multiceps Schrank
Brachionus angularis Gosse
Brachionus calyciflorus Pallas
Brachionus capsuliflorus Pallas
Brachionus falcatus Zacharias
Brachionus forficula Wierzejski
Brachionus urceus Linnaeus
Chromogaster ovalis Bergendal
Collotheca ambigua Hudson
Colurella uncinata Müller
Conochiloides dossuarius Hudson
Conochilus hippocrepis Schrank
Conochilus unicornis Rousselet
Diplois daviesiae Gosse
Diurella sulcata Jennings
Euchlanis dilatata Ehrenberg
Filinia longiseta Ehrenberg
Gastropus hyptopus Ehrenberg
Hexarthra mira (Hudson)
Keratella cochlearis Gosse
Keratella quadrata O.F. Müller
Keratella valga Ehrenberg
Lacinularia flosculosa (O.F. Müller)
Lecane luna O.F. Müller
Lecane ungulata Gosse
Lepadella ovalis Müller
Monostyla bulla Gosse
Monostyla hamata Stokes
Monostyla lunaris Ehrenberg

Mytilina ventralis Ehrenberg
Notommata aurita (Muller)
Philodina erythrophthalma Ehrenberg
Platyias militaris Daday
Ploesoma hudsoni Imhof
Polyarthra spp.
Pompholyx sulcata Hudson
Rotaria neptunia (Ehrenberg)
Rotaria rotatoria (Pallas)
Scaridium longicaudum O.F. Müller
Schizocerca diversicornis Daday
Synchaeta pectinata Ehrenberg
Testudinella patina (Hermann)
Trichocerca cyclindrica Imhof
Trichocera gracilis (Tessin)
Trichocerca longiseta Schrank
Trichocerca pusilla (Lauterborn)
Trichocerca ruttus (O.F. Müller)
Trichotria tetractis Ehrenberg

Cladocera

Alona guttata Sars
Alonella excisa Fischer
Bosmina coregoni Baird
Bosmina longirostris Müller
Bosminopsis deitersi Richard
Ceriodaphnia cornuta Sars
Chydorus gibbus Lilljeborg
Chydorus ovalis Kurz
Chydorus sphaericus Müller
Daphnia longispina (Müller)
Daphnia pulex Leydig emend. Scourfield
Diaphanosoma brachyurum (Liéven)
Diaphanosoma leuchtenbergianum Fischer
Diaphanosoma sarsi Rishard
Grapholeberis testudinaria Fischer
Leptodora kindtii Focke
Macrothrix rosea Jurine
Moina micrura Kurz
Moina rectirostris Leydig
Scapholeberis mucronata Müller
Sida crystallina (Müller)
Simocephalus vetulus Müller

Copepoda

Acanthocyclops viridis Jurine
Cyclops strenuus Fischer
Cyclops vicinus Uljanin
Eodiaptomus sinensis Burckhardt
Eucyclops serrulatus Fischer
Eucyclops speratus Fischer
Limnothora sinensis Burckhardt
Mesocyclops "leuckarti" Claus
Microcyclops javanus Kiefer
Neodiaptomus schmackeri Poppe et Richard
Neutrodiaptomus incongruens Poppe
Paracyclops fimbriatus Fischer
Schmackeria forbesi Poppe et Richard
Schmackeria inopina Burckhardt
Sinocalanus dorrii Brehm
Sinodiaptomus sarsi Rylov
Thermocyclops taihokuensis Harada

C Species of Zoobenthos in Lake Taihu

Arthropoda

Asellus nipponensis Nichols
Caridina nilotica gracilipes de Man
Chironomus spp.
Eriocheir sinensis H. Milne-Edwards
Gammarus sp.
Macrobrachium asperulum (Von Marlens)
Macrobrachium nipponensis (De Haan)
Macrobrachium superbum (Heller)
Neocaridina denticulata (Kemp)
Neorhynchoplax intronersus (Kemp)
Palaemon modestus (Heller)
Palamoretes sinensis Sollaud
Tachaea chinensis Thieleman
Tanypus sp.

Hirudinea

Glossiphonia complanata (Linné)
Hemiclepsis kasimiana Oka

Hirudo nippinia (Whitm.)
Witmania pigra (Whitm.)

Mollusca

Alocinma longicornis (Benson)
Angulagia polyzenata (Fraueufeld)
Anodonta woodiana (Lea)
Anodonta woodiana elliptica (Heude)
Anodonta woodiana pacifica (Heude)
Aronaia lanceolata (Lea)
Bellamya aeruginosa (Reeve)
Bellamya purificata (Heude)
Bellamya quadrata (Benson)
Bithynia fuchsiana (Moéllendorff)
Cipangopaludina cathayensis (Heude)
Cipangopaludina chinensis (Gray)
Corbicula fluminea (Müller)
Cristaria plicata (Leach)
Gyraulus convexiusculus (Hütton)
Hippeutis cantori (Benson)
Hippeutis umbillcalis (Benson)
Hyriopsis cumingii (Lea)
Lamprotula caveata (Heude)
Lamprotula leai (Gray)
Lancecolaria giodiola (Heude)
Lancecolaria grayana (Lea)
Limnoprena lacustris (Martens)
Novaculina chinensis Liu et Zhang
Palypylis hemispherula (Benson)
Parafossarulos sinensis (Neumayr)
Parafossarulos strialulus (Benson)
Radix aurieularia (Linnaeus)
Radix swinhoei (H. Adams)
Schistodesmus lampreyanus (Baird et Adams)
Semisulcospira cacellata (Benson)
Sphaerium lacustre (Müller)
Stenothyra glolbra (A. Adams)
Unio donglasiae (Gray)

Oligochaeta

Aulodrilus prochecatus (Chen)
Branchiodrilus hortensis (Stephenson)

Branchiura sowerbyi Beddard
Dero digitata (O.F. Müller)
Limnodrilus hoffmeistei Claperede
Lumbriculus variegatum (O.F. Müller)
Slavina appendiculata (Udekem)
Stylaria fossularis (Leidy)

Polychaeta

Nephthys oligobranchia Southern

D Fish Species in Lake Taihu

*, **, present species (from 2000 to now); **, newly recorded species

Acipenseridae

Acipenser sinensis Cray

Anabantidae

**Macropodus chinensis* (Bloch)

Anguillidae

Anguilla japonica Temninck et Schlegel

Bagridae

**Pseudobagrus fulvidraco* (Richardson)
Leiocassis L. adiposalis (Qshima)
Leiocassis L. albomarginatus Rendhal
Leiocassis L. taeniatus (Gunther)
Leiocassis longirostris Gunther
Leiocassis ussuriensis (Dybowski)
Pseudobagrus eupogon Boulenger
Pseudobagrus nitidus Sauvoge et Dabry

Catostomidae

Myxocyprixus asiaticus (Bleeker)

Clupeidae

Macrura reetesii (Richardon)

Cobitidae

**Paramisgurnus dabryamus* salvage
**Pararhychobdella sinensis* (Bleeker)
Cobitis laenia Linnaeus
Misgurnus anguillicau datus (Cantor)
Botia wui Tchang
Botia xeaiithi (Gunther)
Misgurnus mizolepis Gunther

Cynoglossidae

Cynoglossus gracilis Gunther
Cynoglossus purpureomaculatus Regan
Cynoglossus trigrammus Gunther

Cyprinidae

**Acheilognathus macropterus* (Bleeker)
**Pseudobrama simoni* (Bleeker)
**Squalidus noteas* (Gunther)
Abbottina rivularis (Basilewsky)
Acanthorhodeus barbatutus Gunther
Acanthorhodeus macropterus Bleeker
Acanthorhodeus tonkinensis Vaillant
Carassius auratus (Linnaeus)
Ctenopharyngodon idellus (Cuvier et Valenciennes)
Culter erythropterus Basilewsky
Cyprinus carpio (Linnaeus)
Elopichthys bambusa (Richardson)
Erythroculter ilishaeformis (Bleeker)
Erythroculter mongolicus (Basilewsky)
Hemibarbus maculatus Bleeker
Hemiculter leucisculus (Basilewsky)
Hypohthalmichthys molitrix Cuvier et Valenciennes

**Megalobrama amblycephala* Yih
**Mylopharyngodon piceus* (Richardson)
**Paracanthobrama guichenoti* Bleeker
**Parasilurus asotus* (Richardson)
**Pseudorasbora parva* Wu
**Rhodeus lighti* (Wu)
**Sarcocheilichtys nigripinnis* (Gunther)
**Sarcocheilichtys sinensis* Bleeker
**Squaliobarbus curriculus* (Richardson)
**Toxabramis swinhonis* Gunther
Acanthobrama simoni Blecker
Acanthorhodeus chanhaensis (Dybowsky)
Acanthorhodeus taenianalis Gunther
Acheilognathus sp.
Aphyocypris chinensis Gunther
Coreius heterodon (Bleeker)
Distoechodon tumirostris Peters
Erythroculter dabryi (Bleeker)
Erythroculter sp.
Gnathopogon argentatus (Sauvag et Dabry)
Gnathopogon sihuensis (Chu)
Gnathopogon wolterstorffi (Regan)
Hemibarbus labeo (Pallas)
Hemiculter bleekeri Warpachowsky
Megalobrama terminalis (Richardson)
Ochetobius elongatus (Kner)
Opsarichthys uncirostris bidens Gunther
Parabramis pekinensis (Basilewsky)
Parapelecus argenteus (Gunther)
Parapelecus engraulis (Nichols)
Pararhodeus fangi Miao
Plagiognathops microlepis (Bleeker)
Rhinogobio typus Bleeker
Rhodeus ocellatus (Ker)
Rhodeus sinensis Gunther
Sarcocheilichtys fukiensis (Nickols)
Saurogobio dabryi Bleeker
Saurogobio dumerili Bleeker
Sinilabeo decorus tungting (Nichols)
Xenocypris argentea Gunther
Xenocypris davidi Bleeker

Cyprinodontidae

Oryzias latipes (Schlegel)

Eleotridae

**Hypseleotris swinhonis* (Gitnther)
**Odontobutis obscure* (Temminck et Schlegel)

Engraulidae

**Coilia brachygnathus* Kreyenberger Pappenheim
**Coilia ectenes* Jordon et Seale
Coilia ectenes taihuensis Yuan et Lin

Gallionymidae

Caltionymus olidus Gunther

Gobiidae

**Ctyenogobius giurinus* (Rutter)
Ctyenogobius cliffordpopei (Nichols)
Odontamblyopus rubicundus (Hamilton Buchanan)
Taenioides cirratus (Blyth)

Gottidae

Trachidermus fasciatus Heckel

Hemirhamphidae

**Hemirhamphus intermedius* (Cantor)

Mastacembelidae

Mastacembelus aculeatus (Basilewsky)

Mugilidae

Mugil cephalus Linnaeus
Mugil soiuy Basilewsky

Ophiocephalidae

**Ophiocephalus argus* (Cantor)
Channa asiatica (Linnaeus)

Salangidae

**Neosalanx oligodontis* Chen
**Neosalanx tanghahkeii taihuensis* Chen
**Protosalanx hyalocranius* (Abbott)
Reganisalanx brachyrostratis (Fang)

Serranidae

**Lateolabrax japonicus* (Cuvier et Valenciennes)
**Siniperca chuatsi* (Basilewsky)
Siniperca kneri German
Siniperca scherzeri Steindachner

Siluridae

**Silurus asotus* Linneus

Symbranchidae

**Monopterus abus* (Zuiew)

Tetraodontidae

Fugu obscurus (Abe)
Fugu ocellatus(Linnaeus)

E Species of Aquatic Vegetation in Lake Taihu

Acorus calarnus L.
Acorus gramineus Soland.
Alternanthera philoxeroides (Mart.) Griseb
Azolla imbricata (Roxb) Nakai

Brasenia schreberi J.F. Gmel
Butomus umbellatus L.
Ceratophyllum demersum L.
Ceratopteris thalictroides (L.) Brongn
Colocasia esculenta (L.) Schott.
Cynodon dactylon (L.) Pers.
Echinochloa crusgalli (L.) Beauv.
Eichhornia crassipes (Mart.) Solms
Eleocharis acicularis (L.) Roem. & Schult
Eleocharis tuberosa (Roxb.) Roem. et Schult
Elodea nuttalli (Planch.) St. John
Euryale ferox Salisb.
Fimbristylis dichotoma (L.) Vahl
Hydrilla verticillata (L.f.) Royle
Hydrocharis dubia (Bl.) Backer
Hygroryza aristata (Retz.) Nees
Juncus setchuensis var. *effusoides* (Juncaceae)
Jussiaea repens L.
Leersia japonica (Makino) Honda
Lemna minor L.
Leptochloa chinensis (L.) Nees
Limnophila sessiliflora (Vahl) Blume
Lobelia chinensis Lour
Marsilea quadrifolia L.
Miscanthus sacchariflorus (Mixim.) Hack
Monochoria korsakowii Regel. et Maack
Monochoria vaginalis (Burm. F.) Presl ex Kunth
Murdannia keisak (Hassk.) Hand.-Mazz.
Myriophyllum spicatum L.
Myriophyllum ussuriensis (Regel) Maxim
Najas graminea Del.
Najas minor All.
Nelutnbo nucifera Gaertn
Nymphaea tetragona Georgi
Nymphoides indica (L.) O. Kuntze
Nymphoides peltatum (Gmel.) O. Kuntze
Nyphar pumilum (Hoffm.) DC.
Oenanthe javanica (Bl.) DC.
Ottelia alismoides (L.) Pers.
Phragmites communis Trin.
Polygonum hydropiper L.
Polygonum lapathifolium L.
Polygonum orientale L.
Polygonum perfoliatum L.
Polygonum sagittatum L.

Potamogeton crispus L.
Potamogeton franchetii A. Benn. et Baag.
Potamogeton maackianus A. Benn.
Potamogeton malaianus Miq.
Sagittaria pygmaea Miq.
Sagittaria sagittifolia L.
Salvinia natans (L.) All
Scirpus lacustris L. var. *tabernaemontani* (C.C. Gmel.)
Scirpus yagara Ohwi
Spirodela polyrhiza (L.) Schleid
Stipa tenacissima L.
Torilis japonica (Houtt) DC.
Trapa incisa Sieb et Zucc var. *quadricaudata* Glück
Trapella sinensis Olive
Typha angustifolia L.
Utricularia aurea Lour
Vallisneria natans (Lour.) Hara
Zizania latifolia (Griseb.) Stapf

Index

B

G

T

U

V

MIX
Papier aus verantwortungsvollen Quellen
Paper from responsible sources
FSC® C105338

If you have any concerns about our products,
you can contact us on
ProductSafety@springernature.com

In case Publisher is established outside the EU,
the EU authorized representative is:
Springer Nature Customer Service Center GmbH
Europaplatz 3, 69115 Heidelberg, Germany

Printed by Libri Plureos GmbH
in Hamburg, Germany